Partial Differential Equations through Examples and Exercises

Kluwer Texts in the Mathematical Sciences

VOLUME 18

A Graduate-Level Book Series

Partial Differential Equations through Examples and Exercises

by

Endre Pap
Arpad Takači

and

Djurdjica Takači

Institute of Mathematics,
University of Novi Sad,
Novi Sad, Yugoslavia

SPRINGER-SCIENCE+BUSINESS MEDIA, B.V.

A C.I.P. Catalogue record for this book is available from the Library of Congress.

ISBN 978-0-7923-4724-8   ISBN 978-94-011-5574-8 (eBook)
DOI 10.1007/978-94-011-5574-8

Printed on acid-free paper

Contents

Preface   ix

List of symbols   xi

1 Introduction   **1**
1.1   Basic Notions 1
     1.1.1   Preliminaries 1
     1.1.2   Examples and Exercises 3
1.2   The Cauchy–Kowalevskaya Theorem 12
     1.2.1   Preliminaries 12
     1.2.2   Examples and Exercises 13
1.3   Equations of Mathematical Physics 15

2 First Order PDEs   **17**
2.1   Quasi–linear PDEs 17
     2.1.1   Preliminaries 17
     2.1.2   Examples and Exercises 18
2.2   Pfaff's Equations 32
     2.2.1   Preliminaries 32
     2.2.2   Examples and Exercises 33
2.3   Nonlinear First Order PDEs 35
     2.3.1   Preliminaries 35
     2.3.2   Examples and Exercises 38

3 Classification of the Second Order PDEs   **49**
3.1   Two Independent Variables 49
     3.1.1   Preliminaries 49
     3.1.2   Examples and Exercises 53
3.2 n Independent Variables 64
     3.2.1   Preliminaries 64
     3.2.2   Examples and Exercises 66

3.3 Wave, Potential and Heat Equation 69

4 Hyperbolic Equations **71**
4.1 Cauchy Problem for the One-dimensional Wave Equation 71
    4.1.1 Preliminaries 71
    4.1.2 Examples and Exercises 72
4.2 Cauchy Problem for the n-dimensional Wave Equation 80
    4.2.1 Preliminaries 80
    4.2.2 Examples and Exercises 82
4.3 The Fourier Method of Separation Variables 89
    4.3.1 Preliminaries 89
    4.3.2 Examples and Exercises 93
4.4 The Sturm–Liouville Problem 106
    4.4.1 Preliminaries 106
    4.4.2 Examples and Exercises 109
4.5 Miscellaneous Problems 129
4.6 The Vibrating String 141

5 Elliptic Equations **143**
5.1 Dirichlet Problem 143
    5.1.1 Preliminaries 143
    5.1.2 Examples and Exercises 144
5.2 The Maximum Principle 163
    5.2.1 Preliminaries 163
    5.2.2 Examples and Exercises 163
5.3 The Green Function 167
    5.3.1 Preliminaries 167
    5.3.2 Examples and Exercises 168
5.4 The Harmonic Functions 173
    5.4.1 Examples and Exercises 173
5.5 Gravitational Potential 182

6 Parabolic Equations **183**
6.1 Cauchy Problem 183
    6.1.1 Preliminaries 183
    6.1.2 Examples and Exercise 184
6.2 Mixed Type Problem 193
    6.2.1 Preliminaries 193
    6.2.2 Examples and Exercises 194
6.3 Heat conduction 223

7 Numerical Methods   **227**
    7.0.1   Preliminaries 227
    7.0.2   Examples and Exercises 230

8 Lebesgue's Integral, Fourier Transform   **249**
   8.1   Lebesgue's Integral and the $L_2(Q)$ Space 249
    8.1.1   Preliminaries 249
    8.1.2   Examples and Exercises 252
   8.2   Delta Nets 256
    8.2.1   Preliminaries 256
    8.2.2   Examples and Exercises 257
   8.3   The Surface Integrals 260
    8.3.1   Preliminaries 260
    8.3.2   Examples and Exercises 261
   8.4   The Fourier Transform 267
    8.4.1   Preliminaries 267
    8.4.2   Examples and Exercises 269

9 Generalized Derivative and Sobolev Spaces   **279**
   9.1   Generalized Derivative 279
    9.1.1   Preliminaries 279
    9.1.2   Examples and Exercises 279
   9.2   Sobolev Spaces 285
    9.2.1   Preliminaries 285
    9.2.2   Examples and Exercises 286

10 Some Elements from Functional Analysis   **303**
   10.1   Hilbert Space 303
    10.1.1   Preliminaries 303
    10.1.2   Examples and Exercises 305
   10.2   The Fredholm Alternatives 313
    10.2.1   Preliminaries 313
    10.2.2   Examples and Exercises 314
   10.3   Normed Vector Spaces 321
    10.3.1   Preliminaries 321
    10.3.2   Examples and Exercises 323

11 Functional Analysis Methods in PDEs   **329**
   11.1   Generalized Dirichlet Problem 329
    11.1.1   Preliminaries 329
    11.1.2   Examples and Exercises 330
   11.2   The Generalized Mixed Problems 355
    11.2.1   Examples and Exercises 355

11.3 Numerical Solutions 366
    11.3.1 Preliminaries 366
    11.3.2 Examples and Exercises 367
11.4 Miscellaneous 368
    11.4.1 Preliminaries 368
    11.4.2 Examples and Exercises 369

12 Distributions in the theory of PDEs   373
12.1 Basic Properties 373
    12.1.1 Preliminaries 373
    12.1.2 Examples and Exercises 376
12.2 Fundamental Solutions 390
    12.2.1 Preliminaries 390
    12.2.2 Examples and Exercises 390

Bibliography   397

Index   401

Preface

The book *Partial Differential Equations through Examples and Exercises* has evolved from the lectures and exercises that the authors have given for more than fifteen years, mostly for mathematics, computer science, physics and chemistry students. By our best knowledge, the book is a first attempt to present the rather complex subject of partial differential equations (PDEs for short) through active reader-participation. Thus this book is a combination of theory and examples.

In the theory of PDEs, on one hand, one has an interplay of several mathematical disciplines, including the theories of analytical functions, harmonic analysis, ODEs, topology and last, but not least, functional analysis, while on the other hand there are various methods, tools and approaches. In view of that, the exposition of new notions and methods in our book is "step by step". A minimal amount of expository theory is included at the beginning of each section *Preliminaries* with maximum emphasis placed on well selected examples and exercises capturing the essence of the material. Actually, we have divided the problems into two classes termed *Examples* and *Exercises* (often containing proofs of the statements from Preliminaries). The examples contain complete solutions, and also serve as a model for solving similar problems, given in the exercises. The readers are left to find the solution in the exercises; the answers, and occasionally, some hints, are still given.

The book is implicitly divided in two parts, classical and abstract. In the first (classical) part, the necessary prerequisites are a standard undergraduate course on ODEs, on Riemann's multiple and surface integrals and, of course, on Fourier series. For the second (abstract) part, it would be desirable that the reader is familiar with the elements of Lebesgue integrals and functional analysis (in particular, Hilbert spaces and operator theory). We tried to make the book as self-contained as possible. For that reason, we also included in the *Preliminaries* and *Examples* some of the mentioned mathematical tools (see, e.g., elementary proofs of the Closed Graph Theorem, Adjoint Theorem and Uniform Boundedness Theorem in Chapter 10).

Many different tools are presented for solving important problems with the basic three partial differential equations: the wave equation, Laplace equation, heat equation and their generalizations. We also give the usual three types of problems with PDEs: initial value problems, boundary value problems and mixed type (eigenvalue) problems. For the solutions of the stated problems, we discuss the three important questions: existence, uniqueness, stability (continuous dependence of solutions upon data). We investigate also three important questions for the solutions of PDEs mostly for applications: construction, regularity and approximation.

We present, among other tools, the three principal methods for solving the stated problems: Fourier method, Green's function and the energy (variational) method. One of the very useful constructive techniques the Fourier method of separation of variables, is applied first in Chapter 4 for hyperbolic equations with respect to the classical Fourier series, where the eigenfunctions are the sine and cosine functions. In the next step, we generalize this method through the Sturm-Liouville problem also with respect to other systems of orthogonal functions, e.g., Legendre polynomials and Bessel functions. The Fourier method is applied also in Chapters 5 and 6 to elliptic and parabolic equations, respectivily. This theoretical background for these methods is obtained in Chapters 10 and 11 in the language of functional analysis through special spaces as, e.g., Sobolev spaces, with generalized eigenvalues and eigenfunctions. The Fourier analysis is completed in Chapter 8 by the Fourier transform.

Most of the book is devoted to second or higher order PDEs. However, for completeness, Chapter 2 treats first order PDEs.

In the last Chapter we present a part of the distribution theory, which also covers the theory of Dirac's delta distribution ("delta function").

The majority of the problems are of mathematical character, though we often give physical interpretations (see sections at the ends of Chapter 1, 3, 4, 5 and 6). The numerical approximations and computation of the stated problems are presented in Chapter 7, with an abstract theoretical background in Chapter 11.

The book is prepared for undergraduate and graduate students in mathematics, physics, technology, economics and everybody with an interest in partial differential equations for modeling complex systems.

We have used *Mathematica* and *Scientific Work Place 2.5.* for some calculations and drawings.

We are grateful to Prof. Olga Hadžić for her numerous remarks and advice on the text, and to Prof. Darko Kapor on his useful suggestions on the physical aspects of PDEs. Dr Dušanka Perišić made some contributions to Subsections 3.2 and 10.2 and has prepared the Figures 4.1-4.4. It is our pleasure to thank the Institute of Mathematics in Novi Sad for working conditions and financial support. We would like to thank Kluwer Academic Publishers, specially to Dr Paul Roos and Ms Angelique Hempel for their encouragement and patience.

Novi Sad, April 1997   ENDRE PAP
                                                                       ARPAD TAKAČI
                                                                    DJURDJICA TAKAČI

List of Symbols

N		set of natural numbers
Z		set of integers
$\mathbf{Z}_+$		set of non–negative integers
R		set of real numbers
$\mathbf{R}^n$		n–dimensional real Euclidean space
C		set of complex numbers
$\Re z$	$=$	real part of a complex number z
$\Im z$	$=$	imaginary part of a complex number z
$\imath$	$=$	$\sqrt{-1}$ imaginary unit
A^c		complement of the set A
χ_A		characteristic function of the set A
$\cap$		intersection of sets
$\cup$		union of sets
$\setminus$		set difference
Σ		σ–algebra of subsets of a set X
Q		region of $\mathbf{R}^n$
∂Q		border of the region Q
$\overline{Q}$		closure of the region Q
$Q' \subset\subset Q$		closure of Q' is a subset of Q
x	$=$	$(x_1, \ldots, x_n) \in \mathbf{R}^n$
$\lvert x \rvert$	$=$	$\sqrt{x_1^2 + \cdots + x_n^2}$
α	$=$	$(\alpha_1, \ldots, \alpha_n)$ multi–index, where $\alpha_i \in \mathbf{Z}_+$, $i = 1, 2, \ldots, n$
$\mathbf{Z}_+^n$		set of multi–indices
$\lvert \alpha \rvert$	$=$	$\alpha_1 + \cdots + \alpha_n$
$\alpha!$	$=$	$\alpha_1! \cdots \alpha_n!$
x^α	$=$	$x_1^{\alpha_1} \cdots x_n^{\alpha_n}$
$\lim\inf$		limes inferior
$\lim\sup$		limes superior

$$D_i \qquad = \quad \frac{\partial}{\partial x_i}$$

$$D = \nabla \quad = \quad (D_1, \ldots, D_n)$$

$$\Delta \qquad = \quad \frac{\partial^2}{\partial x_1{}^2} + \cdots + \frac{\partial^2}{\partial x_n{}^2} \quad \text{(Laplace operator)}$$

$(f|g)$ the scalar product

supp f support of a function or distribution

$C^k(Q)$ space of continuous functions on Q with continuous derivatives of order $\leq k$

$C^\infty(O)$ space of infinitely differentiable functions over an open set O

$C_0^\infty(O)$ space of infinitely differentiable functions over an open set O with compact support

$L_2(Q)$ space of measurable functions f with $\int_Q |f(x)|^2 \, dx < \infty$

$\|f\|_{L_2(Q)} \quad = \quad (\int_Q |f(x)|^2 \, dx)^{1/2}$

$W^k(Q)$ Sobolev space of order k

$$\|f\|_{W^k(Q)} \quad = \quad \left(\sum_{|\alpha| \leq k} \int_Q |D^\alpha f(x)|^2 \, dx \right)^{1/2}$$

$\overset{\circ}{W}{}^k(Q)$ Sobolev space which is $\overline{C_0^\infty(\overline{Q})}$ (closure with respect to $\| \cdot \|_{W^k(Q)}$)

δ delta distribution ("delta function")

$\mathcal{D}(O)$ space of test functions over the open set O

$\mathcal{D}'(O)$ space of distributions over the open set O

Chapter 1

Introduction

1.1 Basic Notions

1.1.1 Preliminaries

A partial differential equation (briefly PDE) for a function $u = u(x_1, \ldots, x_n)$ is a relation of the form

$$F\left(x_1, \ldots, x_n, \frac{\partial u}{\partial x_1}, \ldots, \frac{\partial u}{\partial x_n}, \frac{\partial^2 u}{\partial x_1^2}, \ldots, \frac{\partial^m u}{\partial x_n^m}\right) = 0, \tag{1.1}$$

where F is a given function of the independent variables $x_1, \ldots, x_n$, $n > 1$, and of the (unknown) function u of a finite number of its partial derivatives. The *order* of the PDE (1.1) is the order of the highest derivative that occurs.

A subset Q of $\mathbf{R}^n$ is a *region* if it is open and connected. A function $u = u(x_1, \ldots, x_n)$ is a *solution* of (1.1) on the region $Q \subset \mathbf{R}^n$ if after substitution of u and its partial derivatives in (1.1) the relation (1.1) is satisfied identically for all $(x_1, \ldots, x_n)$ from the region Q.

The vectors (n–tuples) $\alpha = (\alpha_1, \ldots, \alpha_n)$, whose components are non–negative integers α_k, are called *multi–indices*. The set of all multi–indices is $\mathbf{Z}_+^n$. For $\alpha \in \mathbf{Z}_+^n$ and $x \in \mathbf{R}^n$ we define

$$|\alpha| = \alpha_1 + \cdots + \alpha_n, \quad |x| = \sqrt{x_1^2 + \cdots + x_n^2},$$

$$\alpha! = \alpha_1! \cdots \alpha_n!, \quad x^\alpha = x_1^{\alpha_1} \cdots x_n^{\alpha_n}.$$

We are using the convention $0^0 = 1$. A partial order in $\mathbf{Z}_+^n$ is defined by

$$\alpha \geq \beta \text{ whenever } \alpha_k \geq \beta_k \text{ for } k = 1, \ldots, n.$$

We introduce the differentiation symbol $D_k = \dfrac{\partial}{\partial x_k}$, and by a gradient vector of differentiation $D = (D_1, \ldots, D_n)$ we define the gradient $Du = \nabla u$ of a function

1

$u = u(x_1, \ldots, x_n)$ as the vector

$$Du = \nabla u = (D_1 u, \ldots, D_n u).$$

The general partial differential operator D^α of order $m = |\alpha|$ is then

$$D^\alpha = D_1^{\alpha_1} \cdots D_n^{\alpha_n} = \frac{\partial^{|\alpha|}}{\partial x_1^{\alpha_1} \cdots \partial x_n^{\alpha_n}}.$$

A PDE is *linear* if it is linear in the unknown function and its derivatives, with coefficients depending on the independent variables $x_1, \ldots, x_n$. A linear PDE of order m is of the form

$$\sum_{|\alpha| \le m} a_\alpha D^\alpha u = F,$$

where a_α and F are given functions depending on $x = (x_1, \ldots, x_n)$.

A PDE of order m is said to be *quasi–linear* if it is linear in the derivatives of order m with coefficients depending on the independent variables $x_1, \ldots, x_n$, and on the unknown function and its derivatives of orders strictly smaller than m.

Classification of problems with PDEs.

1) *Cauchy problem or initial–value problem.* For a given PDE on a region Q, some additional initial–values for the unknown function and its derivatives are given on some subsets of Q.

2) *Boundary value problem.* For a given PDE on a region Q some additional boundary conditions for the unknown function and its derivatives are given on the boundary ∂Q.

3) *Mixed type problem.* For a given PDE on a region Q both some additional initial–values and boundary conditions for the unknown function and its derivatives are given.

A problem with a PDE is *well–posed* in a class of functions $\mathcal{C}$, if the following three conditions are satisfied

(i) there exists a solution in $\mathcal{C}$;

(ii) the solution is unique;

(iii) the solution is continuously dependent on the given conditions, e.g., initial–values, boundary conditions, coefficients, etc.

1.1.2 Examples and Exercises

Exercise 1.1 *Show that the function $u(x, y) = y^2 + x$ is the solution of the Cauchy problem on $\mathbf{R}^2$*

$$\frac{\partial u}{\partial x} = 1 \quad and \quad u(0, y) = y^2 \quad (y \in \mathbf{R}).$$

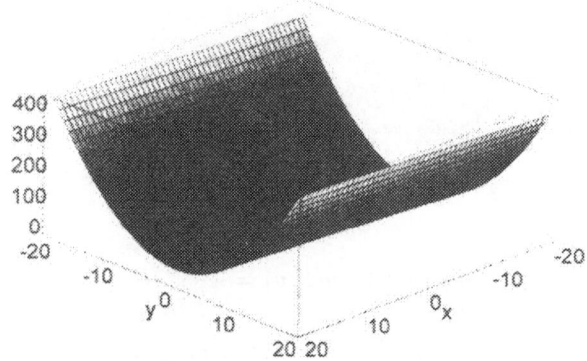

Figure 1.1 $u(x, y) = y^2 + x$

Exercise 1.2 *Show that the function $u(x, y) = \cos\sqrt{x^2 + y^2}$ is the solution of the Cauchy problem*

$$\left(\frac{\partial u}{\partial x}\right)^2 + \left(\frac{\partial u}{\partial y}\right)^2 = 1 - u^2 \quad and \quad u(0, y) = \cos y.$$

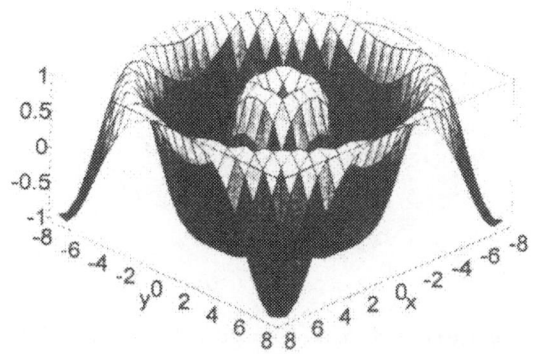

Figure 1.2 $u(x, y) = \cos\sqrt{x^2 + y^2}$

Exercise 1.3 *Show that the function $u(x, y) = -x^2/2 + y^2/2$ is the solution of the equation*

$$\frac{\partial^2 u}{\partial x^2}\frac{\partial^2 u}{\partial y^2} - \left(\frac{\partial^2 u}{\partial x \partial y}\right)^2 = -1.$$

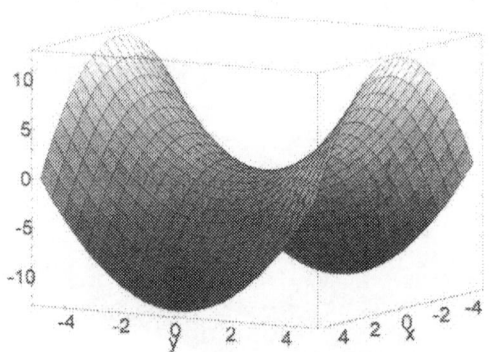

Figure 1.3 $u(x,y) = -\dfrac{1}{2}x^2 + \dfrac{1}{2}y^2$

Exercise 1.4 *Show that the function $u(x,t) = (x+ct)^4 + (x-ct)^3$, where c is a positive constant (Figure 1.4 for $c = 3$), is a solution of the wave equation*

$$\frac{\partial^2 u}{\partial t^2} - c^2 \frac{\partial^2 u}{\partial x^2} = 0.$$

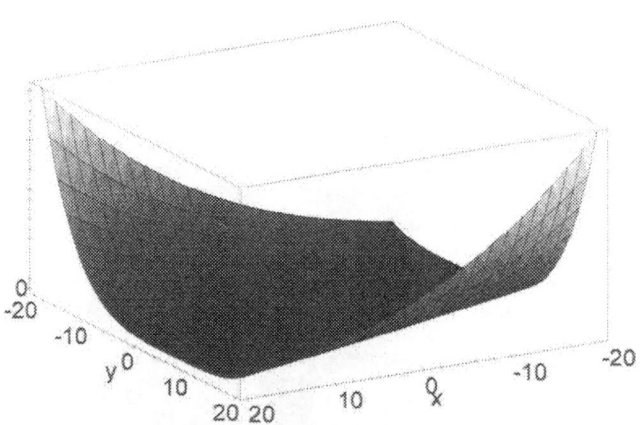

Figure 1.4 $u(x,t) = (x+3t)^4 + (x-3t)^3$

Exercise 1.5 *Show that the function $u(x_1, \ldots, x_n, t) = (x_1 + \cdots + x_n + \sqrt{n}ct)^{10}$, where c is a positive constant, is a solution of the $n-$dimensional wave equation*

$$\frac{\partial^2 u}{\partial t^2} - c^2 \left(\frac{\partial^2 u}{\partial x_1^2} + \cdots + \frac{\partial^2 u}{\partial x_1^2} \right) = 0,$$

which, using the Laplace operator Δ given by

$$\Delta = \frac{\partial^2}{\partial x_1^2} + \cdots + \frac{\partial^2}{\partial x_1^2},$$

can be written in the shorter form

$$\frac{\partial^2 u}{\partial t^2} - c^2 \Delta u = 0.$$

Example 1.6 *Let $f : Q \to \mathbf{C}$ be an analytic function, where Q is a region in $\mathbf{C}$. Prove that the functions $u(x,y) = \Re f(x+\imath y)$ and $v(x,y) = \Im f(x+\imath y)$ for $x+\imath y \in Q$ (where $\imath$ is the imaginary unit, $\imath^2 = -1$) satisfy*

a) the Cauchy–Riemann equations

$$\frac{\partial u}{\partial x} = \frac{\partial v}{\partial y} \quad and \quad \frac{\partial u}{\partial y} = -\frac{\partial v}{\partial x};$$

b) the Laplace equation

$$\frac{\partial^2 u}{\partial x^2} + \frac{\partial^2 u}{\partial y^2} = 0,$$

if, additionally, u and v are from the class $C^2(Q)$.

Solution.

a) Since the function f is analytic, there exists

$$f'(z) = \lim_{h \to 0} \frac{f(z+h) - f(z)}{h}$$

for $z = x + \imath y \in Q$. We evaluate the preceding limit in two ways. First let $h \to 0$ for real h. We have for $h \neq 0$

$$\frac{f(z+h) - f(z)}{h} = \frac{f(x+h+\imath y) - f(x+\imath y)}{h}$$

$$= \frac{u(x+h,y) - u(x,y)}{h} + \imath \frac{v(x+h,y) - v(x,y)}{h}.$$

Taking $h \to 0$ we obtain

$$f'(z) = \frac{\partial u(x,y)}{\partial x} + \imath \frac{\partial v(x,y)}{\partial x}. \tag{1.2}$$

Now let $\imath h \to 0$ for real h. We have for $h \neq 0$

$$\frac{f(z+\imath h) - f(z)}{\imath h} = \frac{f(x+\imath(h+y)) - f(x+\imath y)}{\imath h}$$

$$= -\imath \frac{u(x, y+h) - u(x, y)}{h} + \frac{v(x, y+h) - v(x, y)}{h}.$$

Letting $h \to 0$ we obtain

$$f'(z) = -\imath \frac{\partial u(x, y)}{\partial y} + \frac{\partial v(x, y)}{\partial y}. \tag{1.3}$$

Since both the real and imaginary parts of (1.2) and (1.3) must be equal, we get the Cauchy–Riemann equations.

b) Differentiating the first Cauchy–Riemann equation with respect to x and the second one with respect to y we obtain

$$\frac{\partial^2 u}{\partial x^2} = \frac{\partial^2 v}{\partial x \partial y} \quad \text{and} \quad \frac{\partial^2 u}{\partial y^2} = -\frac{\partial^2 v}{\partial y \partial x}.$$

Adding the obtained equalities we get that the function u satisfies the Laplace equation. Differentiating now the first Cauchy–Riemann equation with respect to y and the second one with respect to x and repeating the preceding procedure we get that the function v satisfies the Laplace equation, too.

Example 1.7 *Show that the function $u = u(x, t)$, given by*

$$u(x, t) = ae^{-t/a^2} \sin \frac{x}{a} \qquad \left((x, t) \in \mathbf{R}^2\right),$$

for a real constant $a \neq 0$, is a solution of the one–dimensional heat equation $\frac{\partial u}{\partial t} = \frac{\partial^2 u}{\partial x^2}$, which for $t \geq 0$ converges to zero as $a \to 0$, but does not for $t < 0$.

Solution. For the given function it holds

$$\frac{\partial u(x, t)}{\partial t} = -\frac{1}{a} e^{-t/a^2} \sin \frac{x}{a}$$

and

$$\frac{\partial u(x, t)}{\partial x} = e^{-t/a^2} \cos \frac{x}{a}, \quad \frac{\partial^2 u(x, t)}{\partial x^2} = -\frac{1}{a} e^{-t/a^2} \sin \frac{x}{a}.$$

Putting the derivatives $\dfrac{\partial u(x, t)}{\partial t}$ and $\dfrac{\partial^2 u(x, t)}{\partial x^2}$ in the heat equation, we obtain that it is satisfied identically for all $x, t \in \mathbf{R}$.

We have for $t \geq 0$

$$\lim_{a \to 0} u(x, t) = \lim_{a \to 0} e^{-t/a^2} \frac{\sin \frac{x}{a}}{\frac{x}{a}} \cdot x = 0,$$

see Figures 1.5 and 1.6.

For $t < 0$ we have $\lim_{a \to 0} u(x,t) = +\infty$, see Figures 1.7 and 1.8. Thus the last result shows that the Cauchy problem for the heat equation is not well–posed.

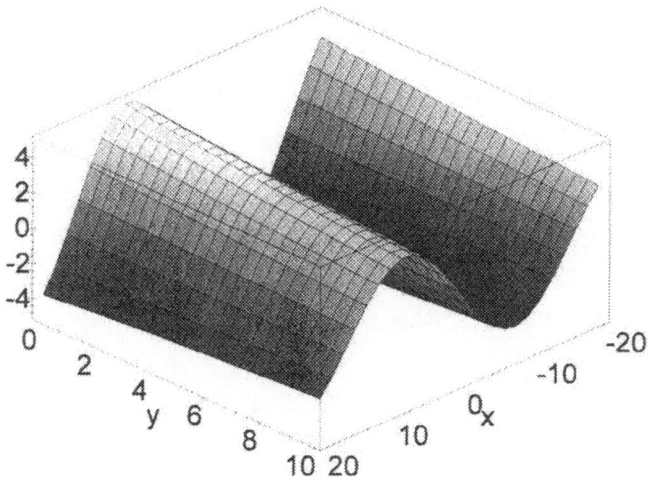

Figure 1.5 Case $a = 5$, $u(x,t) = 5e^{-t/25} \sin \dfrac{x}{5}$, $t > 0$

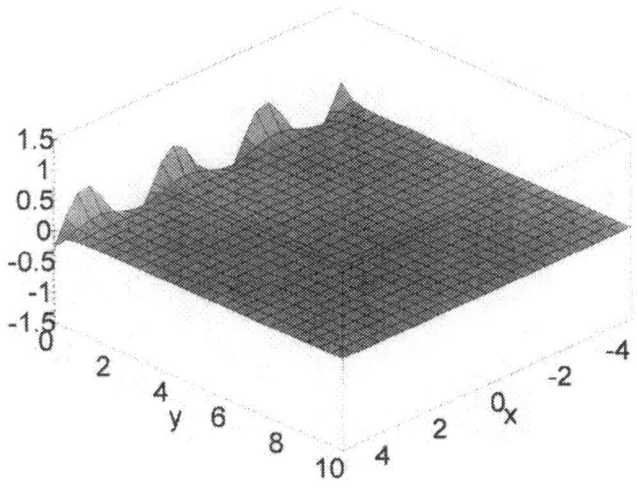

Figure 1.6 Case $a = 0.5$, $u(x,t) = 0.5e^{-t/0.25} \sin 2x$, $t > 0$

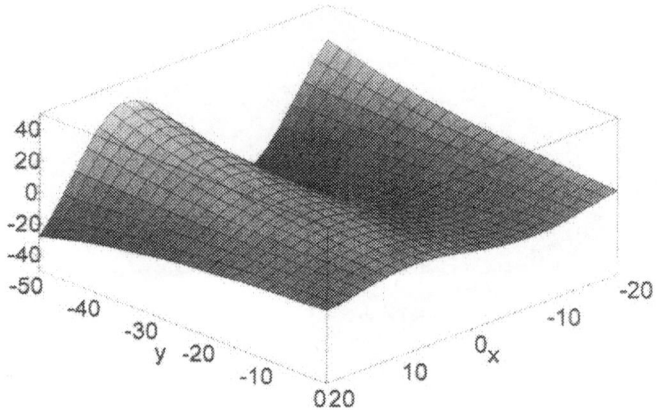

Figure 1.7 Case $a = 5$, $u(x,t) = 5e^{-t/25} \sin \dfrac{x}{5}$, $t < 0$

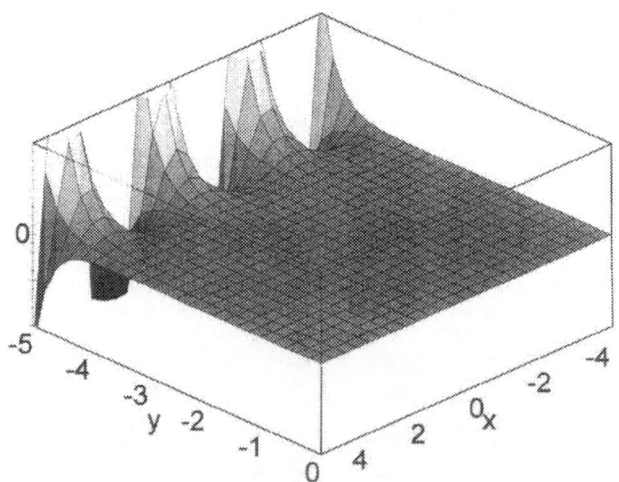

Figure 1.8 Case $a = 0.5$, $u(x,t) = 0.5e^{-t/0.25} \sin 2x$, $t < 0$,
$10^{-8} \le u \le 10^8$

Example 1.8 (Hadamard) *Show that the function* $u = u(x,y)$ *given by*

$$u(x,y) = \frac{e^{ny} - e^{-ny}}{2n^2} \sin nx$$

for $n \in \mathbf{N}$ *is a solution on* $D = \{(x,y) \mid x^2 + y^2 < 1\}$ *of the Cauchy problem for the Laplace equation*

$$\frac{\partial^2 u}{\partial x^2} + \frac{\partial^2 u}{\partial y^2} = 0,$$

$$u(x,0) = 0, \quad \frac{\partial u(x,0)}{\partial y} = \frac{\sin nx}{n}.$$

Prove that this problem is not well–posed.

Solution. It easy to check that the function u_n for an arbitrary but fixed $n \in \mathbf{N}$ given by

$$u_n(x,y) = \frac{e^{ny} - e^{-ny}}{2n^2} \sin nx$$

is a solution of the given Cauchy problem. Letting $n \to \infty$ we obtain for $x, y \in D$, $x \neq 0, y \neq 0$

$$|u_n(x,y)| \to \infty.$$

On the other side, the given Cauchy problem for $n \to \infty$ reduces to the problem

$$\frac{\partial^2 u}{\partial x^2} + \frac{\partial^2 u}{\partial y^2} = 0,$$

$$u(x,0) = 0, \quad \frac{\partial u(x,0)}{\partial y} = 0,$$

which has only a trivial solution $u = 0$. Therefore the considered solution of the given Cauchy problem does not depend continuously of the initial condition. Hence it is not well–posed (it is ill–posed).

Remark 1.8.1. In the contrast with this simple problem with a PDE which is not well–posed, let us remind some well–known results from the theory of ordinary differential equations where for general classes of problems with ordinary differential equations the well–posedness can be ensured. For example, the well–posedness of the Cauchy problem

$$y' = f(x,y), \quad y(x_0) = y_0$$

is ensured supposing that the function f is continuous and satisfies the Lipschitz condition in some region which contains the point (x_0, y_0).

Example 1.9 *Find the solution $u = u(x, y)$ of the following problem on the set* $\mathbf{R}^2$.

$$\frac{\partial^2 u}{\partial x \partial y} - \frac{\partial u}{\partial y} = 4, \tag{1.4}$$

$$\frac{\partial u(0, y)}{\partial y} = 3y^2, \quad u(x, 0) = 0. \tag{1.5}$$

Solution. The equation (1.4) can be written in the form

$$\frac{\partial}{\partial y}\left(\frac{\partial u}{\partial x} - u\right) = 4,$$

which after integration with respect to y reduces on the equation

$$\frac{\partial u}{\partial x} - u = 4y + F_1(x),$$

where F_1 is an arbitrary differentiable function. The solution of this equation is

$$u(x, y) = e^x F_2(y) - 4y + e^x \int_{x_0}^x e^{-t} F_1(t)\, dt, \tag{1.6}$$

where F_2 is an arbitrary differentiable function. Putting u from (1.6) in the first initial condition in (1.5) we obtain

$$\frac{\partial u(0, y)}{\partial y} = F_2'(y) - 4 = 3y^2,$$

which implies

$$F_2(y) = y^3 + 4y + C, \tag{1.7}$$

where C is an arbitrary real constant. Putting now u from (1.6) in the second initial condition in (1.5) and using (1.7) we obtain

$$u(x, 0) = Ce^x + e^x \int_{x_0}^x e^{-t} F_1(t)\, dt = 0.$$

Hence the solution of equation (1.4) with the initial–conditions (1.4) is

$$u(x, y) = e^x(y^3 + 4y) - 4y.$$

Example 1.10 *Find the general solutions of the equations*

a) $\dfrac{\partial u}{\partial y} = 0$;

b) $\dfrac{\partial^2 u}{\partial x \partial y} = 0$;

c) $\dfrac{\partial^2 u}{\partial x \partial y} + x \dfrac{\partial u}{\partial y} = 0.$

Solution.

a) $u(x,y) = f(x)$, where f is an arbitrary function from $C^1(\mathbf{R})$.

b) $u(x,y) = p(x) + q(y)$ for $p, q \in C^2(\mathbf{R})$.

c) Taking the second variable fixed, say $y = y_0$, and introducing a function of one variable $w(x) = \dfrac{\partial u(x, y_0)}{\partial y}$, the given equation reduces on an ordinary differential equation

$$w' + xw = 0,$$

which can easily be solved by the method of separation of variables. Hence $w(x) = Ce^{-x^2/2}$, where C depends from y_0. Therefore for arbitrary y we have $C = q(y)$, and

$$\frac{\partial u}{\partial y} = q(y) e^{-x^2/2}$$

Applying the integral with respect to y we obtain

$$u(x,y) = e^{-x^2/2} \int q(y)\, dy + f(x) = f(x) + e^{-x^2/2} g(y),$$

where f and g are functions from $C^1(\mathbf{R})$.

Exercise 1.11 *Prove that the following problem on the set $\mathbf{R}^2$*

$$\frac{\partial^2 u}{\partial x \partial y} = 0,$$

$$u(x,0) = f(x), \quad \frac{\partial u(x,0)}{\partial y} = g(x),$$

has a solution if and only if $f \in C^2(\mathbf{R})$ and $g(x) = const$. Then the solution is given by

$$u(x,y) = f(x) + F(y) - F(0) + y(g(0) - F'(0)),$$

where F is an arbitrary function from $C^2(\mathbf{R})$.

Exercise 1.12 *Prove that the equation*

$$\frac{\partial^2 u}{\partial x^2} - \frac{\partial^2 u}{\partial y^2} + \frac{a}{x} \frac{\partial u}{\partial x} = 0,$$

taking the substitution $v(x,y) = x^{a/2} u(x,y)$, reduces on the equation

$$\frac{\partial^2 v}{\partial x^2} - \frac{\partial^2 v}{\partial y^2} + \frac{a(2-a)}{4x^2} v = 0.$$

Exercise 1.13 *Prove that the heat equation*

$$\frac{\partial u}{\partial t} = \frac{a}{2}\frac{\partial^2 u}{\partial x^2} \quad (x \in \mathbf{R}, t > 0),$$

a real constant, taking the substitution $u(x,t) = \exp(-v(x,t)/a)$ reduces to the Burgers equation

$$\frac{\partial v}{\partial t} + \frac{1}{2}\left(\frac{\partial v}{\partial x}\right)^2 - \frac{a}{2}\frac{\partial^2 u}{\partial x^2} = 0.$$

1.2   The Cauchy–Kowalevskaya Theorem

1.2.1   Preliminaries

Definition 1.1 *A function $f : O \to \mathbf{R}$, where O is an open set in $\mathbf{R}^n$, is called real analytic at $x^0 \in O$ if there exist $c_\alpha \in \mathbf{R}$ ($\alpha \in \mathbf{Z}_+^n$) and a neighbourhood $V(x^0)$ of x^0 such that*

$$f(x) = \sum_{\alpha \in \mathbf{Z}_+^n} c_\alpha(x - x^0)^\alpha$$

for all $x \in V(x^0 \cap O)$. The function f is called real analytic in O if it is real analytic at every point x^0 from O.

Theorem 1.2 (Cauchy–Kowalevskaya) *Let a_{jk}^i and b_j be real analytic functions depending on $z = (x_1, \ldots, x_{n-1}, u_1, \ldots, u_s)$ at the point*

$$z^0 = (x_1^0, \ldots, x_{n-1}^0, u_1^0, \ldots, u_s^0) \in \mathbf{R}^{n+s-1}, \quad i = 1, \ldots, n-1; \quad k, j = 1, \ldots, s,$$

and let $\varphi_j = \varphi_j(x_1, \ldots, x_{n-1})$ be real analytic functions at $(x_1^0, \ldots, x_{n-1}^0)$ for $j = 1, \ldots, s$. Then the system of quasi–linear PDEs

$$\frac{\partial u_j}{\partial x_n} = \sum_{i=1}^{n-1}\sum_{k=1}^{s} a_{jk}^i \frac{\partial u_k}{\partial x_i} + b_j$$

for $j = 1, \ldots, s$, with the initial conditions

$$u_j(x_1, \ldots, x_{n-1}, x_n^0) = \varphi_j(x_1, \ldots, x_{n-1})$$

for $j = 1, \ldots, s$, has a unique solution $(u_1, \ldots, u_s)$ which is real analytic at the point x^0.

1.2.2 Examples and Exercises

Exercise 1.14 *For $\alpha = (\alpha_1, \dots, \alpha_n) \in \mathbf{Z}_+^n$ and $x = (x_1, \dots, x_n) \in \mathbf{R}^n$ prove*

a) $\alpha! \leq |\alpha|! \leq n^{|\alpha|}\alpha!$;

b) $(x_1 + \dots + x_n)^m = \sum\limits_{|\alpha|=m} \dfrac{m!}{\alpha!} x^\alpha$ *for* $m \in \mathbf{Z}_+$.

Exercise 1.15 *Prove that for $f, g \in C^m(\mathbf{R}^n)$ it holds*

$$D^\alpha(f \cdot g) = \sum_{\beta,\gamma,\beta+\gamma=\alpha} \frac{\alpha!}{\beta!\gamma!} D^\beta f D^\gamma g,$$

where $|\alpha| = \alpha_1 + \dots + \alpha_n = m$.

Exercise 1.16 *Let O be an open subset of the set $\mathbf{R}^n$. Prove that a function f is real analytic on O if and only if $f \in C^\infty(O)$ and for every compact set $K \subset O$ there exist $M > 0$ and $r > 0$ such that for every $y \in K$ we have*

$$|D^\beta f(y)| \leq M|\beta|! r^{-|\beta|} \quad (\beta \in \mathbf{Z}_+^n).$$

Example 1.17 *Prove that there is no real analytic solution in $(0,0)$ of the equation*

$$\frac{\partial^2 u}{\partial x \partial y} - \frac{\partial u}{\partial y} = 0,$$

which would also satisfy the initial–condition

$$u(x,0) = \frac{1}{1+x^2}.$$

Solution. If there were a real analytic (in $(0,0)$) solution

$$u(x,y) = \sum_{i=1}^\infty \sum_{j=1}^\infty a_{i,j} x^i y^j, \tag{1.8}$$

then the coefficients $a_{i,j}$ would be given by

$$a_{2s,k} = \frac{(2s+2k)!}{(s)!k!}(-1)^{k+s} \quad \text{and} \quad a_{2s+1,k} = 0 \quad (s, k \in \mathbf{Z}_+).$$

But then the series (1.8) would not converge in any neighborhood of $(0,0)$, since it is divergent for all $(0,y), y \neq 0$.

Remark 1.17.1. The reason for this is the fact that $y = 0$ is a characteristic of the given equation (see more in Chapter 3.).

Example 1.18 *Prove that the equation*

$$\frac{\partial u}{\partial x} + i\frac{\partial u}{\partial y} + 2i(x+iy)\frac{\partial u}{\partial z} = f(z),$$

for a given function f which is not analytic, has no C^1-solution $u = u(x,y,z)$ in any neighborhood of $(0,0,0)$ in $\mathbf{R}^3$.

Solution. We give only the sketch of the proof. Suppose the contrary, i.e., that the given equation has a solution $u \in C^1(\overline{S})$, where

$$S = \{(x,y,z) \,|\, x^2 + y^2 < R^2, |z| < M\}$$

for some $R > 0$ and $M > 0$ and f is a non–analytical function on the interval $(-M, M)$. Introducing the function u_1 by $u_1(\rho, \varphi, z) = u(\rho\cos\varphi, \rho\sin\varphi, z)$, we have that the function v defined by

$$v(r,z) = \sqrt{r}\int_0^{2\pi} u_1(\sqrt{r}, \varphi, z)e^{i\varphi}\,d\varphi$$

is a solution of the equation

$$\frac{\partial v}{\partial r} + i\frac{\partial v}{\partial z} = \pi f(z)$$

on the set $P = \{(r,z)| 0 < r < R^2, |z| < M\}$. Therefore the function w given by

$$w(r,z) = v(r,z) = v(r,z) + i\pi\int_0^z f(t)\,dt$$

is an analytic function on P and continuous on $\overline{P}$ with respect to $r + iz$. By the Schwarz Reflection Principle (see [21]) the function w as a function of complex variable $r+iz$ can be analytically extended on $\{(r,z)| \,|r| < R^2, |z| < M\}$. Therefore the function

$$w(0,z) = i\pi\int_0^z f(t)\,dt$$

is analytic for $|z| < M$. Hence the function f is also analytic for $|z| < M$. A contradiction.

Remark 1.18.1. It is interesting that the given linear equation of the first order (Levy example from 1957) has no solution even in the space of distributions $\mathcal{D}'$ (see Chapter 12). On the other side this equation has a weak solution in the Colombeau space $\mathcal{G}(\mathbf{R}^n)$ of the so called new generalized functions (see [5]).

1.3 Equations of Mathematical Physics

This section is devoted to a brief presentation of some problems in mathematical physics which are modeled by some partial differential equations.

We start with some equations of motion from mechanics.

The equation of *vibrating string* is given by

$$\frac{\partial^2 u}{\partial x^2} - \frac{\rho}{T}\frac{\partial^2 u}{\partial t^2} = -\frac{q(x,t)}{T},$$

where $u = u(x,t)$ is the displacement of the point of the string with the abscissa x at the time t, T is the tension, ρ is the linear density and $q = q(x,t)$ is the external load per unit length.

The equation of *longitudinal oscillations of a rod* of constant cross section is given by

$$\frac{\partial^2 u}{\partial x^2} - \frac{\rho}{E}\frac{\partial^2 u}{\partial t^2} = -\frac{q(x,t)}{T},$$

where $u = u(x,t)$ is the displacement of the cross section of the rod with abscissa x at the time t, E is Young's modulus and ρ is the density.

The equation of a *vibrating membrane* is given by

$$\frac{\partial^2 u}{\partial x^2} + \frac{\partial^2 u}{\partial y^2} - \frac{\rho}{T}\frac{\partial^2 u}{\partial t^2} = -\frac{q(x,y,t)}{T},$$

where $u = u(x,y,t)$ is the displacement of the point (x,y) of the membrane at the time t, T is the tension per unit length of the boundary of the membrane, ρ is the surface density and $q = q(x,y,t)$ is the external load per unit area.

The equation of *the flow of heat in a body* is given by

$$\Delta u = \frac{c\rho}{k}\frac{\partial u}{\partial t} - \frac{Q}{k},$$

where $u = u(x,y,z,t)$ is the temperature at the point (x,y,z) at the time t, k the thermal conductivity of the body, c is the specific heat, ρ is the heat density and Q is the density of heat sources within the body.

The equations of *the electrostatics* are given by

$$\Delta u = -\frac{4\pi\rho}{\varepsilon}, \quad \mathbf{E} = -\nabla u,$$

where u is the potential of the electrostatic field $\mathbf{E}$, ρ is the volume density of charge at the considered point, ε is the dielectric constant of the medium.

The Maxwell equations are given by

$$\mathrm{div}(\varepsilon \mathbf{E}) \quad = \quad 4\pi\rho, \quad \mathrm{div}(\mu \mathbf{H}) = 0,$$

$$\mathrm{rot}\mathbf{E} \quad = \quad -\frac{1}{c}\frac{\partial(\mu\mathbf{H})}{\partial t},$$

$$\mathrm{rot}\mathbf{H} \quad = \quad \frac{1}{c}\frac{(\varepsilon\mathbf{E})}{\partial t} + \frac{4\pi}{c}\mathbf{I},$$

where $\mathbf{E} = \mathbf{E}(x,t)$ is the electric field vector, $\mathbf{H} = \mathbf{H}(x,t)$ is the magnetic field vector, ε is the dielectric constant, μ is the magnetic permeability, c is the velocity of the light, ρ is the charge and

$$\mathrm{div} = \frac{\partial}{\partial x_1} + \frac{\partial}{\partial x_2} + \frac{\partial}{\partial x_3},$$

for the definition of rot see Section 2.2.1. *The Schrödinger equation* in quantum mechanics is given by

$$i\frac{h}{2\pi}\frac{\partial\psi}{\partial t} = -\frac{h^2}{4\pi m_0}\Delta\psi + V\psi,$$

where $\psi = \psi(x,t)$ is the wave function of a particle of mass m_0 in the field of the potential V and h is the Planck constant.

The equations of *hydrodynamics* are given by

$$\frac{\partial\rho}{\partial t} + \mathrm{div}(\rho\mathbf{V}) = f \quad \text{(continuity equation)},$$

$$\frac{\partial\mathbf{V}}{\partial t} + (\mathbf{V}|\nabla\mathbf{V})\mathbf{V} + \frac{1}{\rho}\nabla p = \mathbf{F} \quad \text{(Euler movement equation)},$$

where $\mathbf{V} = \mathbf{V}(x,t)$, $x = (x_1, x_2, x_3)$ is the velocity vector of the fluid movements, $\rho = \rho(x,t)$ is the fluid density, $f = f(x,t)$ the source density and where $p = p(x,t)$ is the pressure and $\mathbf{F} = \mathbf{F}(x,t)$ is the massforce.

Chapter 2

First Order PDEs

2.1   Quasi–linear PDEs

2.1.1   Preliminaries

A *first order PDE* is an equation of the form

$$F\left(x_1, x_2, \ldots, x_n, u, \frac{\partial u}{\partial x_1}, \frac{\partial u}{\partial x_2}, \ldots, \frac{\partial u}{\partial x_n}\right) = 0, \qquad (2.1)$$

where F is a given and $u = u(x_1, x_2, \ldots, x_n)$ is the unknown function of n independent variables x_1, $x_2, \ldots, x_n$. A solution of (2.1) is called *integral surface*.

Let L be a smooth curve in the Euclidean space $\mathbf{R}^{n+1}$ given by

$$x_1 = x_1(\tau), \ x_2 = x_2(\tau), \ldots, \ x_n = x_n(\tau), \ u = u(x_n) \qquad (\tau \in I), \qquad (2.2)$$

where I is a real interval. *Cauchy's problem* for the first order PDE (2.1) is a problem to determine its solution (i.e., its integral surface) u which passes through the curve given by (2.2). Such a solution is then called *Cauchy's integral* for the problem (2.1) and (2.2).

Quasi–linear first order PDE is an equation of the form

$$P_1 \frac{\partial u}{\partial x_1} + P_2 \frac{\partial u}{\partial x_2} + \cdots + P_n \frac{\partial u}{\partial x_n} = R, \qquad (2.3)$$

where $u = u(x_1, x_2, \ldots, x_n)$ is the unknown function, while $P_i = P_i(x_1, x_2, \ldots, x_n, u)$, $i = 1, 2, \ldots, n$, and $R = R(x_1, x_2, \ldots, x_n, u)$ are given continuously differentiable functions on a region in $\mathbf{R}^{n+1}$. If, moreover, for every $i = 1, 2, \ldots, n$ the functions P_i depend only on the independent variables $x_1, x_2, \ldots, x_n$, then (2.3) is a *linear first order PDE*.

We associate to (2.3) the following symmetric system of ordinary differential equations (shortly ODEs):

$$\frac{dx_1}{P_1} = \frac{dx_2}{P_2} = \cdots = \frac{dx_n}{P_n} = \frac{du}{R}. \qquad (2.4)$$

17

If

$$\psi_i = \psi_i(x_1, x_2, \ldots, x_n, u) = C_i \quad (i = 1, 2, \ldots, n), \qquad (2.5)$$

are the functionally independent solutions of (2.4), the so called *first integrals,* where C_i, $i = 1, 2, \ldots, n$, are arbitrary constants, then the general solution of (2.3) is

$$\Psi(\psi_1, \psi_2, \ldots, \psi_n) = 0, \qquad (2.6)$$

where Ψ is an arbitrary function from the class $C^1(\mathbf{R}^n)$. (See Example 2.1 below for a relation between an arbitrary system of first order ODEs and its first integrals.)

The *characteristic curve* of a first order PDE is a curve L with the property that infinitely many integral surfaces pass through it. In the case of quasi–linear equation (2.3), the characteristic curves are given by the first integrals (2.5) of the system of ODEs (2.4). One sees at once that in the geometrically most interesting case $n = 2$, any integral surface is constituted of characteristic curves.

Throughout this chapter, in the case $n = 2$, we shall always denote by x and y the independent variables, and by $z = z(x, y)$ the dependent variable. Also, it is usual to put

$$p = \frac{\partial z}{\partial x} \quad \text{and} \quad q = \frac{\partial z}{\partial y}$$

for the first order partial derivatives of the function z.

2.1.2   Examples and Exercises

In this subsection, C, C_1, $C_2, \ldots$ denote arbitrary real constants.

Example 2.1 *Let Q be a region in $\mathbf{R}$ and suppose the following system of ODEs*

$$\frac{dY}{dx} = f(x, Y) \quad (x \in Q) \qquad (2.7)$$

is given. In (2.7), $Y = (y_1(x), y_2(x), \ldots, y_n(x))$ is the unknown vector–function and $f = (f_1(x, Y), f_2(x, Y), \ldots, f_n(x, Y))$ is the given vector function such that every f_i, $i = 1, 2, \ldots, n$, is a continuous function which also satisfies the Lipschitz condition. Then the first integral of (2.7) is any relation between the variables x and Y, provided it is not identically equal to a constant, but is equal to a constant if $Y = Y(x)$ is a solution of (2.7). Prove

a) *A continuously differentiable function $\psi = \psi(x, y_1, \ldots, y_n) \not\equiv 0$ is a first integral of the system (2.7) iff it holds*

$$\frac{\partial \psi}{\partial x} + \frac{\partial \psi}{\partial y_1} f_1 + \frac{\partial \psi}{\partial y_2} f_2 + \cdots + \frac{\partial \psi}{\partial y_n} f_n = 0 \qquad (2.8)$$

identically on Q, when $(y_1, \ldots, y_n) = (y_1(x), \ldots, y_n(x))$ is a solution of (2.7).

b) *The general solution of (2.7) is determined with n first integrals, say $\psi_1, \psi_2, \ldots, \psi_n$, satisfying the condition*

$$\frac{\partial(\psi_1, \ldots, \psi_n)}{\partial(y_1, \ldots, y_n)} \neq 0,$$

   i.e., if they are functionally independent.

Solution.

a) Suppose $\psi(x, y_1, y_2, \ldots, y_n) = C$ is a first integral of the system (2.7). Then it follows from the chain rule

$$0 = \frac{d}{dx}\psi(x, y_1, \ldots, y_n) = \frac{\partial\psi}{\partial x} + \frac{\partial\psi}{\partial y_1}y_1'(x) + \cdots + \frac{\partial\psi}{\partial y_n}y_n'(x).$$

Since $(y_1, y_2, \ldots, y_n)$ is a solution of (2.7), the relation (2.8) follows. Conversely, if (2.8) holds, then

$$\frac{d}{dx}\psi(x, y_1, \ldots, y_n) = \frac{\partial\psi}{\partial x} + \frac{\partial\psi}{\partial y_1}y_1'(x) + \cdots + \frac{\partial\psi}{\partial y_n}y_n'(x) = 0.$$

Hence if $(y_1, y_2, \ldots, y_n) = (y_1(x), y_2(x), \ldots, y_n(x))$ is a solution of (2.7), then it holds

$$\psi(x, y_1, y_2, \ldots, y_n) = \text{const.}$$

b) The statement immediately follows from the Implicit Function Theorem.

Example 2.2 *Solve the following Cauchy problems for the linear first order PDEs:*

a) $y\dfrac{\partial z}{\partial x} - x\dfrac{\partial z}{\partial y} = 0, \quad z(x, 0) = x^2;$

b) $x\dfrac{\partial z}{\partial x} + y\dfrac{\partial z}{\partial y} = 0, \quad z(x, 1) = |x|.$

Solutions.

a) Let us determine firstly the general solution of the given equation. One of the equations of the associated system of ODEs (2.4) is

$$\frac{dx}{y} = \frac{dy}{-x}, \quad \text{or} \quad -\int x\,dx = \int y\,dy,$$

which gives the first integral

$$x^2 + y^2 = C_1 \quad (C_1 \text{ an arbitrary real constant}). \tag{2.9}$$

Since clearly $z = C_2$ is another first integral of the given PDE, the general solution is

$$z = \varphi(x^2 + y^2),$$

where φ is an arbitrary function from $C^1(\mathbf{R})$. The initial condition gives

$$x^2 = z(x, 0) = \varphi(x^2), \quad \text{hence} \quad \varphi(x) = x.$$

Thus the Cauchy's integral is $z(x, y) = x^2 + y^2$.

b) In this case, one of the equations of the system (2.4) is

$$\frac{dx}{x} = \frac{dy}{y}, \quad \text{or} \quad \ln|x| = \ln|y| + \ln C$$

for an arbitrary constant C. Hence $|x|/|y| = C_1$, C_1 a real constant, is a first integral of the associated symmetric system of ODEs. As in part a), the general solution is $z = \varphi(|x|/|y|)$, where $\varphi \in C^1(\mathbf{R})$ is an arbitrary function and $y \neq 0$. The initial condition implies

$$|x| = z(x, 1) = \varphi(|x|/1), \quad \text{which gives} \quad \varphi(t) = t.$$

Thus the Cauchy's integral is

$$z(x, y) = \frac{|x|}{|y|} \quad (y \neq 0).$$

Example 2.3 *Find the equation of the surface which satisfies the linear first order PDE*

$$4yzp + q + 2y = 0,$$

and passes through the ellipse

$$y^2 + z^2 = 1, \quad x + z = 2. \tag{2.10}$$

Solution. *The corresponding system of ODEs is*

$$\frac{dx}{4yz} = \frac{dy}{1} = \frac{dz}{-2y},$$

whose first integrals are

$$z + y^2 = C_1, \quad x + z^2 = C_2. \tag{2.11}$$

Thus the general solution is

$$\Phi(z + y^2, x + z^2) = 0,$$

where Φ is an arbitrary function of the class $C^1(\mathbf{R}^2)$. Eliminating the variables x, y and z from (2.10) and (2.11), gives $C_1 + C_2 = 3$, and thus the Cauchy integral is

$$x + y^2 + z + z^2 = 3.$$

Exercise 2.4 *Let the first order linear PDE*

$$xp + yq = z - x^2 - y^2$$

be given.

a) *Find the general solution of this equation.*

b) *Find the Cauchy integral through the curve* $x = 1,\ z = -y^2 + 2y - 1$.

Answers.

a) $z = \varphi(x/y)\sqrt{x^2 + y^2} - x^2 - y^2 \quad (\varphi \in C^1(\mathbf{R}))$.

b) $z = -x^2 - (y - 1)^2 + 1$.

Exercise 2.5 *Find the solution of the PDE*

$$4x^3yp + (y^4 + 4x^2y^2 - x^4)q = 4yz(x^2 + y^2),$$

which passes through the parabola $y = 0,\ x^2 = z/2$.

Answer. The general solution of the given PDE is

$$z = \frac{(x^2 + y^2)^2}{x^2}\varphi\left(\frac{x^2 - y^2}{x(x^2 + y^2)}\right) \quad (\varphi \in C^1(\mathbf{R})),$$

and the Cauchy integral is

$$2(x^2 + y^2)^2 = x^2 z.$$

Exercise 2.6 *Solve the Cauchy problem*

$$x(y^2 + z)p - yq(x^2 + z) = z(x^2 - y^2), \qquad x + y = 0, \quad z = 1.$$

Answer. The general solution is

$$\Phi(xyz, x^2 + y^2 - 2z) = 0 \quad \left(\Phi \in C^1(\mathbf{R}^2)\right),$$

and the Cauchy integral is

$$2xyz + x^2 + y^2 + 2 - 2z = 0.$$

Example 2.7 *Find the surfaces whose each tangent plane meets the* x-*axis in a point whose* x-*coordinate is one half of the* x-*coordinate of the point in which the tangent plane was set.*

Solution. The equation of the tangent plane of the surface $z = z(x, y)$ in its point $M(x, y, z)$ is

$$p(X - x) + q(Y - y) = Z - z,$$

where (X, Y, Z) are the coordinates of a point on the tangent plane. The given condition implies $X = x/2$, $Y = 0$ and $Z = 0$, hence the PDE of these surfaces is

$$px + 2qy = 2z.$$

Note that it is a linear PDE of first order. Thus the required surfaces are of the form

$$z = y\varphi(x^2/y),$$

for an arbitrary function $\varphi \in C^1(\mathbf{R})$.

Example 2.8 *Find the surfaces whose tangent planes pass through the point* $(0, 0, 1)$.

Solution. Firstly, a plane

$$Ax + By + Cz = D \tag{2.12}$$

passes through the point $(0, 0, 1)$ iff it holds

$$A \cdot 0 + B \cdot 0 + C \cdot 1 = D, \quad \text{hence} \quad C = D.$$

Secondly, the plane (2.12) is a tangent plane of a surface $z = z(x, y)$ iff the vectors $\mathbf{n}(A, B, C)$ and $(p, q, -1)$ are parallel, where the first order partial derivatives p and q are evaluated in the meeting point of the surface and its tangent plane. Thus we can take $p = -A/C$ and $q = -B/C$, provided that $C \neq 0$. So we obtain the PDE of these surfaces:

$$px + qy = z - 1,$$

whose general solution is

$$\Phi\left(\frac{y}{x}, \frac{z-1}{x}\right) = 0, \quad \text{or} \quad z = 1 + x\varphi\left(\frac{y}{x}\right),$$

where $\Phi \in C^1(\mathbf{R}^2)$ and $\varphi \in C^1(\mathbf{R})$ are arbitrary functions of their variables.

Example 2.9 *Find the general equation of the surfaces S which are perpendicular to each surface from the family*

$$z^2 = kxy \quad (k \in \mathbf{R}). \tag{2.13}$$

Solution. If we denote by $\mathbf{n} = (p, q, -1)$ and $\mathbf{n_1} = (p_1, q_1, -1)$ the normal vectors of the surface S and one from the given family, then by the given condition they should be perpendicular, i.e., $\mathbf{n} \perp \mathbf{n_1}$. The last condition holds if their scalar product is zero, i.e.,

$$pp_1 + qq_1 + 1 = 0. \tag{2.14}$$

Now differentiating in x and y, respectively, equation (2.13), gives

$$p_1 = \frac{ky}{2z} \quad \text{and} \quad q_1 = \frac{kx}{2z}.$$

Hence from (2.14) it follows

$$k(py + qx) = -2z,$$

and replacing the parameter k from (2.13) gives the PDE

$$pyz^2 + qxz^2 + 2zxy = 0. \tag{2.15}$$

The first integrals of the associated system of ODEs (2.4) are

$$x^2 - y^2 = C_1 \quad \text{and} \quad 2x^2 + z^2 = C_2,$$

and the general solution of (2.15) is

$$\Phi(x^2 - y^2, 2x^2 + z^2) = 0 \quad \left(\Phi \in C^1(\mathbf{R}^2)\right).$$

Example 2.10 *Find the integral surface of the PDE*

$$pxz + qyz = -xy,$$

which passes through the curve

$$x = t, \quad y = t^2, \quad z = t^3 \quad (t \in \mathbf{R}). \tag{2.16}$$

Solution. The first integrals of the given PDE are

$$x/y = C_1 \quad \text{and} \quad z^2 + xy = C_2, \tag{2.17}$$

hence its general solution is $\Phi\left(\dfrac{x}{y}, z^2 + xy\right) = 0$. Replacing x, y and z from (2.16) into (2.17) gives

$$C_1 = \frac{1}{t} \quad \text{and} \quad C_2 = t^6 + t^3.$$

Eliminating the parameter t from these equations gives the first integral

$$x^6 z^2 + x^7 y = y^6 + x^3 y^3.$$

Example 2.11 *Let M be a point on a surface, P its orthogonal projection to the xy–plane and N the intersection of the xy–plane and the perpendicular line of the surface at the point M.*

a) *Find those surfaces for which it holds $\angle NOP = 45°$, where $O(0,0,0)$ is the origin.*

b) *From the surfaces obtained in a), find the one which passes through the $x-$axis.*

Solutions.

a) The perpendicular line of a surface $z = z(x, y)$ passing through its point $M(x, y, z)$ has the equation

$$\frac{X - x}{p} = \frac{Y - y}{q} = \frac{Z - z}{-1},$$

where X, Y and Z are the coordinates on the perpendicular line. Hence the coordinates of the point N are $(x + pz, y + qz, 0)$. Denote by α and β, respectively, the angles $\angle NOx_+$ and $\angle POx_+$, where x_+ is the positive direction of the $x-$axis; by assumption, it should hold $\alpha = \beta + 45°$. Since it holds

$$\tan \alpha = \frac{Y}{X} = \frac{y + qz}{x + pz}, \quad \tan \beta = \frac{y}{x},$$

the required PDE is

$$p(zx + zy) + q(zy - zx) = -x^2 - y^2. \tag{2.18}$$

One first integral of (2.18) is

$$\sqrt{x^2 + y^2} \, e^{\arctan(y/x)} = C_1,$$

while the second first integral we get from the ODE

$$-\frac{dz}{x^2 + y^2} = \frac{x \, dx + y \, dy}{z(x^2 + y^2)},$$

and it is $C_2 = x^2 + y^2 + z^2$. Thus the general solution of (2.18) is

$$x^2 + y^2 + z^2 = \Psi \left(\sqrt{x^2 + y^2} \, e^{\arctan(y/x)} \right), \tag{2.19}$$

or, in polar coordinates, $x = \rho \cos \phi$, $y = \rho \sin \phi$,

$$\rho^2 + z^2 = \Psi \left(\rho \, e^\phi \right) \quad \left(\Psi \in C^1(\mathbf{R}) \right).$$

b) Since the equation of the $x-$axis is $y = z = 0$, from (2.19) it follows $x^2 = \Psi(x)$. Hence the required surface has the equation

$$x^2 + y^2 + z^2 = (x^2 + y^2) \cdot e^{2 \arctan(y/x)}.$$

Example 2.12 *Find the equation of the surfaces which have the property that in arbitrary point (x, y, z) their tangent planes determine on the $x-$axis a segment which depends on a prescribed way from y/x. Compile a second order PDE which these surfaces satisfy.*

Solution. Similarly as in Example 2.7, the tangent plane of the surface $z = z(x, y)$ at the point (x, y, z) is given by

$$p(X - x) + q(Y - y) = Z - z,$$

where (X, Y, Z) is a point on the tangent plane. For $X = 0$ and $Y = 0$ by supposition it follows $Z = f(y/x)$, where f is a given function. Thus we obtain

$$z - px - qy = f\left(\frac{y}{x}\right). \tag{2.20}$$

The general solution of (2.20) is $z = x\varphi(x) + f(y/x)$, where φ is an arbitrary function of class $C^1(\mathbf{R})$. If we differentiate (2.20) in x and in y, we get

$$rx^2 + 2sxy + ty^2 = 0,$$

where

$$r = \frac{\partial^2 z}{\partial x^2}, \quad s = \frac{\partial^2 z}{\partial x \partial y} \quad \text{and} \quad \frac{\partial^2 z}{\partial y^2}.$$

Example 2.13 *Let $f \in C^2(\mathbf{R})$ be a given function.*

a) *Form a PDE satisfied by the function $z = z(x, y)$, if the expression*

$$\left(x\frac{\partial z}{\partial x} + y\right) dx + \left(x\frac{\partial z}{\partial y} + x - 2xy + 2yf'(z - x)\right) dy \tag{2.21}$$

is a total differential.

b) *Find the solution of this equation which for $x = 0$ becomes $z = y^2$. For such function z find the function ζ whose total differential is the expression (2.21).*

Solutions.

a) From the condition

$$\frac{\partial}{\partial y}\left(x\frac{\partial z}{\partial x} + y\right) = \frac{\partial}{\partial x}\left(x\frac{\partial z}{\partial y} + x - 2xy + 2yf'(z - x)\right)$$

it follows

$$2yf''(z - x)\frac{\partial z}{\partial x} + \frac{\partial z}{\partial y} = 2y(1 + f''(z - x)).$$

The general solution of the last PDE is

$$\Phi\left(z - x - y^2, f'(z - x) - x\right) = 0 \quad \left(\Phi \in C^1(\mathbf{R}^2)\right).$$

b) The solution is

$$z = x + y^2$$

and the function ζ is given by

$$\zeta(x,y) = \frac{x^2}{2} + xy + f(y^2) + C,$$

where C is an arbitrary constant.

Example 2.14 *Let a family of surfaces be given with the equation*

$$ax^2 + by^2 - \lambda z^2 = c \quad (\lambda \in \mathbf{R}),$$

where a, b and c are given real numbers. Determine all the orthogonal trajectories for the given family of surfaces. Then, in particular, find those trajectories that pass through the curve $x = 1$, $z^2 = y^a$.

Solution. Put

$$f(x,y,z) = ax^2 + by^2 - \lambda z^2 = c.$$

If $z = \varphi(x,y)$ is the sought after trajectory, then it holds

$$p\frac{\partial f}{\partial x} + q\frac{\partial f}{\partial y} - \frac{\partial f}{\partial z} = 0.$$

This gives the PDE

$$axzp + byzq = c - ax^2 - by^2,$$

whose general solution is

$$\Phi\left(\frac{y^a}{x^b}, a(x^2 + y^2 + z^2) - 2c\ln x\right) = 0,$$

where $\Phi \in C^1(\mathbf{R})$ is an arbitrary function of the class $C^1(\mathbf{R}^2)$. The sought after particular solution is

$$a(x^2 + y^2 + z^2) - 2c\ln x = a \cdot \left(1 + \frac{y^2}{x^{2b/a}} + \frac{y^a}{x^b}\right).$$

Example 2.15 *Find the general solution of the PDE*

$$(1 + \sqrt{z - x - y})\,\frac{\partial z}{\partial x} + \frac{\partial z}{\partial y} = 2,$$

and check whether there exits a solution which is not included in the obtained general solution.

Solution. The associated symmetric system of PDEs is

$$\frac{dx}{1 + \sqrt{z - x - y}} = \frac{dy}{1} = \frac{dz}{2}.$$

Then one of the first integrals is obtained as follows:

$$\frac{dy}{1} = \frac{dz}{2}, \quad \text{hence} \quad y = \frac{z}{2} + C_1,$$

where C_1 is an arbitrary constant. Further on we have

$$\frac{dx + dy - dz}{\sqrt{z - x - y}} = \frac{dy}{1}, \quad \text{which gives} \quad \frac{-d(z - x - y)}{\sqrt{z - x - y}} = dy.$$

Thus another first integral is

$$2(z - x - y)^{1/2} = -y + C_2,$$

where C_2 is also an arbitrary constant.

The general solution of the given PDE is

$$\Phi\left(2y - z, y + 2(z - x - y)^{1/2}\right), \tag{2.22}$$

where Φ is an arbitrary function of the class $C^1(\mathbf{R}^2)$. However, note that

$$z = x + y$$

is also a solution of the given PDE, though it is not included in (2.22).

Remark 2.15.1 In general, the surfaces on which a partial derivative of some of the coefficients P or Q in the linear PDE

$$Pp + Qq = R$$

is not bounded, are good candidates for such "special" solutions.

Example 2.16 *Prove that the integrability condition for the ODE*

$$\mu \cdot (P(x, y)\, dx + Q(x, y)\, dy) = 0$$

is a first order linear PDE with the unknown function $\mu = \mu(x, y)$. *Then find some cases when the integrating factor* μ *can be easily found.*

Solution. The integrability condition gives the following PDE:

$$\mu\left(\frac{\partial Q}{\partial x} - \frac{\partial P}{\partial y}\right) = P\frac{\partial \mu}{\partial y} - Q\frac{\partial \mu}{\partial x}.$$

It is well known that this equation has a solution in μ if at least one of the functions P or Q is not identically equal to zero and both P and Q are of the class $C^1(\mathbf{R}^2)$. Let us point out the following two cases:

a) If the expression $X = \dfrac{-1}{Q}\left(\dfrac{\partial Q}{\partial x} - \dfrac{\partial P}{\partial y}\right)$ depends only on x, i.e., $X = X(x)$, then the integration factor also depends only on x and is equal to

$$\mu(x) = \exp\left(\int X(x)\,dx\right).$$

b) Similarly, if the expression $Y = \dfrac{1}{P}\left(\dfrac{\partial Q}{\partial x} - \dfrac{\partial P}{\partial y}\right)$ depends only on y, i.e., $Y = Y(y)$, then

$$\mu(y) = \exp\left(\int Y(y)\,dy\right).$$

Exercise 2.17 *Find the integration factor* $\mu = \mu(x,y)$ *for the PDE*

$$\left(2x^3 y - y^2\right) dx - \left(2x^4 + xy\right) dy = 0.$$

Answer. $\mu(x,y) = x^{-2}y^{-3}.$

Example 2.18 *Find the solutions* $u = u(x,y,z)$ *of the following linear PDEs:*

a) $x\dfrac{\partial u}{\partial x} + (z+u)\dfrac{\partial u}{\partial y} + (y+u)\dfrac{\partial u}{\partial x} = y+z;$

b) $xy\dfrac{\partial u}{\partial x} - \sqrt{1-y^2}\left(y\dfrac{\partial u}{\partial y} - z\dfrac{\partial u}{\partial z}\right) = xy\dfrac{\partial u}{\partial z};$

c) $(u-x)\dfrac{\partial u}{\partial x} + (u-y)\dfrac{\partial u}{\partial y} - z\dfrac{\partial u}{\partial z} = x+y.$

Solutions.

a) We start from the system of ODEs:

$$\frac{dx}{x} = \frac{dy}{z+u} = \frac{dz}{y+u} = \frac{du}{y+z}.$$

Firstly we have

$$\frac{dz-du}{u-z} = \frac{dy-dz}{z-y}, \quad \text{which gives} \quad \frac{z-u}{y-z} = C_1.$$

Secondly we have

$$\frac{dx}{x} = \frac{d(y+z+u)}{2(y+z+u)}, \quad \text{which gives} \quad \frac{x^2}{y+z+u} = C_2,$$

and, finally,

$$\frac{dx}{x} = \frac{dz - du}{u - z}, \quad \text{hence} \quad x(u - z) = C_3.$$

Thus the general solution of the given PDE is

$$\Phi\left(\frac{z - u}{y - z}, \frac{x^2}{y^2 + z + u}, x(u - z)\right) = 0,$$

where Φ is an arbitrary function of the class $C^1(\mathbf{R}^3)$.

b) $u = \varphi\left(x \exp\left(\arcsin y\right), 2yz + x\left(y + \sqrt{1 - y^2}\right)\right) \quad (\varphi \in C^1(\mathbf{R}^2))$.

Hint. The second first integral can be found from the following ODE.

$$\frac{\left(y + \sqrt{1 - y^2}\right) dx}{xy\left(y + \sqrt{1 - y^2}\right)} + \frac{\left(2z + x - \dfrac{xy}{\sqrt{1 - y^2}}\right) dy}{\left(2z + x - \dfrac{xy}{\sqrt{1 - y^2}}\right)\left(-y\sqrt{1 - y^2}\right)} = \frac{2y\, dz}{2y\left(z\sqrt{1 - y^2} - xy\right)}.$$

c) $\Phi\left(\dfrac{x - y}{z}, z(2u + x + y), \dfrac{u - x - y}{z^2}\right), \quad \Phi \in C^1(\mathbf{R}^3)$.

Exercise 2.19 *Solve by* $u = u(x_1, \ldots, x_n)$ *the first order linear PDE*

$$\sum_{i=1}^{n} x_i \frac{\partial u}{\partial x_i} = x_1 x_2 \cdots x_n.$$

Answer. The general solution is

$$\Phi\left(\frac{x_1}{x_2}, \frac{x_1}{x_3}, \ldots, \frac{x_1}{x_n}, x_1 x_2 \cdots x_n - n u\right) \quad \left(\Phi \in C^1(\mathbf{R}^n)\right).$$

Example 2.20 *Solve by* $u = u(x, y)$ *the system of PDEs*

$$\frac{\partial u}{\partial x} = ay^2, \quad \frac{\partial u}{\partial y} = \frac{b}{2y^2} + \frac{2u}{y} - ay^2,$$

where a and b are given constants.

Solution. Let us solve (as an ODE) the first equation by the unknown function u observed as a function of the variable x and treating y as a parameter. This solution is

$$u(x, y) = axy^2 + G(y), \tag{2.23}$$

where G is an arbitrary function from the class $C^1(\mathbf{R})$. Putting the right–hand side of (2.23) in the second given PDE, we get the ODE

$$G'(y) - \frac{2}{y}G(y) = \frac{b}{2y^2} - ay^2.$$

This linear ODE by the unknown function $G = G(y)$ has the solution

$$G(y) = -\frac{b}{6}y^{-1} - ay^3 + Cy^2,$$

where C is an arbitrary constant. Thus the solution of the given system is

$$u(x,y) = axy^2 - \frac{b}{6y} - ay^3 + Cy^2.$$

Example 2.21 *Solve the following system of PDEs by the functions* $u = u(x,y)$ *and* $v = v(x,y)$:

$$y\frac{\partial u}{\partial x} + x\frac{\partial v}{\partial y} = 4xy, \qquad x\frac{\partial u}{\partial y} + y\frac{\partial v}{\partial x} = 8xy,$$

and find the solution that satisfies the initial condition

$$(u,v)\big|_{y=2x} = (9x^2, 6x^2).$$

Solution. Adding the first PDE to the second one gives us

$$y\frac{\partial(u+v)}{\partial x} + x\frac{\partial(u+v)}{\partial y} = 12xy, \qquad (2.24)$$

while subtracting the second PDE from the first one gives

$$y\frac{\partial(u-v)}{\partial x} - x\frac{\partial(u-v)}{\partial y} = -4xy. \qquad (2.25)$$

From (2.24) it follows

$$u + v = 6x^2 + \Phi(x^2 - y^2),$$

while from (2.25) it follows

$$u - v = -2x^2 + \Psi(x^2 + y^2),$$

where Φ and Ψ are arbitrary functions of the class $C^1(\mathbf{R})$. Hence

$$u(x,y) = 2x^2 + \frac{1}{2}\Phi(x^2 - y^2) + \frac{1}{2}\Psi(x^2 + y^2)$$

and

$$v(x,y) = 4x^2 + \frac{1}{2}\Phi(x^2 - y^2) - \frac{1}{2}\Psi(x^2 + y^2).$$

The initial conditions give

$$u(x,2x) = 9x^2 = 2x^2 + \frac{1}{2}\left(\Phi(-3x^2) + \Psi(5x^2)\right)$$

and

$$v(x, 2x) = 6x^2 = 4x^2 + \frac{1}{2}\left(\Phi(-3x^2) - \Psi(5x^2)\right).$$

Solving by Φ and Ψ the last system we obtain

$$\Phi(x) = -3x \quad \text{and} \quad \Psi(x) = x,$$

hence, finally,

$$u(x, y) = x^2 + 2y^2 \quad \text{and} \quad v(x, y) = 2x^2 + y^2$$

Exercise 2.22 *Find the characteristic curves of the following PDEs:*

a) $z = px + qy$;

b) $(mz - ny)p + (nx - lz)q = ly - mz.$

Answers.

a) Since the first integrals of the given equation are $z/x = C_1$ and $y/x = C_2$, the two families of characteristic curves are obtained for different constants C_1 and C_2.

b) The families of concentric spheres $x^2 + y^2 + z^2 = C_1$ and of parallel planes $lx + my + nz = C_2$ are the two families of characteristic curves. The intersections of these spheres and planes are the circles that constitute each integral surface.

Example 2.23 *Show that the Cauchy problem*

$$2\cos x \cdot p + 2y \sin x \cdot q = z \sin x, \qquad z = 0, \ y = \alpha \cos x,$$

for a given constant α, does not have a unique solution.

Solution. One first integral is $z^2/y = C_1$, while the other is $y/\cos x = C_2$, where C_1 and C_2 are arbitrary constants; these equations define the characteristic curves of the given PDE. In view of the given condition, we see that the given Cauchy problem does not have a unique solution. In fact, for any function $\varphi \in C^1(\mathbf{R})$ such that $\varphi(\alpha) = 0$ the function $z = z(x, y)$ given by

$$z^2 = y\varphi\left(\frac{y}{\cos x}\right)$$

is a solution of the given problem.

2.2   Pfaff's Equations

2.2.1   Preliminaries

Pfaff's equation is an equation of the form

$$P\,dx + Q\,dy + R\,dz = 0, \tag{2.26}$$

where $z = z(x, y)$ is the unknown function, while P, Q and R are given continuously differentiable functions of x, y and z in a region in $\mathbf{R}^3$. Equation (2.26) can be easily solved in the following two cases.

1. The following *integrability conditions* hold:

$$\frac{\partial P}{\partial y} = \frac{\partial Q}{\partial x}, \quad \frac{\partial P}{\partial z} = \frac{\partial R}{\partial x} \quad \text{and} \quad \frac{\partial Q}{\partial z} = \frac{\partial R}{\partial y}. \tag{2.27}$$

   Then there is a function $u = u(x, y, z)$ such that

$$du = P\,dx + Q\,dy + R\,dz\,,$$

   i.e., the left–hand side of (2.26) is a total differential of a function $u = u(x, y, z)$. Hence for an arbitrary constant C the function

$$u(x, y, z) = C$$

   is an implicit solution of (2.26).

2. The following *complete integrability condition* holds:

$$P\left(\frac{\partial Q}{\partial z} - \frac{\partial R}{\partial y}\right) + Q\left(\frac{\partial R}{\partial x} - \frac{\partial P}{\partial z}\right) + R\left(\frac{\partial P}{\partial y} - \frac{\partial Q}{\partial x}\right) = 0. \tag{2.28}$$

   Then there are functions $u = u(x, y, z)$ and $v = v(x, y, z)$ such that

$$v\,du = P\,dx + Q\,dy + R\,dz$$

(existence of an integration factor!).

It is convenient to put $\mathbf{A} = (P, Q, R)$. Then (2.28) can be written in the form

$$\text{rot}\,\mathbf{A} \cdot \mathbf{A} = 0 \tag{2.29}$$

where

$$\text{rot}\,\mathbf{A} = \begin{vmatrix} \mathbf{i} & \mathbf{j} & \mathbf{k} \\ \dfrac{\partial}{\partial x} & \dfrac{\partial}{\partial y} & \dfrac{\partial}{\partial z} \\ P & Q & R \end{vmatrix}.$$

If neither of the upper two cases hold, then there exist functions u, v and w such that

$$P\,dx + Q\,dy + R\,dz = du + v\,dw, \tag{2.30}$$

where $v\,dw = P_1 dx + Q_1 dy + R_1 dz = 0$, and for the vector $\mathbf{A}_1 = (P_1, Q_1, R_1)$ the complete integrability condition (2.28) holds.

2.2.2 Examples and Exercises

Example 2.24 *Solve the following Pfaff's equations.*

a) $(6x + yz)\,dx + (xz - 2y)\,dy + (xy + 2z)\,dz = 0$;

b) $yz\,dx + (xz - yz^3)\,dy = 2xy\,dz$.

Solutions.

a) Since the equalities in (2.27) hold (integrability conditions), it follows that the left–hand side of the given PDE is a total differential of some function $u = u(x, y, z)$. It can be found as follows:

$$u(x, y, z) = \int_{x_0}^{x}(6x + yz)\,dx + \int_{y_0}^{y}(x_0 z - 2y)\,dy + \int_{z_0}^{z}(x_0 y_0 + 2z)\,dz,$$

or

$$u(x, y, z) = 3x^2 + xyz - y^2 + z^2 + C,$$

where C is an arbitrary constant.

b) Let us observe first that the complete integrability condition (2.28) (or (2.29)) is satisfied, where $\mathbf{A} = (P, Q, R) = (yz, xz - yz^3, -2xy)$ and

$$\text{rot}\,\mathbf{A} = \begin{vmatrix} \mathbf{i} & \mathbf{j} & \mathbf{k} \\[4pt] \dfrac{\partial}{\partial x} & \dfrac{\partial}{\partial y} & \dfrac{\partial}{\partial z} \\[8pt] yz & xz - yz^3 & -2xy \end{vmatrix}.$$

The given PDE can be solved in the following way. Let us assume for a moment that one of the variables, say y, is a constant; then we get

$$yz\,dx - 2xy\,dz = 0, \quad \text{hence} \quad \frac{x}{z^2} = \varphi(y).$$

Differentiation of the last equation gives

$$\frac{1}{z^2}\,dx - \frac{2x}{z^3}\,dz = \varphi'(y)\,dy. \tag{2.31}$$

Comparing the given PDE and (2.31) we see that the corresponding functions (with the same differentials) must be proportional, i.e.,

$$\frac{\frac{1}{z^2}}{yz} = \frac{-\frac{2x}{z^3}}{-2xy} = \frac{-\varphi'(y)}{xz - yz^3}.$$

The complete integrability condition (2.28) implies that now we should get an equation containing only the variable y and the functions $\varphi(y)$ and $\varphi'(y)$. In fact, we have

$$\frac{\varphi'(y)}{yz^3 - xz} = \frac{1}{yz^3}, \quad \text{hence} \quad \varphi'(y) = 1 - \frac{x}{z^2} \cdot \frac{1}{y} = 1 - \frac{\varphi(y)}{y},$$

and thus

$$\varphi(y) = \frac{C}{y} + \frac{y}{2} \quad (C \in \mathbf{R}).$$

So we obtain $\dfrac{x}{z^2} - \dfrac{C}{y} + \dfrac{y}{2} = 0,$ or

$$2xy - 2Cz^2 + y^2z^2 = 0,$$

where C is an arbitrary constant.

Exercise 2.25 *Prove that for the following PDE*

$$y\,dx + z\,dy + x\,dz = 0$$

the complete integrability condition does not hold, and find the functions u, v and w from (2.30).

Answer. It holds

$$y\,dx + z\,dy + x\,dz = \frac{1}{2}d(xy + yz + zx) + \frac{1}{2}(x - y)^2 d\left(\frac{z - x}{x - y}\right),$$

and thus

$$u(x,y,z) = \frac{1}{2}(xy + yz + zx), \quad v(x,y,z) = \frac{1}{2}(x - y)^2 \quad \text{and} \quad w(x,y,z) = 1\frac{z - x}{x - y}.$$

Exercise 2.26 *Solve the following Pfaff's equations.*

a) $yz\,dx + 2xz\,dy + xy\,dz = 0$;

b) $(e^x y + e^z)\,dx + (e^y z + e^x)\,dy + (e^y - e^x y - e^y z)\,dz = 0$;

c) $(2x^2 + 2xy + 2xz^2 + 1)\,dx + dy + 2z\,dz = 0$;

d) $2x(y+z)\,dx + (2yz - x^2 + y^2 - z^2)\,dy + (2yz - x^2 - y^2 + z^2)\,dz = 0;$

e) $(2x^3y + 1)\,dx + x^4dy + x^2\tan z\,dz = 0;$

f) $2(y+z)dx - (x+z)\,dy + (2y - x + z)\,dz = 0;$

Answers. *In the solutions given below, C is an arbitrary constant.*

a) $xy^2z = C.$

    Hint.   *Put $y = $ const and solve the obtained ODE; then it follows $\varphi(y) = C/y^2$, for some $\varphi \in C^1(\mathbf{R})$. Then continue as in Example 2.24 b).*

b) $e^xy + e^yz + e^zx = Ce^z.$

c) $e^{x^2}(x + y + z^2) = C.$

d) $x^2 + y^2 + z^2 = C(y + z).$

e) $x^2y - \dfrac{1}{x} - \ln\cos z = C.$

    Hint. $\mu(x) = \dfrac{1}{x^2}.$

f) $y + z = C(x + z)^2.$

    Hint.   *Put $x = x_1z$ and $y = y_1z$. Then it holds*

$$\frac{2\,dx_1}{x_1 + 1} - \frac{2\,dy_1}{y_1 + 1} + \frac{dz}{z} = 0.$$

2.3   Nonlinear First Order PDEs

2.3.1   Preliminaries

A *nonlinear first order PDE* is an equation of the form

$$F(x, y, z, p, q) = 0, \tag{2.32}$$

where $z = z(x, y)$ is the unknown function of two independent variables x and y, while $p = \dfrac{\partial u}{\partial z}$ and $q = \dfrac{\partial z}{\partial y}$.

Equation (2.32) has exactly three types of solutions, namely the complete, singular and general solution.

The *complete solution* is a solution of (2.32) which depends on two arbitrary and independent constants, say a and b, and thus can be written as

$$V(x, y, z, a, b) = 0. \tag{2.33}$$

The *singular solution* is a solution of (2.32) which depends neither from arbitrary constants nor from arbitrary functions. When the singular solution exists, it is obtained from the complete solution (2.33) by elimination of a and b from the following system of three equations:

$$V(x,y,z,a,b) = 0, \quad \frac{\partial V}{\partial a}(x,y,z,a,b) = 0 \quad \text{and} \quad \frac{\partial g}{\partial b}(x,y,a,b) = 0.$$

The *general solution* is a solution of (2.32) which depends from an arbitrary function. If the complete solution of (2.32) is given by (2.33), then its general solution is given by the system of two equations, namely

$$V(x,y,z,a,b(a)) = 0 \quad \text{and} \quad \frac{\partial V}{\partial a}(x,y,z,a,b(a)) + \frac{\partial V}{\partial b}(x,y,z,a,b(a)) \cdot b'(a) = 0,$$

where $b = b(a)$ is an arbitrary function.

As we shall see in Example 2.27, both the singular (if any) and the general solution of a nonlinear first order PDE can be obtained from the complete solution. Thus our main task in solving (2.32) is to find a complete solution. It can be obtained from the Lagrange–Charpite method exposed below.

The Lagrange–Charpite Method

Let a nonlinear first order PDE of the form (2.32) be given. The main idea of the *Lagrange–Charpite method* for obtaining its complete solution, is to find a function Φ, functionally independent of F, of the form

$$\Phi(x,y,z,p,q) = a, \tag{2.34}$$

where a is an arbitrary constant. The function Φ should have the property that the system of PDEs

$$F = 0 \quad \text{and} \quad \Phi = a \tag{2.35}$$

can be solved in p and q :

$$p = \varphi(x,y,z,a) \quad \text{and} \quad q = \psi(x,y,z,a). \tag{2.36}$$

Clearly, the functions φ and ψ from (2.36) should satisfy the equality

$$\frac{\partial \varphi}{\partial y} = \frac{\partial \psi}{\partial x}. \tag{2.37}$$

Then Pfaff's equation

$$dz = p\,dx + q\,dy = \varphi(x,y,z,a)\,dx + \psi(x,y,z,a)\,dy$$

satisfies the complete integrability condition (2.28), and its general solution depends on (another) arbitrary constant b. Thus it can be written in the form

$$V(x, y, z, a, b) = 0.$$

This equation is then the complete solution of the given nonlinear first order PDE (2.32).

We still have to find the function Φ from (2.34). To that end, differentiating both equations in (2.35) and putting p and q from (2.36) in the obtained equations, we obtain a linear first order PDE with the unknown function Φ. Its associated symmetric system of ODEs (2.4) is

$$\frac{dx}{\dfrac{\partial F}{\partial p}} = \frac{dy}{\dfrac{\partial F}{\partial q}} = \frac{dz}{p\dfrac{\partial F}{\partial p} + q\dfrac{\partial F}{\partial q}} = \frac{-dp}{\dfrac{\partial F}{\partial x} + p\dfrac{\partial F}{\partial z}} = \frac{-dq}{\dfrac{\partial F}{\partial y} + q\dfrac{\partial F}{\partial z}}. \tag{2.38}$$

One can show that equality (2.36) holds iff the functions F and Φ are in *involution*. By definition, this means that it holds

$$[F, \Phi] = 0.$$

where

$$[F, \Phi] = \begin{vmatrix} \dfrac{\partial F}{\partial p} & \dfrac{\partial F}{\partial x} + p\dfrac{\partial F}{\partial z} \\[2ex] \dfrac{\partial \Phi}{\partial p} & \dfrac{\partial \Phi}{\partial x} + p\dfrac{\partial \Phi}{\partial z} \end{vmatrix} + \begin{vmatrix} \dfrac{\partial F}{\partial q} & \dfrac{\partial F}{\partial y} + q\dfrac{\partial F}{\partial z} \\[2ex] \dfrac{\partial \Phi}{\partial q} & \dfrac{\partial \Phi}{\partial y} + q\dfrac{\partial \Phi}{\partial z} \end{vmatrix}. \tag{2.39}$$

In particular, if the function F from (2.32) does not depend on z, i.e., it has the form

$$F(x, y, p, q) = 0$$

then (2.39) can be written as

$$(F, \Phi) = \begin{vmatrix} \dfrac{\partial F}{\partial p} & \dfrac{\partial F}{\partial x} \\[2ex] \dfrac{\partial \Phi}{\partial p} & \dfrac{\partial \Phi}{\partial x} \end{vmatrix} + \begin{vmatrix} \dfrac{\partial F}{\partial q} & \dfrac{\partial F}{\partial y} \\[2ex] \dfrac{\partial \Phi}{\partial q} & \dfrac{\partial \Phi}{\partial y} \end{vmatrix}. \tag{2.40}$$

The expression (F, Φ) given by (2.40) is called *Poisson bracket*.

2.3.2 Examples and Exercises

Example 2.27 *Assume equation (2.32) can be written in the form*

$$\frac{\partial z}{\partial x} = f\left(x, y, z, \frac{\partial z}{\partial y}\right) \tag{2.41}$$

(for example, in a neighbourhood of a point where it holds $\frac{\partial F}{\partial p} \neq 0$.) Then using the method of variation of constants, prove that all the solutions of (2.32) can be obtained from the complete solution.

Hint. Firstly take $a = a(x, y)$ and $b = b(x, y)$ in the complete solution

$$z = g(x, y, z, a, b).$$

Then, by the definition of the complete solution, replacing z, a and b with the functions $g(x, y, z, a, b)$, $\frac{\partial g}{\partial x}(x, y, z, a, b)$ and $\frac{\partial g}{\partial y}(x, y, z, a, b)$, respectively, one gets an identity in (2.41) for arbitrary functions a and b.

Conversely, a function $z = g(x, y, z, a, b)$ is a solution of (2.41) only if

$$\frac{\partial z}{\partial x} = \frac{\partial g}{\partial x} \quad \text{and} \quad \frac{\partial z}{\partial y} = \frac{\partial g}{\partial y}.$$

Show that this will hold only if

$$\frac{\partial g}{\partial a}(x, y, a, b)\frac{\partial a}{\partial x} + \frac{\partial g}{\partial b}(x, y, a, b)\frac{\partial b}{\partial x} = 0,$$

$$\frac{\partial g}{\partial a}(x, y, a, b)\frac{\partial a}{\partial y} + \frac{\partial g}{\partial b}(x, y, a, b)\frac{\partial b}{\partial y} = 0. \tag{2.42}$$

Putting $\frac{\partial g}{\partial a}(x, y, a, b) = 0$ and $\frac{\partial g}{\partial b}(x, y, a, b) = 0$, and, if possible, eliminating a and b from equations (2.42), we obtain the *singular solution*. Note that it depends neither from arbitrary constants nor from arbitrary functions.

Next, assume it holds $\frac{\partial(a, b)}{\partial(x, y)} = 0$, where

$$\frac{\partial(a, b)}{\partial(x, y)} = \begin{vmatrix} \frac{\partial a}{\partial x} & \frac{\partial a}{\partial y} \\ \frac{\partial b}{\partial x} & \frac{\partial b}{\partial y} \end{vmatrix},$$

and let a and b be some nonconstant functions. Then the functions a and b are *not* functionally independent, i.e., there exists a differentiable function w such that $b = w(a)$. This leads to the *general solution* of (2.41).

Finally, check that *any* solution of (2.41) reduces to one of the following three ones: singular, complete or general.

Example 2.28 *Determine a PDE whose complete solution is a family of spheres with the same radius R and with centers in the $xy-$plane, i.e.,*

$$(x - a)^2 + (y - b)^2 + z^2 = R^2, \tag{2.43}$$

where a and b are arbitrary constants. Find then the general and the singular solution of the obtained PDE.

Solution. Starting from the complete solution (2.43), we can reconstruct the PDE as follows. Differentiating equation (2.43) in x and y, respectively, we get

$$2(x - a) + 2zp = 0 \quad \text{and} \quad 2(y - b) + 2zq = 0,$$

which after the elimination of the constants a and b gives the PDE

$$z^2(p^2 + q^2 + 1) = R^2. \tag{2.44}$$

The general solution of (2.44) is given by the system of equations

$$(x - a)^2 + (y - b(a))^2 + z^2 = R^2 \quad \text{and} \quad (x - a) + b'(a)(y - b(a)) = 0, \tag{2.45}$$

where $b = b(a)$ is an arbitrary function of class $C^1(\mathbf{R})$. (The last equation was obtained by putting $b = b(a)$ in (2.43) and differentiation in a.) The geometric interpretation of (2.45) is that it represents an envelope of the one–parameter family of spheres given by (2.43) for $b = b(a)$.

The singular solution is given by the system

$$(x - a)^2 + (y - b)^2 + z^2 = R^2, \quad 2(x - a) = 0, \quad 2(y - b) = 0$$

which, after the elimination of the constants a and b, gives simply $z^2 = R^2$. The geometric interpretation of the singular solution is that it represents a union of two parallel planes, which are the envelopes of a family of surfaces (spheres) that make the complete solution.

Example 2.29 *Find the complete and singular solution(s), if any, of the following nonlinear PDEs of first order:*

a) $1 + p^2 = zq$;

b) $z = px + qy + p^2 + pq + q^2$;

c) $z^2(p^2 + q^2 + 1) = 1$.

Solutions.

a) Putting

$$F(x, y, z, p, q) = 1 + p^2 - zq,$$

we obtain

$$\frac{\partial F}{\partial x}\frac{\partial F}{\partial x} + p\frac{\partial F}{\partial z} = -pq \quad \text{and} \quad \frac{\partial F}{\partial y} + q\frac{\partial F}{\partial z} = -q^2.$$

Then from system 2.38 we get

$$\frac{-dp}{-pq} = \frac{-dq}{-q^2}$$

and thus we can take $q = ap$ for an arbitrary constant a. Solving in p the quadratic equation $1 + p^2 - z \cdot (ap) = 0$, we obtain

$$p_{1,2} = \frac{-az \pm \sqrt{a^2 z^2 - 4}}{2}. \tag{2.46}$$

Let us take the sign " $+$ " in (2.46); then it holds

$$\int \frac{2\,dz}{-az + \sqrt{a^2 z^2 - 4}} = x + ay + b,$$

where b is another arbitrary constant, and thus the complete solution is

$$x + ay + b + \frac{z}{4}\sqrt{a^2 z^2 - 4} - \frac{1}{a}\ln\left|az + \sqrt{a^2 z^2 - 4}\right| + \frac{az^2}{4} = 0.$$

There is no singular solution.

Remark 2.29.1 The reader should check that *any* nonlinear first order PDE of the form $F(p, q) = 0$ has a first integral $q = ap$, for an arbitrary constant a.

Remark 2.29.2 If we took the sign " $-$ " in (2.46), we would get another complete solution. This shows that the complete solution of a nonlinear first order PDE is not unique; in fact, it even might have infinitely many complete solutions.

b) One easily gets the complete solution

$$z = ax + by + a^2 + ab + b^2 \quad (a, b \text{ arbitrary constants}).$$

Taking the partial derivatives in a and b of the complete solution, respectively, we obtain the system

$$x + 2a + b = 0, \quad y + a + 2b = 0.$$

Solving it in a and b, we get the singular solution

$$z = \frac{1}{3}(xy - x^2 - y^2).$$

Remark 2.29.3 The equation of the form $z = px + qy + f(p,q)$, for some given function f is called "generalized Clairaut's equation" .

c) The first integral is $p = \dfrac{\sqrt{1-z^2}}{z\sqrt{1+a^2}}$, and thus the complete integral is

$$-\sqrt{1-z^2} = \frac{1}{\sqrt{1+a^2}}(x+ay) + b.$$

The singular integrals are $z = 1$ and $z = -1$.

Example 2.30

a) *Let M be a point on a surface, and let the point N be the intersection of the xy−plane and surface's perpendicular line through M. Find the surfaces which have the property that for every point M on the surface the segment MN is equal to a given number $a > 0$. .*

b) *Find the Cauchy integral through the circle $z = b$, $x^2 + y^2 = R^2$, for given b, $0 < b < a$, and $R > 0$.*

Solutions.

a) Let $z = z(x,y)$ be the sought after surface and $M(x,y,z)$ its arbitrary point. Then the equation of the surface's perpendicular line through M is

$$\frac{X-x}{p} = \frac{Y-y}{q} = \frac{Z-z}{-1}. \tag{2.47}$$

Here X, Y and Z are the coordinates on the perpendicular line. If $N(X_0, Y_0, 0)$ is its intersection with the xy−plane, then by supposition it holds

$$\overline{MN}^2 = (x - X_0)^2 + (y - Y_0)^2 + (z - 0)^2 = a^2.$$

From (2.47) we get $X_0 = pz + x$ and $Y_0 = qz + y$, which gives the PDE

$$a^2 = z^2(p^2 + q^2 + 1).$$

As in Example 2.29 a), we obtain that a first integral is $q = sp$, for an arbitrary constant s. Thus the complete integral is

$$s \mp \sqrt{a^2 - z^2} = x\cos t + y\sin t,$$

where t is another arbitrary constant.

Clearly, the singular integrals are $z = a$ and $z = -a$ (give a geometric explanation).

Finally, putting $s = f(t)$, where f is an arbitrary function of the class $C^1(\mathbf{R})$, we obtain the general solution given with the system of equations

$$f(t) \mp \sqrt{a^2 - z^2} = x\cos t + y\sin t, \quad f'(t) = -x\sin t + y\cos t.$$

b) By supposition we have

$$f(t) = \sqrt{a^2 - b^2} + R\sin(s + t),$$

hence the Cauchy integral is

$$\left(R \pm \sqrt{a^2 - b^2} \mp \sqrt{a^2 - z^2}\right)^2 = x^2 + y^2.$$

Example 2.31 *Find the surfaces which satisfy the PDE*

$$z = px + qy + pq$$

and pass through the parabola $x = 0$, $z = y^2$.

Solution. One easily finds the complete solution:

$$z = xa + yb + ab \quad (a, b \text{ arbitrary constants}).$$

If we put $y = t$, then for $x = 0$ we have $z = t^2$, hence

$$t^2 = bt + ab. \tag{2.48}$$

Differentiating this equation in t we get $b = 2t$, hence from (2.48) we obtain $a = -1/2$. Thus

$$z = -\frac{t}{2}x + 2ty - t^2,$$

which after another differentiation gives $t = y - \frac{x}{4}$. Thus the solution is

$$z = \left(\frac{x}{4} - y\right)^2.$$

Example 2.32 *Find the complete, general and singular solution of the PDE*

$$\left(\frac{p}{x}\right)^2 + \left(\frac{q}{y}\right)^2 = \frac{2}{z^2}.$$

Moreover, find the surface which passes through the line $y = 0$, $z = x - 1$.

Solution. Firstly, we shall introduce a new unknown function $u = u(x, y)$ by $u = z^2$. Then we have

$$\frac{1}{x^2}\left(\frac{\partial u}{\partial x}\right)^2 + \frac{1}{y^2}\left(\frac{\partial u}{\partial y}\right)^2 = 8,$$

and the complete solution is

$$u = z^2 = ax^2 + y^2 + \sqrt{2 - a^2} + b.$$

There is no singular solution.
The general solution is

$$z^2 = ax^2 + y^2\sqrt{2 - a^2} + f(a), \quad x^2 = \frac{ay^2}{\sqrt{2 - a^2}} + f'(a).$$

where f is an arbitrary function of the class $C^1(\mathbf{R})$.
   Let us find the sought after surface. By assumption we have

$$(x - 1)^2 = ax^2 + f(a),$$

or

$$2(x - 1) = 2ax, \quad \text{or} \quad x = \frac{1}{1 - a}.$$

In view of the equation $f'(a) + x^2 = 0$, we obtain

$$f(a) = -\frac{1}{a - 1}.$$

Finally, the parametric equation of the Cauchy integral is

$$z^2 = a^2 + y^2\sqrt{2 - a^2} - \frac{1}{a - 1}, \quad x^2 = \frac{ay^2}{2 - a^2} - \frac{1}{(a - 1)^2}.$$

Exercise 2.33 *Using an appropriate change of variable(s), solve the following PDEs.*

a) $x^2 p^2 + y^2 q^2 = z$;   **b)** $4xyz = pq + 2px^2 y + 2qxy^2.$

Answers.

a) The complete solution is

$$2\sqrt{z} = a \ln x + \sqrt{1 - a^2} \ln y + b,$$

while the singular one is $z + x^2 + y^2 = 0$.
   Hint. Put $X = \ln x$, $Y = \ln y$ and $Z = 2\sqrt{z}$. Then the given PDE transforms
to $P^2 + Q^2 = 1$, where $P = \dfrac{\partial Z}{\partial X}$ and $Q = \dfrac{\partial Z}{\partial Y}$.

b) $z = ax^2 + by^2 + ab$ (complete solution), $z = -x^2 y^2$ (singular solution).
   Hint. Put $x = \sqrt{X}$, $y = \sqrt{Y}$.

Exercise 2.34 *Let the PDE $pq = z^m xy$ be given, $m \in \mathbf{Z}$.*

a) *Find the complete, singular and the general solution of this PDE.*

b) *For $m = 0$ find the Cauchy integral through the curve $x = 1$, $z = \sqrt{1 + y^2}$.*

Answers.

a) If $m = 2$, then put $Z = \ln z$, and show that the transformed equation becomes $PQ = xy$ (P, Q are as in Exercise 2.33 a). The complete solution is

$$2 \ln z = ax^2 + \frac{y^2}{2a} + b.$$

If $m \neq 2$, then put $Z = \frac{2}{2 - m} z^{1-m/2}$, which gives again the PDE $PQ = xy$. Hence

$$\frac{2}{2 - m} z^{1-m/2} = \frac{ax^2}{2} + \frac{y^2}{2a} + b.$$

b) In a) we saw that for $m = 0$ the complete solution is

$$z = \frac{ax^2}{2} + \frac{y^2}{2a} + b.$$

The Cauchy integral is

$$z = x\sqrt{1 + y^2}.$$

Example 2.35 *Find the complete solution of the PDE*

$$yzp^2 = q.$$

Solution. From the system (2.38) we have

$$\frac{dz}{2p^2yz - q} = -\frac{dp}{yp^3}, \quad \text{hence} \quad \frac{dz}{p^2yz} = -\frac{dp}{yp^3}.$$

The solution is $a = zp$, hence $p = a/z$ and $q = yz \cdot (a/z)^2$. Thus we obtain Pfaff's equation

$$dz = \frac{a}{z}dx + \frac{ya^2}{z}dy,$$

and the complete solution

$$z^2 = 2ax + a^2y + b.$$

Example 2.36

a) *Find and solve the PDE of the form $z = F(x, y, p, q)$, satisfied by the conic surfaces with a peak in a fixed point $V(a, b, c)$.*

b) *Assume now $a = b = c = 0$. Then find a function $\Phi = \Phi(x, y, p, q)$, such that it is in involution with F, see (2.39).*

Solution.

a) The given condition can be written as a scalar product

$$(x - a, y - b, z - c) \cdot (p, q, -1) = 0,$$

or

$$p(x - a) + q(y - b) = z - c. \tag{2.49}$$

The solution of this linear first order PDE is

$$G\left(\frac{z - c}{x - a}, \frac{y - b}{x - a}\right) = 0,$$

where G is an arbitrary function of the class $C^1(\mathbf{R}^2)$.

b) For $a = b = c = 0$ we get from (2.49)

$$F(x, y, p, q) = px + qy.$$

We are looking for a function Φ which is in involution with F, i.e., the Poisson bracket (F, Φ), see (2.40), is equal to zero:

$$(F, \Phi) = \begin{vmatrix} x & p \\ \dfrac{\partial \Phi}{\partial p} & \dfrac{\partial \Phi}{\partial x} \end{vmatrix} + \begin{vmatrix} y & q \\ \dfrac{\partial \Phi}{\partial q} & \dfrac{\partial \Phi}{\partial y} \end{vmatrix}$$

$$= x\frac{\partial \Phi}{\partial x} - p\frac{\partial \Phi}{\partial p} + y\frac{\partial \Phi}{\partial y} - q\frac{\partial \Phi}{\partial q}$$

$$= 0.$$

Using this equality, we get a linear first order PDE in Φ, whose solution is easily found to be

$$\Phi(x, y, p, q) = \Psi(xp, yq, x/y) = 0 \quad \left(\Psi \in C^1(\mathbf{R}^3)\right).$$

Exercise 2.37 *Let the following first order PDE be given:*

$$z^2(2 + p^2 + q^2) = x^2 + y^2 + 2z(xp + yq).$$

a) *If* $Z = \dfrac{z^2}{2}$, $P = \dfrac{\partial Z}{\partial x}$ *and* $Q = \dfrac{\partial Z}{\partial y}$, *then prove that the transformed equation (in Z) has the following two first integrals in involution:*

$$F = y + Q + a \quad and \quad \Phi = x + P + b, \tag{2.50}$$

where a and b are arbitrary constants.

b) *Using a), find the complete integrals of the transformed and the given equation.*

c) *Find the general and, if any, the singular solution of the given equation.*

Answers.

a) Using the transformation $Z = z^2/2$, check that one obtains

$$4Z + P^2 + Q^2 - x^2 - y^2 - 2(xP + yQ) = 0,$$

and hence its first integrals from (2.50) are in involution.

b) The complete solution is

$$V(x, y, z, a, b) = 2(x^2 + y^2 + z^2) - 4(ax + by) + a^2 + b^2 = 0.$$

c) The general solution is given by the system

$$V(x, y, z, a, w(a)) = 0, \quad \frac{\partial V}{\partial a} + \frac{\partial V}{\partial b} b'(a) = 0,$$

where $b \in C^1(\mathbf{R})$ is an arbitrary function.

From the system

$$V = 0, \quad \frac{\partial V}{\partial a} = 0, \quad \frac{\partial V}{\partial b} = 0,$$

we obtain the singular solution

$$z = x^2 + y^2.$$

Example 2.38 *Determine all surfaces $z = z(x, y)$ with the property that in every surface point $M(x, y, z)$ the scalar product of the unit normal vector and the vector $\vec{OM}$ is equal to*

$$\frac{z - H(p, q)}{(1 + p^2 + q^2)^{1/2}},$$

where H is a homogeneous function of degree $n \neq 1$, i.e.,

$$H(tp, tq) = t^n H(p, q) \quad \text{for all} \quad t > 0,$$

and, moreover, it is assumed that the direction of the normal vector has been chosen so that it makes an acute angle with the positive direction of the $x-$axis.

Solution. If we denote by **i**, **j** and **k**, respectively, the unit vectors on the $x-$, $y-$ and $z-$axis, then by assumption it holds

$$(x\mathbf{i} + y\mathbf{j} + z\mathbf{k}) \cdot \frac{-p\mathbf{i} - q\mathbf{j} + \mathbf{k}}{\sqrt{1 + p^2 + q^2}} = \frac{z - H(p, q)}{(1 + p^2 + q^2)^{1/2}}.$$

Thus we get

$$px + qy = H(p, q).$$

Using the homogeneity of H, we obtain the complete solution

$$z = \frac{n-1}{n} \frac{(x+ay)^{n/(n-1)}}{(H(1,b))^{n/(n-1)}} + b$$

It is left to the reader to find the general solution; clearly, there is no singular solution.

Example 2.39 *Solve the PDE*

$$p^2 + q^2 + pq - qx - py - 2z + xy = 0.$$

Answer.

I mode. From the system (2.38) we have

$$\frac{dx}{2p+q-y} = \frac{dy}{2q+p-x} = \frac{dz}{2p^2+2q^2+2pq-py-qx}.$$

$$= \frac{-dp}{-q+y-2p} = \frac{-dq}{-p+x-2q}$$

Now we easily find one first integral, namely

$$x + a = p, \quad \text{and thus} \quad \Psi_1 = p - x.$$

Another first integral is

$$\Psi_2 = q - y.$$

Since these two first integrals are in involution (check that!), it follows that the complete solution is

$$z = \frac{1}{2}\left((x+a)^2 + (y+b)^2 + ab\right).$$

II mode. As in mode I, we find that $p = x + a$ is a first integral. From the given equation it follows

$$q = -\frac{a}{2} + \sqrt{2z + ay - (x+a)^2 + \frac{a^2}{4}}.$$

Thus we obtain Pfaff's equation

$$dz = (x+a)\,dx + \left(-\frac{a}{2} + \sqrt{2z + ay - (x+a)^2 + \frac{a^2}{4}}\right)dy,$$

or

$$\frac{dz + \dfrac{a}{2}dy - (x + a)dx}{\sqrt{2z + ay - (x + a)^2 + \dfrac{a^2}{4}}} = 0.$$

Hence the complete solution is

$$y + b + \frac{a}{2} = \sqrt{2z + ay - (x + a)^2 + \frac{a^2}{4}}.$$

Example 2.40 *Find a one–parameter family of the first order PDE*

$$pq = x + y + z.$$

Solution. One easily finds that

$$p - q + x - y = 2a \quad \text{and} \quad \frac{p+1}{q+1} = b$$

are two first integrals which are *not* in involution. Solving in p and q gives

$$p = \frac{1}{b-1}\left(b(y - x) + 2ab - b + 1\right), \quad q = \frac{1}{b-1}\left(y - x + 2a - b + 1\right).$$

Since necessarily $\dfrac{\partial p}{\partial y} = \dfrac{\partial q}{\partial x}$, it follows that $b = -1$ and

$$2p = y - x + 2a - 2, \quad 2q = x - y - 2a - 2.$$

Hence the sought after one–parameter family is

$$z = pq - x - y$$

$$= -\frac{(x - y)^2}{4} + a(x - y) - (x + y) + 1 - a^2.$$

Chapter 3

Classification of the Second Order PDEs

3.1 Two Independent Variables

3.1.1 Preliminaries

The *quasi–linear* second order PDE on some region $Q \subset \mathbf{R}^2$ is given by

$$A(x,y)\frac{\partial^2 u}{\partial x^2} + 2B(x,y)\frac{\partial^2 u}{\partial x \partial y} + C(x,y)\frac{\partial^2 u}{\partial y^2} = F\left(x,y,u,\frac{\partial u}{\partial x},\frac{\partial u}{\partial y}\right). \qquad (3.1)$$

The functions A, B, C and F are the *coefficients* of (3.1). If, additionally, the right–hand side function F has the form

$$F\left(x,y,u,\frac{\partial u}{\partial x},\frac{\partial u}{\partial y}\right) = F_1(x,y)u + F_2(x,y)\frac{\partial u}{\partial x} + F_3(x,y)\frac{\partial u}{\partial y} + F_4(x,y),$$

for some functions F_j, $j = 1,\ldots,4$, then (3.1) is *linear*.

The unknown function $u = u(x,y)$ is looked for in the set $C^2(Q)$. Thus we assume that the functions A, B and C are continuously differentiable on the region Q, while F is assumed to be a continuous function on the set $Q \times \mathbf{R}^3$.

The *type* of the second order quasi–linear PDE (3.1) depends on the sign of the function

$$D(x,y) = B^2(x,y) - A(x,y)\,C(x,y), \quad (x,y) \in Q. \qquad (3.2)$$

By definition, the second order quasi–linear PDE (3.1) is

1. *hyperbolic* at $(x,y) \in Q$ if $D(x,y) > 0$;

2. *elliptic* at $(x,y) \in Q$ if $D(x,y) < 0$;

3. *parabolic* at $(x,y) \in Q$ if $D(x,y) = 0$.

The equation is hyperbolic (respectively elliptic, parabolic) on the region Q if it is hyperbolic (resp. elliptic, parabolic) at every point $(x, y) \in Q$.

Note that the three most important second order PDEs: the wave, the Poisson (hence also Laplace's) and the heat equation are respectively hyperbolic, elliptic and parabolic in the whole plane $\mathbf{R}^2$ (see Section 3.3 below).

Suppose now that the functions

$$\xi = \xi(x, y), \quad \eta = \eta(x, y), \tag{3.3}$$

are from the class $C^2(Q)$, and, moreover, that the Jacobian

$$\frac{\partial(\xi, \eta)}{\partial(x, y)} = \begin{vmatrix} \dfrac{\partial \xi}{\partial x} & \dfrac{\partial \xi}{\partial y} \\ \dfrac{\partial \eta}{\partial x} & \dfrac{\partial \eta}{\partial y} \end{vmatrix} \tag{3.4}$$

is nonzero for every $(x, y) \in Q$. We also put

$$v(\xi, \eta) = u(x(\xi, \eta), y(\xi, \eta)) \tag{3.5}$$

Then from the chain rule it follows

$$\frac{\partial u}{\partial x} = \frac{\partial v}{\partial \xi}\frac{\partial \xi}{\partial x} + \frac{\partial v}{\partial \eta}\frac{\partial \eta}{\partial x}, \quad \frac{\partial u}{\partial y} = \frac{\partial v}{\partial \xi}\frac{\partial \xi}{\partial y} + \frac{\partial v}{\partial \eta}\frac{\partial \eta}{\partial y}, \tag{3.6}$$

$$\frac{\partial^2 u}{\partial x^2} = \frac{\partial^2 v}{\partial \xi^2}\left(\frac{\partial \xi}{\partial x}\right)^2 + 2\frac{\partial^2 v}{\partial \xi \partial \eta}\frac{\partial \xi}{\partial x}\frac{\partial \eta}{\partial x} + \frac{\partial^2 v}{\partial \eta^2}\left(\frac{\partial \eta}{\partial x}\right)^2 + \frac{\partial v}{\partial \xi}\frac{\partial^2 \xi}{\partial x^2} + \frac{\partial v}{\partial \eta}\frac{\partial^2 \eta}{\partial x^2},$$

$$\frac{\partial^2 u}{\partial x \partial y} = \frac{\partial^2 v}{\partial \xi^2}\frac{\partial \xi}{\partial x}\frac{\partial \xi}{\partial y} + \frac{\partial^2 v}{\partial \xi \partial \eta}\frac{\partial \xi}{\partial x}\frac{\partial \eta}{\partial y} + \frac{\partial^2 v}{\partial \eta \partial \xi}\frac{\partial \eta}{\partial x}\frac{\partial \xi}{\partial y} + \frac{\partial^2 v}{\partial \eta^2}\frac{\partial \eta}{\partial x}\frac{\partial \eta}{\partial y}$$

$$\qquad + \frac{\partial v}{\partial \xi}\frac{\partial^2 \xi}{\partial x \partial y} + \frac{\partial v}{\partial \eta}\frac{\partial^2 \eta}{\partial x \partial y}, \tag{3.7}$$

$$\frac{\partial^2 u}{\partial y^2} = \frac{\partial^2 v}{\partial \xi^2}\left(\frac{\partial \xi}{\partial y}\right)^2 + 2\frac{\partial^2 v}{\partial \xi \partial \eta}\frac{\partial \xi}{\partial y}\frac{\partial \eta}{\partial y} + \frac{\partial^2 v}{\partial \eta^2}\left(\frac{\partial \eta}{\partial y}\right)^2 + \frac{\partial v}{\partial \xi}\frac{\partial^2 \xi}{\partial y^2} + \frac{\partial v}{\partial \eta}\frac{\partial^2 \eta}{\partial y^2}.$$

Then the nonsingular transformation $T : Q \to \mathbf{R}^2$ of the independent variables x and y, given by $(x, y) \mapsto (\xi, \eta)$, transforms (3.1) into the quasi–linear second order PDE

$$\alpha(\xi, \eta)\frac{\partial^2 v}{\partial \xi^2} + 2\beta(\xi, \eta)\frac{\partial^2 v}{\partial \xi \partial \eta} + \gamma(\xi, \eta)\frac{\partial^2 v}{\partial \eta^2} = \Phi(\xi, \eta, v, \frac{\partial v}{\partial \xi}, \frac{\partial v}{\partial \eta}) \tag{3.8}$$

with new unknown function $v = v(\xi, \eta)$ from (3.5). The reader should check that the coefficients of (3.8) are:

$$\alpha(\xi, \eta) = A\left(\frac{\partial \xi}{\partial x}\right)^2 + 2B\frac{\partial \xi}{\partial x}\frac{\partial \xi}{\partial y} + C\left(\frac{\partial \xi}{\partial y}\right)^2, \tag{3.9}$$

$$\beta(\xi,\eta) = A\frac{\partial\xi}{\partial x}\frac{\partial\eta}{\partial x} + B\left(\frac{\partial\xi}{\partial x}\frac{\partial\eta}{\partial y} + \frac{\partial\xi}{\partial y}\frac{\partial\eta}{\partial x}\right) + C\frac{\partial\xi}{\partial y}\frac{\partial\eta}{\partial y}, \tag{3.10}$$

$$\gamma(\xi,\eta) = A\left(\frac{\partial\eta}{\partial x}\right)^2 + 2B\frac{\partial\eta}{\partial x}\frac{\partial\eta}{\partial y} + C\left(\frac{\partial\eta}{\partial y}\right)^2 \tag{3.11}$$

and

$$\Phi\left(\xi,\eta,v,\frac{\partial v}{\partial\xi},\frac{\partial v}{\partial\eta}\right) = F\left(x(\xi,\eta),y(\xi,\eta),v,\frac{\partial v}{\partial\xi}\frac{\partial\xi}{\partial x} + \frac{\partial v}{\partial\eta}\frac{\partial\eta}{\partial x},\frac{\partial v}{\partial\xi}\frac{\partial\xi}{\partial y} + \frac{\partial v}{\partial\eta}\frac{\partial\eta}{\partial y}\right)$$
$$- \left(a\frac{\partial^2\xi}{\partial x^2} + 2b\frac{\partial^2\xi}{\partial x\partial y} + c\frac{\partial^2\xi}{\partial y^2}\right)\frac{\partial v}{\partial\xi} - \left(a\frac{\partial^2\eta}{\partial x^2} + 2b\frac{\partial^2\eta}{\partial x\partial y} + c\frac{\partial^2\eta}{\partial y^2}\right)\frac{\partial v}{\partial\eta}. \tag{3.12}$$

The essential property of the quasi–linear second order PDE is the following equality, which immediately follows from equations (3.9), (3.10) and (3.11):

$$D(x,y) = B^2(x,y) - A(x,y)\,C(x,y) = \beta^2(\xi,\eta) - \alpha(\xi,\eta)\,\gamma(\xi,\eta),$$

where on the right–hand side the variables ξ and η are given by (3.3). The above equality shows that a nonsingular transformation *preserves* the type of the equation.

Our main goal is to show that, depending on the sign of the function D from (3.2), an appropriate change of variables reduces the quasi–linear equation (3.1) to one of the following three *canonical forms*: :

$$\frac{\partial^2 v}{\partial\xi\partial\eta} = \Phi\left(\xi,\eta,v,\frac{\partial v}{\partial\xi},\frac{\partial v}{\partial\eta}\right) \qquad \text{(hyperbolic equation)}; \tag{3.13}$$

$$\frac{\partial^2 v}{\partial\xi^2} + \frac{\partial^2 v}{\partial\eta^2} = \Phi\left(\xi,\eta,v,\frac{\partial v}{\partial\xi},\frac{\partial v}{\partial\eta}\right) \qquad \text{(elliptic equation)}; \tag{3.14}$$

$$\frac{\partial^2 v}{\partial\xi^2} - \frac{\partial v}{\partial\eta} = \Phi\left(\xi,\eta,v,\frac{\partial v}{\partial\xi},\frac{\partial v}{\partial\eta}\right) \qquad \text{(parabolic equation)}. \tag{3.15}$$

To that end, let us analyze the quadratic equation in λ :

$$A(x,y)\,\lambda^2 + 2B(x,y)\,\lambda + C(x,y) = 0. \tag{3.16}$$

In the following, C_1 and C_2 are arbitrary constants.

1. If the function D from (3.2) satisfies $D(x,y) > 0$ for all $(x,y) \in Q$, then equation (3.16) has two real and distinct solutions, say $\lambda_1 = \lambda_1(x,y)$ and $\lambda_2 = \lambda_2(x,y)$. If we put

$$\frac{dy}{dx} = -\lambda_1(x,y) \quad \text{and} \quad \frac{dy}{dx} = -\lambda_2(x,y),$$

and assume the upper two ODEs have solutions of the form

$$\xi(x,y) = C_1 \quad \text{and} \quad \eta(x,y) = C_2, \tag{3.17}$$

then reduces (3.1) to (3.13) (check that).

2. If D from (3.2) is negative on Q, then the solutions of (3.16) are complex conjugate numbers:
$$\lambda_{1,2} = \Re\lambda \pm \imath\Im\lambda.$$

Then assume the equations

$$\frac{dy}{dx} = -\Re\lambda(x,y) \quad \text{and} \quad \frac{dy}{dx} = -\Im\lambda(x,y),$$

have implicit solutions

$$\xi(x,y) = C_1 \quad \text{and} \quad \eta(x,y) = C_2. \tag{3.18}$$

Then the change of variables $(x,y) \mapsto (\xi,\eta)$ reduces (3.1) to (3.14).

3. In the parabolic case, $D(x,y) = 0$ for $(x,y) \in Q$, the solutions of (3.16) are equal and real, say $\lambda = \lambda(x,y)$. Then assume the ODE

$$\frac{dy}{dx} = -\lambda(x,y)$$

has an implicit solution

$$\xi(x,y) = C_1. \tag{3.19}$$

Then take $\xi = \xi(x,y)$; for the other independent variable η one can choose any function $\eta = \eta(x,y)$, provided that the Jacobian (3.4) is nonzero on Q. Then this change of variables reduces (3.1) to (3.15).

The two families of curves given by (3.17) are the *characteristic curves* for the hyperbolic second order PDE. Also, the curves from the (single) family (3.19) are the *characteristics curves* for the parabolic second order PDE. However, the elliptic second order PDEs have no characteristic curves (why?).

Cauchy's Problem

Let a curve L be given in parametric form

$$x = \varphi(\tau), \quad y = \psi(\tau) \quad (\tau \in I), \tag{3.20}$$

where φ and ψ are C^1 functions on some interval I in $\mathbf{R}$. Then *Cauchy's problem* for (3.1) is finding a solution of (3.1) that passes through the curve L and satisfies the initial conditions

$$u|_L = p(\tau), \quad \left.\frac{\partial u}{\partial x}\right|_L = q(\tau) \quad \text{and} \quad \left.\frac{\partial u}{\partial y}\right|_L = r(\tau), \tag{3.21}$$

where p, q and r are given functions of the real parameter τ.

3.1.2   Examples and Exercises

Example 3.1 *Determine the type of the following second order linear PDEs and find their canonical forms.*

a) $\dfrac{\partial^2 u}{\partial x^2} + 3\dfrac{\partial^2 u}{\partial x \partial y} - 4\dfrac{\partial^2 u}{\partial y^2} + \dfrac{\partial u}{\partial x} + 4\dfrac{\partial u}{\partial y} = 0;$

b) $\dfrac{\partial^2 u}{\partial x^2} + 6\dfrac{\partial^2 u}{\partial x \partial y} + 10\dfrac{\partial^2 u}{\partial y^2} + \dfrac{\partial u}{\partial x} + 3\dfrac{\partial u}{\partial y} = 0;$

c) $\dfrac{\partial^2 u}{\partial x^2} - 2\dfrac{\partial^2 u}{\partial x \partial y} + \dfrac{\partial^2 u}{\partial y^2} + 3\dfrac{\partial u}{\partial x} + \dfrac{\partial u}{\partial y} + 2u = 0$

Solutions.

a) The function D from (3.2) is equal to $(3/2)^2 - 1 \cdot (-4) = 25/4 > 0$, and thus it follows that the given PDE is hyperbolic in the whole xy–plane. In this case, the quadratic equation (3.16) is

$$\lambda^2 + 3\lambda - 4 = 0,$$

and its solutions are $\lambda_1 = 1$ and $\lambda_2 = -4$. Thus we have the differential equations

$$\frac{dy}{dx} = -1, \quad \frac{dy}{dx} = 4,$$

whose solutions are

$$x + y = C_1 \quad \text{and} \quad 4x - y = C_2,$$

respectively. In order to obtain the canonical form (3.13) we introduce the new variables ξ and η by $\xi = x + y$ and $\eta = 4x - y$, see equation (3.17), and we put $v(\xi, \eta) = u(x, y)$. Note that the above transformation $(x, y) \mapsto (\xi, \eta)$ is nonsingular, since the corresponding Jacobian is nonzero:

$$\frac{\partial(\xi, \eta)}{\partial(x, y)} = \begin{vmatrix} 1 & 1 \\ 4 & -1 \end{vmatrix} = -5.$$

Thus we obtain the canonical form

$$\frac{\partial^2 v}{\partial \xi \partial \eta} + \frac{1}{5}\frac{\partial v}{\partial \xi} = 0.$$

b) The solutions of the quadratic equation $\lambda^2 + 6\lambda + 10 = 0$ are conjugate complex numbers, namely $\lambda_{1,2} = -3 \pm \imath$, hence the given PDE is elliptic in the whole plane. The characteristics are

$$y = 3x - \imath x + C_1, \quad y = 3x + \imath x + C_2,$$

where C_1 and C_2 are arbitrary constants. Thus we can put $\xi = 3x - y$ and $\eta = x$ (the "real" and the "imaginary" part of the characteristics). Now the transformation $(x, y) \mapsto (\xi, \eta)$ is nonsingular, and applying the equations (3.6) and (3.7), we get the canonical form of the given elliptic PDE

$$\frac{\partial^2 v}{\partial \xi^2} + \frac{\partial^2 v}{\partial \eta^2} + \frac{\partial v}{\partial \eta} = 0.$$

c) The given PDE is parabolic in the whole xy-plane, since both solutions of the corresponding quadratic equation are equal to 1. Thus we can use the change of variables $\xi = x + y$ and $\eta = x$ in order to obtain the canonical form of parabolic equations. In our case we get

$$\frac{\partial^2 v}{\partial \eta^2} + 4\frac{\partial v}{\partial \xi} + 3\frac{\partial v}{\partial \eta} + v = 0.$$

Remark 3.1.1 In case c), once we have put $\xi = x + y$, for the new variable η we could have chosen any function of x and y of the class $C^2(\mathbf{R}^2)$, provided that the transformation $(x, y) \mapsto (\xi, \eta)$ is nonsingular.

Exercise 3.2 *Determine the type of the following second order linear PDEs, depending on the points in xy-plane, then use a suitable change of variables, $\xi = \xi(x, y)$, $\eta = \eta(x, y)$ which, after putting $v(\xi, \eta) = u(x, y)$, gives the canonical form of the transformed equation.*

a) $\dfrac{\partial^2 u}{\partial x^2} + 2\dfrac{\partial^2 u}{\partial x \partial y} + (1 - \operatorname{sgn} y)\dfrac{\partial^2 u}{\partial y^2} = 0;$

b) $\dfrac{\partial^2 u}{\partial x^2} + x^2\dfrac{\partial^2 u}{\partial y^2} = 0 \quad (x \neq 0);$

c) $x^2\dfrac{\partial^2 u}{\partial x^2} - y^2\dfrac{\partial^2 u}{\partial y^2} = 0;$

d) $y^2\dfrac{\partial^2 u}{\partial x^2} + 2xy\dfrac{\partial^2 u}{\partial x \partial y} + 2x^2\dfrac{\partial^2 u}{\partial y^2} + y\dfrac{\partial u}{\partial y} = 0;$

e) $x\dfrac{\partial^2 u}{\partial x^2} - y\dfrac{\partial^2 u}{\partial y^2} + \dfrac{1}{2}\dfrac{\partial u}{\partial x} - \dfrac{1}{2}\dfrac{\partial u}{\partial y} = 0 \quad (x, y > 0);$

f) $y\dfrac{\partial^2 u}{\partial x^2} + x\dfrac{\partial^2 u}{\partial y^2} = 0;$

g) $4y^2\dfrac{\partial^2 u}{\partial x^2} - e^{2x}\dfrac{\partial^2 u}{\partial y^2} - 4y^2\dfrac{\partial u}{\partial x} = 0;$

h) $\sin^2 x \dfrac{\partial^2 u}{\partial x^2} - 2y \sin x \dfrac{\partial^2 u}{\partial x \partial y} + y^2 \dfrac{\partial^2 u}{\partial y^2} = 0;$

i) $(1 - x^2)\dfrac{\partial^2 u}{\partial x^2} - 2xy \dfrac{\partial^2 u}{\partial x \partial y} + (1 - y^2)\dfrac{\partial^2 u}{\partial y^2} - 2x \dfrac{\partial u}{\partial x} - 2y \dfrac{\partial u}{\partial y} = 0.$

Answers.

a) For $y > 0$ the given equation is hyperbolic. Putting the new variables $\xi = y - 2x$, $\eta = y$, we obtain the canonical form $\dfrac{\partial^2 v}{\partial \xi \partial \eta} = 0.$

However, for $y < 0$ the given equation is elliptic. Putting $\xi = y - x$, $\eta = x$, it becomes the Laplace equation $\dfrac{\partial^2 v}{\partial \xi^2} + \dfrac{\partial^2 v}{\partial \eta^2} = 0.$

b) For $x \neq 0$ the equation is elliptic. Putting $\xi = y$, $\eta = x^2/2$, it becomes

$$\frac{\partial^2 v}{\partial \xi^2} + \frac{\partial^2 v}{\partial \eta^2} + \frac{1}{2\eta}\frac{\partial v}{\partial \eta} = 0.$$

c) For $x \neq 0$, $y \neq 0$, the equation is hyperbolic. Putting $\xi = xy$, $\eta = y/x$, it becomes

$$\frac{\partial^2 v}{\partial \xi \partial \eta} - \frac{1}{2\xi}\frac{\partial v}{\partial \eta} = 0.$$

Note that on the $x-$ and on the $y-$ axis the given equation is parabolic.

d) The PDE is elliptic in the whole $xy-$plane. Putting $\xi = x^2 - y^2$, $\eta = x^2$, it becomes

$$\frac{\partial^2 v}{\partial \xi^2} + \frac{\partial^2 v}{\partial \eta^2} + \frac{1}{\xi - \eta}\frac{\partial v}{\partial \xi} + \frac{1}{2\eta}\frac{\partial v}{\partial \eta} = 0.$$

e) The PDE is parabolic in the first quadrant $x > 0$, $y > 0$. Putting $\xi = \sqrt{x} + \sqrt{y}$, $\eta = \sqrt{x} - \sqrt{y}$, we get

$$\frac{\partial^2 v}{\partial \xi \partial \eta} = 0.$$

f) The PDE is elliptic in the first and in the third quadrant. Using the new variables

$$\xi = x^{3/2}, \quad \eta = y^{3/2} \quad \text{for } x > 0, y > 0,$$

and

$$\xi = (-x)^{3/2}, \quad \eta = (-y)^{3/2} \quad \text{for } x < 0, y < 0,$$

we get the canonical form

$$\frac{\partial^2 v}{\partial \xi^2} + \frac{\partial^2 v}{\partial \eta^2} + \frac{1}{3\xi}\frac{\partial v}{\partial \xi} + \frac{1}{3\eta}\frac{\partial v}{\partial \eta} = 0.$$

However, this equation is hyperbolic in the second and in the fourth quadrant. Then we put

$$\xi = (-x)^{3/2} - y^{3/2}, \quad \eta = (-x)^{3/2} + y^{3/2} \quad \text{for } x < 0, \ y > 0,$$

and

$$\xi = x^{3/2} - (-y)^{3/2}, \quad \eta = x^{3/2} + (-y)^{3/2} \quad \text{for } x > 0, \ y < 0.$$

So we obtain

$$\frac{\partial^2 v}{\partial \xi \partial \eta} + \frac{1}{3} \frac{1}{\eta^2 - \xi^2} \left(\eta \frac{\partial v}{\partial \xi} - \xi \frac{\partial v}{\partial \eta} \right) = 0.$$

Finally, on the $x-$ and on the $y-$axis the equation is parabolic.

g) The PDE is hyperbolic in the $xy-$plane. Putting $\xi = e^x + y^2$, $\eta = -e^x + y^2$, it reduces to the canonical form

$$\frac{\partial^2 v}{\partial \xi \partial \eta} = \frac{1}{8(\xi + \eta)^2} \left(\frac{\partial v}{\partial \xi} + \frac{\partial v}{\partial \eta} \right).$$

h) The PDE is parabolic for every x and y. Putting $\xi = y \tan(x/2)$ and $\eta = y$, we get

$$\frac{\partial^2 v}{\partial \eta^2} - \frac{2\xi}{\xi^2 + \eta^2} \frac{\partial v}{\partial \xi} = 0.$$

i) The quadratic equation

$$(1 - x^2)\lambda^2 - 2xy\lambda + (1 - y^2) = 0 \tag{3.22}$$

has a discriminant equal to $4(x^2 + y^2 - 1)$, hence the given PDE is elliptic inside the unit circle, i.e., for $x^2 + y^2 < 1$, and hyperbolic outside it. On the circle $x^2 + y^2 = 1$ it is parabolic. From (3.22) we get the differential equation of the characteristics

$$\left(xy \pm \sqrt{x^2 + y^2 - 1} \right) dx + (1 - x^2) dy = 0,$$

which can be most easily solved using the new independent variable t given by $t = \sqrt{1 - x^2}$ and the new dependent variable z given by $z = y/t$. Thus putting

$$\xi = \frac{y}{x - 1}, \quad \eta = \frac{\sqrt{1 - x^2 - y^2}}{x - 1} \quad \text{for } x^2 + y^2 < 1,$$

we obtain the canonical form

$$\frac{\partial^2 v}{\partial \xi^2} + \frac{\partial^2 v}{\partial \eta^2} = 0.$$

If $x^2 + y^2 > 1$, then we put

$$\xi = \frac{y}{x-1}, \quad \eta = \frac{\sqrt{x^2 + y^2 - 1}}{x - 1}$$

and obtain

$$\frac{\partial^2 v}{\partial \xi^2} - \frac{\partial^2 v}{\partial \eta^2} = 0.$$

Exercise 3.3 *Transform the given PDEs with the given changes of variables.*

a) $\dfrac{\partial^2 u}{\partial x^2} - 2\sin x \dfrac{\partial^2 u}{\partial x \partial y} - \cos^2 x \dfrac{\partial^2 u}{\partial y^2} - \cos x \dfrac{\partial u}{\partial y} = 0,$
     $\xi = x + y - \cos x, \ \eta = x - y + \cos x;$

b) $y^2 \dfrac{\partial^2 u}{\partial x^2} + x^2 \dfrac{\partial^2 u}{\partial y^2} = 0, \quad \xi = y^2, \ \eta = x^2;$

c) $x \dfrac{\partial^2 u}{\partial x^2} + 2xy \dfrac{\partial^2 u}{\partial x \partial y} - y \dfrac{\partial^2 u}{\partial y^2} = 0, \quad \xi = y/x, \ \eta = y.$

Answers. In all exercises we put $v(\xi, \eta) = u(x, y)$.

a) $\dfrac{\partial^2 v}{\partial \xi \partial \eta} = 0.$

b) $\dfrac{\partial^2 v}{\partial^2 \xi} + \dfrac{\partial^2 v}{\partial \eta^2} + \dfrac{1}{2\xi} \dfrac{\partial v}{\partial \xi} + \dfrac{1}{2\eta} \dfrac{\partial v}{\partial \eta} = 0.$

c) $\dfrac{\partial^2 v}{\partial \eta^2} = 0.$

Exercise 3.4 *Prove that using suitable changes of variables the given PDEs can be reduced to the given canonical forms.*

a) $\dfrac{\partial^2 u}{\partial x^2} - 2\cos x \dfrac{\partial^2 u}{\partial x \partial y} - (3 + \sin^2 x) \dfrac{\partial^2 u}{\partial y^2} - y \dfrac{\partial u}{\partial y} = 0,$

     $\dfrac{\partial^2 v}{\partial \xi \partial \eta} + \dfrac{1}{32}(\eta - \xi)\left(\dfrac{\partial v}{\partial \xi} - \dfrac{\partial v}{\partial \eta}\right) = 0;$

b) $\dfrac{\partial^2 u}{\partial x^2} - 2x \dfrac{\partial^2 u}{\partial x \partial y} + x^2 \dfrac{\partial^2 u}{\partial y^2} - 2 \dfrac{\partial u}{\partial y} = 0, \quad \dfrac{\partial^2 v}{\partial \eta^2} - \dfrac{\partial v}{\partial \xi} = 0;$

c) $(1 + x^2) \dfrac{\partial^2 u}{\partial x^2} + (1 + y^2) \dfrac{\partial^2 u}{\partial y^2} + x \dfrac{\partial u}{\partial x} + y \dfrac{\partial u}{\partial y} = 0, \quad \dfrac{\partial^2 v}{\partial \xi^2} + \dfrac{\partial^2 v}{\partial \eta^2} = 0;$

d) $x \dfrac{\partial^2 u}{\partial x^2} - y \dfrac{\partial^2 u}{\partial y^2} = 0 \ \ (x, y, > 0), \quad \dfrac{\partial^2 v}{\partial \xi^2} - \dfrac{1}{2\xi} \dfrac{\partial v}{\partial \eta} = 0 \ \ (\xi, \eta > 0).$

Example 3.5 *Prove that if the second order PDE with constant coefficients*

$$A\frac{\partial^2 u}{\partial x^2} + B\frac{\partial^2 u}{\partial x \partial y} + C\frac{\partial^2 u}{\partial y^2} = 0 \qquad (A^2 + B^2 + C^2 > 0), \tag{3.23}$$

is either hyperbolic or elliptic, (i.e., $B^2 - 4AC \neq 0$), then its solution has the form

$$u(x, y) = F(m_1 x + y) + G(m_2 x + y),$$

where F and G are functions of the class $C^2(\mathbf{R})$, while m_j, $j = 1, 2$, are are the solutions of the quadratic equation

$$Am^2 + Bm + C = 0. \tag{3.24}$$

Solution. Let us assume $A \neq 0$ and $B^2 - 4AC > 0$; the other cases are handled similarly. Then the solutions m_1 and m_2 of (3.24) are real and different. Putting $\xi = m_1 x + y$, $\eta = m_2 x + y$, we obtain

$$(Am_1^2 + Bm_1 + C)\frac{\partial^2 v}{\partial \xi^2} + (Am_2^2 + Bm_2 + C)\frac{\partial^2 v}{\partial \eta^2} + (2Am_1 m_2 + B(m_1 + m_2) + 2C)\frac{\partial^2 v}{\partial \xi \partial \eta} = 0.$$

Since $m_1 + m_2 = -B/A$, $m_1 m_2 = C/A$, we have

$$\frac{1}{A}(4AC - B^2)\frac{\partial^2 v}{\partial \xi \partial \eta} = 0, \quad \text{which implies} \quad \frac{\partial^2 v}{\partial \xi \partial \eta} = 0.$$

The last PDE can be written as

$$\frac{\partial}{\partial \xi}\left(\frac{\partial v}{\partial \eta}\right) = 0,$$

hence for some function G_1 from $C^1(\mathbf{R})$ it holds

$$\frac{\partial v}{\partial \eta} = G_1(\eta).$$

This gives us that for some function F_1 from $C^2(\mathbf{R})$ it holds

$$v(\xi, \eta) = F(\xi) + \int_{\eta_0}^{\eta} G_1(\tau)\, d\tau,$$

or $v(\xi, \eta) = F(\xi) + G_2(\eta)$, and finally we get

$$u(x, y) = F(m_1 x + y) + G(m_2 x + y)$$

for some C^2 functions F and G.

Example 3.6 *Prove that if the second order PDE with constant coefficients (3.23) satisfies $A^2 + B^2 + C^2 > 0$ and $B^2 - 4AC = 0$, i.e., is parabolic, in the xy-plane, then its solution has the form*

$$u(x, y) = F(mx + y) + xG(mx + y),\qquad (3.25)$$

where F and G are functions of the class $C^2(\mathbf{R})$, while $m = -B/A$ is the double solution of (3.22).

Solution. Again we assume $A \neq 0$; the other case $A = 0$ is left to the reader. By supposition, it holds $B^2 - 4AC = 0$, hence in order to obtain the canonical form of (3.23), we can put $\xi = mx + y$, $\eta = x$. This gives

$$(Am^2 + Bm + C)\frac{\partial^2 v}{\partial \xi^2} + A\frac{\partial^2 v}{\partial \eta^2} + (2Am + B)\frac{\partial^2 v}{\partial \xi \partial \eta} = 0,$$

or $\dfrac{\partial^2 v}{\partial \eta^2} = 0$. This gives (3.25).

Exercise 3.7 *Find the general solution of the PDE*

$$\frac{\partial^2 u}{\partial x^2} - 2\frac{\partial^2 u}{\partial x \partial y} + \frac{\partial^2 u}{\partial y^2} = 0.$$

Answer. Since $m_1 = m_2 = 1$, it follows from Example 3.6 that the general solution is

$$u(x, y) = F(x + y) + xG(x + y) \qquad (F, G \in C^2(\mathbf{R})).$$

Exercise 3.8 *Let α, β and γ be some real constants. Simplify then the following PDEs using the change of dependent variable $v(x, y) = e^{\lambda x + \mu y}u(x, y)$ and choosing suitable values for the parameters λ and μ.*

a) $\quad \dfrac{\partial^2 u}{\partial x^2} + \dfrac{\partial^2 u}{\partial y^2} + \alpha\dfrac{\partial u}{\partial x} + \beta\dfrac{\partial u}{\partial y} + \gamma u = 0;$

b) $\quad \dfrac{\partial^2 u}{\partial x^2} = \dfrac{1}{a^2}\dfrac{\partial u}{\partial y} + \alpha u + \beta\dfrac{\partial u}{\partial x};$

c) $\quad \dfrac{\partial^2 u}{\partial x^2} - \dfrac{1}{a^2}\dfrac{\partial^2 u}{\partial y^2} = \alpha\dfrac{\partial u}{\partial x} + \beta\dfrac{\partial u}{\partial y} + \gamma u;$

d) $\quad \dfrac{\partial^2 u}{\partial x \partial y} = \alpha\dfrac{\partial u}{\partial x} + \beta\dfrac{\partial u}{\partial y}.$

Answers.

a)   For $\lambda = \alpha/2$ and $\mu = \beta/2$ we obtain

$$\frac{\partial^2 v}{\partial x^2} + \frac{\partial^2 v}{\partial y^2} + \left(\gamma - \frac{\alpha^2}{4} - \frac{\beta^2}{4}\right) v = 0.$$

b) $\dfrac{\partial^2 v}{\partial x^2} = \dfrac{1}{a^2}\dfrac{\partial v}{\partial y}.$

c) $\dfrac{\partial^2 v}{\partial x^2} - \dfrac{1}{a^2}\dfrac{\partial^2 v}{\partial y^2} = \left(\alpha + \beta + \gamma - \dfrac{\alpha^2}{4} - \dfrac{\beta^2 a^4}{4}\right) v.$

d) $\dfrac{\partial^2 v}{\partial x \partial y} = (\alpha + \beta - \alpha\beta) v.$

Example 3.9 *Find the general solution of the second order PDE with constant coefficients*

$$A\frac{\partial^2 u}{\partial x^2} + B\frac{\partial^2 u}{\partial x \partial y} + C\frac{\partial^2 u}{\partial y^2} + D\frac{\partial u}{\partial x} + E\frac{\partial u}{\partial y} = 0,$$

provided that $\dfrac{B + \sqrt{B^2 - 4AC}}{2A} = -\dfrac{E}{D}$ *and* $B^2 - 4AC > 0.$

Solution. Let m_j, $j = 1, 2$, be the solutions of the quadratic equation (3.24). Then putting $\xi = m_1 x + y$, $\eta = m_2 x + y$ and $v(\xi, \eta) = u(x, y)$, we obtain

$$\frac{4AC - B^2}{A}\frac{\partial^2 v}{\partial \xi \partial \eta} + (Dm_2 + E)\frac{\partial v}{\partial \eta} = 0.$$

Following Exercise 3.8, let $w(\xi, \eta) = \exp(\lambda\xi)\, v(\xi, \eta)$; then we obtain

$$\frac{\partial^2 w}{\partial \xi \partial \eta} = 0, \quad \text{provided that} \quad \lambda = \frac{A(Dm_2 + E)}{B^2 - 4AC}.$$

Thus the general solution of the given PDE is of the form

$$u(x, y) = \exp\left(\frac{A(Dm_2 + E)}{B^2 - 4AC}\right)(m_2 + y) + G(m_2 x + y).$$

Exercise 3.10 *Find the general solution of the PDE*

$$\frac{\partial^2 u}{\partial x \partial y} = \frac{1}{x - y}\left(\frac{\partial u}{\partial x} - \frac{\partial u}{\partial y}\right).$$

Answer. Putting $v(x, y) = (x - y)u(x, y)$ we get

$$\frac{\partial^2 v}{\partial x \partial y} = 0,$$

hence

$$u(x, y) = \frac{f(x) + g(y)}{x - y}.$$

3.1. TWO INDEPENDENT VARIABLES

Example 3.11 *Solve the following Cauchy problems.*

a) $4y^2 \dfrac{\partial^2 u}{\partial x^2} + 2(1 - y^2)\dfrac{\partial^2 u}{\partial x \partial y} - \dfrac{\partial^2 u}{\partial y^2} - \dfrac{2y}{1+y^2}\left(2\dfrac{\partial u}{\partial x} - \dfrac{\partial u}{\partial y}\right) = 0,$

$u(x,0) = f(x), \ \dfrac{\partial u}{\partial y}(x,0) = g(x) \quad (x \in \mathbf{R}),$

where $f \in C^2(\mathbf{R})$ and $g \in C^1(\mathbf{R})$ are given functions.

b) $\dfrac{\partial^2 u}{\partial x^2} - 2\sin x \dfrac{\partial^2 u}{\partial x \partial y} - (3 + \cos^2 x)\dfrac{\partial^2 u}{\partial y^2} + \dfrac{\partial u}{\partial x} + (2 - \sin x - \cos x)\dfrac{\partial u}{\partial y} = 0,$

$u(x, \cos x) = 0, \ \dfrac{\partial u}{\partial y}(x, \cos x) = e^{-x/2} \cos x;$

c) $\dfrac{\partial^2 u}{\partial x^2} + 2\cos x \dfrac{\partial^2 u}{\partial x \partial y} - \sin^2 x \dfrac{\partial^2 u}{\partial y^2} - \sin x \dfrac{\partial u}{\partial y} = 0,$

$u(x, \sin x) = x + \cos x, \ \dfrac{\partial u}{\partial y}(x, \sin x) = \sin x.$

Solutions.

a) In our case, the solutions of the quadratic equation (3.16) are $\lambda_1 = 1/2$ and $\lambda_2 = -1/(2y^2)$, which gives

$$\frac{dy}{dx} = -\lambda_1 = -\frac{1}{2}, \quad \text{hence} \quad x + 2y = C_1,$$

and

$$\frac{dy}{dx} = -\lambda_2 = \frac{1}{2y^2}, \quad \text{hence} \quad x - \frac{2y^3}{3} = C_2$$

for arbitrary constants C_1 and C_2. Thus the new variables $\xi = x + 2y$ and $\eta = x - 2y^3/3$ lead to the PDE

$$\frac{\partial^2 v}{\partial \xi \partial \eta} = 0, \quad v(\xi, \eta) = u(x, y).$$

Thus the general solution of the given PDE is

$$v(\xi, \eta) = F(\xi) + G(\eta), \quad \text{hence} \quad u(x,y) = F(x + 2y) + G\left(x - 2y^3/3\right),$$

where F and G are arbitrary (but appropriately smooth) functions. The initial conditions give

$$F(x) + G(x) = f(x) \quad \text{and} \quad 2F'(x) = g(x),$$

hence the solution is

$$u(x,y) = f\left(x - \frac{2y^3}{3}\right) + \frac{1}{2}\int_{x - 2y^3/3}^{x+2y} g(\tau)\, d\tau.$$

b) The change of variables $\xi = y - \cos x + 2x$, $\eta = y - \cos x - 2x$, $v(\xi, \eta) = u(x, y)$, gives the PDE

$$4\frac{\partial^2 v}{\partial \xi \partial \eta} = \frac{\partial v}{\partial \xi},$$

whose general solution is

$$v(\xi, \eta) = e^{\eta/4} F(\xi) + G(\eta),$$

where F and G are arbitrary functions from $C^2(\mathbf{R}^2)$. Thus

$$u(x, y) = e^{(y - \cos x - 2x)/4} F(y - \cos x + 2x) + G(y - \cos x - 2x). \qquad (3.26)$$

The first initial condition gives the equation

$$e^{-x/2} F(2x) + G(-2x) = 0$$

Differentiating the last equation in x gives

$$e^{-x/2}\left(-\frac{1}{2}F(2x) + 2F'(2x)\right) - 2G'(-2x) = 0 \qquad (3.27)$$

The second initial condition gives

$$e^{-x/2}\left(-\frac{1}{4}F(2x) + F'(2x)\right) + G'(-2x) = e^{-x/2}\cos x,$$

and replacing $G'(-2x)$ from (3.27) we obtain $F(x) = \sin(x/2) + C$. Hence $G(x) = -e^{x/4}\sin(x/2) + C$, and the final solution is

$$\begin{aligned}
u(x, y) &= e^{(y - \cos x - 2x)/4}\left(\sin\frac{y - \cos x - 2x}{2} - \sin\frac{y - \cos x + 2x}{2}\right)\\
&= 2e^{(y - \cos x - 2x)/4}\cos x \cdot \sin\frac{1}{2}(y - \cos x).
\end{aligned}$$

c) Using the change of variables $\xi = y - x - \sin x$, $\eta = y + x - \sin x$ we come to the solution

$$u(x, y) = x + \cos(x - y + \sin x).$$

Example 3.12 *Find the solution $u = u(x, t)$ of the equation*

$$\frac{\partial^2 u}{\partial x^2} + 2\cos x\frac{\partial^2 u}{\partial x \partial t} - \sin^2 x\frac{\partial^2 u}{\partial t^2} - \sin x\frac{\partial u}{\partial t} = 0, \qquad (3.28)$$

on the set $\mathbf{R} \times [0, \infty)$, satisfying the conditions

$$u(x, \sin x) = \varphi_0(x), \qquad \frac{\partial u}{\partial t}(x, \sin x) = \varphi_1(x) \quad (x \in \mathbf{R}), \qquad (3.29)$$

where $\varphi_0 \in C^2(\mathbf{R})$ and $\varphi_1 \in C^1(\mathbf{R})$ are given functions.

Solution. The quadratic equation (3.16) is

$$\lambda^2 + 2\cos x\,\lambda - \sin^2 x = 0,$$

whose solutions are

$$\lambda_{1,2} = -\cos x \pm 1.$$

So we have

$$\left(\frac{dt}{dx}\right)_{1,2} = \cos x \mp 1, \quad \text{hence} \quad t_{1,2} = \sin x \mp x \pm C_{1,2},$$

and thus we introduce the independent variables ξ and η, and the dependent variable $v = v(\xi, \eta)$ by

$$\xi = t - x - \sin x, \quad \eta = t + x - \sin x \quad \text{and} \quad v(\xi, \eta) = u(x, y).$$

Then equation (3.28) becomes

$$\frac{\partial^2 v}{\partial \eta^2} - \frac{\partial^2 v}{\partial \xi^2} = 0. \tag{3.30}$$

The solution of the equation (3.30) is

$$v(\xi, \eta) = A(\xi) + B(\eta),$$

or

$$u(x, t) = A(t - x - \sin x) + B(t + x - \sin x),$$

where A and B functions from $C^2(\mathbf{R})$ are to be determined from the initial conditions (3.29). Using them we have

$$\begin{aligned} u(x, \sin x) &= A(-x) + B(x) = \varphi_0(x) \quad \text{and} \\ \frac{\partial u}{\partial t}(x, \sin x) &= A'(-x) + B'(x) = \varphi_1(x). \end{aligned} \tag{3.31}$$

Differentiating the first equation in (3.31) and adding it to the second one gives

$$2B'(x) = \varphi_0'(x) + \varphi_1(x), \quad \text{hence}$$
$$B(x) = \frac{1}{2}\left(\varphi_0(x) - \varphi_0(x_0) + \int_{x_0}^{x} \varphi_1(\zeta)\,d\zeta\right),$$

for some $x_0 \in \mathbf{R}$. Then we have

$$A(-x) = \varphi_0(x) - B(x),$$

which gives us

$$A(x) \;=\; \varphi_0(-x) - \frac{1}{2}\left(\varphi_0(-x) - \varphi_0(x_0) + \int_{x_0}^{x}\varphi_1(\zeta)\,d\zeta\right)$$

$$=\; \frac{1}{2}\varphi_0(-x) + \frac{1}{2}\varphi_0(x_0) - \frac{1}{2}\int_{x_0}^{x}\varphi_1(\zeta)\,d\zeta.$$

Therefore the solution of the problem (3.28), (3.29) has the form

$$u(x,t) \;=\; \frac{1}{2}\left(\frac{\varphi_0(x + \sin x - t) + \varphi_0(x - \sin x + t)}{2}\right) + \frac{1}{2}\int_{x+\sin x - t}^{x-\sin x + t}\varphi_1(d\zeta)\,d\zeta.$$

3.2 n Independent Variables

3.2.1   Preliminaries

In Section 3.1 we exposed the classification and reduction to canonical forms of quasi–linear second order PDEs, see (3.1), in two independent variables ($n = 2$). The general case ($n > 2$) to be exposed below, is essentially analogous, but technically more involved.

Let Q be a region in R^n, $n > 2$. The quasi–linear second order PDE is an equation of the form

$$\sum_{i=1}^{n}\sum_{j=1}^{n} a_{ij}(x)\frac{\partial^2 u}{\partial x_i \partial x_j} + \Phi(x, u, \mathrm{grad}\; u) = 0 \quad (x \in Q). \tag{3.32}$$

In (3.32), $u = u(x_1, x_2, \ldots, x_n)$ is the unknown function, sought after in $C^2(\mathbf{R}^n)$, the coefficients $a_{i,j}$ are continuously differentiable functions on Q such that for all $i, j = 1, 2, \ldots, n$ and $x \in Q$ it holds $a_{ij}(x) = a_{ji}(x)$. Finally, the function Φ is assumed to be a continuous function of its variables. Note that (3.32) reduces to (3.1) if $n = 2$.

Let us transform equation (3.32) by a change of variables $\xi = Tx$ ($x \in Q$); as in Subsection 3.1, we put for the new dependent variable $v(\xi) = u(Tx)$. Our goal is to show that we can choose T so that it is a nonsingular transformation such that equation (3.32) can be transformed at every point $x^0 = (x_1^0, x_2^0, \ldots, x_n^0) \in Q$ to a canonical form, i.e., to an equation of the form

$$\sum_{i=1}^{n} b_{ii}(\xi^0)\frac{\partial^2 v}{\partial \xi_i^2} + \Psi(\xi^0, v, \mathrm{grad}\; v) = 0, \tag{3.33}$$

for some function Ψ, where $b_{ii} \in \{1, -1, 0\}$, $i = 1, 2, \ldots, n$, and the function $v(\xi) = u(Tx)$ and its derivatives are calculated at the point $\xi^0 = Tx^0$.

To that end, let us quote some well known facts from the matrix theory. One proves there that for every symmetric quadratic form

$$\sum_{i=1}^{n}\sum_{j=1}^{n} a_{ij}(x^0)y_iy_j \tag{3.34}$$

there exists a nonsingular linear transformation $y \mapsto \eta$, $y = (y_1,\ldots,y_n)$, $\eta = (\eta_1,\ldots,\eta_n)$, which reduces (3.34) to the canonical quadratic form

$$\sum_{i=1}^{n}\sum_{j=1}^{n} b_{ij}\eta_i\eta_j,$$

where $b_{ii} \in \{1,-1,0\}$ for $i = 1,2,\ldots,n$, and $b_{i,j} = 0$ for every pair (i,j) such that $i \neq j$, $i,j = 1,2,\ldots,n$. Then it is well known that there exists a matrix $\mathbf{B}$ such that $y = \mathbf{B}^T\eta$, where $\mathbf{B}^T$ is the transposed matrix of $\mathbf{B}$. In fact, $\mathbf{B}$ can be obtained as a product of the following matrices: $\mathbf{E}_{ij}$, $\mathbf{E}_{i(\alpha)}$ and $\mathbf{E}_{ij(\alpha)}, i,j \in \{1,2,...,n\}$, $\alpha \in \mathbf{R}$. These matrices are respectively obtained by transforming the identity matrix so that the i−th and the j−th row of the identity matrix change places, the elements of the i−th row is multiplied by α, and finally the elements of the j−th row are multiplied and added to the elements of the i−th row.

Let now $x^0 \in Q$ and denote by $\mathbf{A} = [a_{ij}(x^0)]_{i,j=1}^n$ the matrix determined by equation (3.32). Then choose matrix $\mathbf{B}$ as above; the product $\mathbf{BAB}^T$ gives the diagonal matrix

$$\begin{bmatrix} i_1 & 0 & \ldots & 0 \\ 0 & i_2 & \ldots & 0 \\ \multicolumn{4}{c}{\ldots\ldots\ldots\ldots} \\ 0 & 0 & \ldots & i_n \end{bmatrix},$$

$i_k \in \{-1,0,1\}$, $k \in \{1,2,...,n\}$. Let us denote by r and s the number of positive and negative, respectively, ones in the set $\{i_1,i_2,...,i_n\}$; hence the number of zeros on the diagonal of the above matrix is $n - (r+s)$. In fact, the numbers r and s determine the type of equation (3.32) at a point $x \in Q$ in the following way.

The equation (3.32) is:

1. *elliptic* in x^0 if $(r,s) = (n,0)$, or $(r,s) = (0,n)$;

2. *ultrahyperbolic* in x^0 if $r > 0$, $s > 0$ and $r + s = n$;
   in particular, it is *hyperbolic* in x^0 if $(r,s) = (n-1,1)$ or $(r,s) = (1,n-1)$;

3. *ultraparabolic* in x^0 if $r + s < n$;
   in particular, it is *parabolic* in x^0 if $(r,s) = (n-1,0)$ or $(r,s) = (0,n-1)$.

3.2.2   Examples and Exercises

Example 3.13 *Determine the canonical forms of the following equations.*

a) $\dfrac{\partial^2 u}{\partial x^2} + 2\dfrac{\partial^2 u}{\partial x \partial y} - 2\dfrac{\partial^2 u}{\partial x \partial z} + 2\dfrac{\partial^2 u}{\partial y \partial y} + 6\dfrac{\partial^2 u}{\partial z^2}, \quad u = u(x,y,z);$

b) $4\dfrac{\partial^2 u}{\partial x^2} - 4\dfrac{\partial^2 u}{\partial x \partial y} - 2\dfrac{\partial^2 u}{\partial y \partial z} + \dfrac{\partial u}{\partial y} + \dfrac{\partial u}{\partial z} = 0, \quad u = u(x,y,z);$

c) $\dfrac{\partial^2 u}{\partial x \partial y} - \dfrac{\partial^2 u}{\partial x \partial z} + \dfrac{\partial u}{\partial x} + \dfrac{\partial u}{\partial y} - \dfrac{\partial u}{\partial z} = 0, \quad u = u(x,y,z);$

a) $\dfrac{\partial^2 u}{\partial x^2} + 2\dfrac{\partial^2 u}{\partial x \partial y} + 2\dfrac{\partial^2 u}{\partial y^2} + 2\dfrac{\partial^2 u}{\partial y \partial z} + 2\dfrac{\partial^2 u}{\partial y \partial t} + 2\dfrac{\partial^2 u}{\partial y \partial z} + 2\dfrac{\partial^2 u}{\partial t^2} = 0, \quad u = u(x,y,z,t).$

Solutions.

a) The matrix determined (in an arbitrary point $x^0 \in \mathbf{R}^3$) by the given equation is

$$\mathbf{A} = \begin{bmatrix} 1 & 1 & -1 \\ 1 & 2 & 0 \\ -1 & 0 & 6 \end{bmatrix}$$

Since $\mathbf{BAB}^T = \mathbf{E}$, where $\mathbf{B} = \mathbf{E}_{3(\frac{1}{2})}\mathbf{E}_{3,2(-1)}\mathbf{E}_{3,1(1)}\mathbf{E}_{2,1(-1)}$, the equation is elliptic on $\mathbf{R}^3$. By the change of variables

$$\begin{bmatrix} \xi \\ \eta \\ \zeta \end{bmatrix} = \mathbf{B}\begin{bmatrix} x \\ y \\ z \end{bmatrix} = \begin{bmatrix} x \\ y - x \\ x - \dfrac{y}{z} + \dfrac{z}{2} \end{bmatrix},$$

the equation transforms to

$$\frac{\partial^2 u}{\partial \xi^2} + \frac{\partial^2 u}{\partial \eta^2} + \frac{\partial^2 u}{\partial \zeta^2} = 0.$$

b) The matrix determined by the given equation is

$$\mathbf{A} = \begin{bmatrix} 4 & -2 & 0 \\ -2 & 0 & -1 \\ 0 & -1 & 0 \end{bmatrix}.$$

Since

$$\mathbf{BAB}^T = \begin{bmatrix} 1 & 0 & 0 \\ 0 & -1 & 0 \\ 0 & 0 & 1 \end{bmatrix}$$

where $\mathbf{B} = \mathbf{E}_{1(\frac{1}{2})}\mathbf{E}_{3,2(-1)}\mathbf{E}_{2,1(\frac{1}{2})}$, the equation is hyperbolic on $\mathbf{R}^3$, and by the change of variables

$$\xi = \frac{x}{2}, \quad \eta = \frac{x}{2} + y, \quad \zeta = -\frac{x}{2} - y + z,$$

it transforms to the canonical form

$$\frac{\partial^2 u}{\partial \xi^2} - \frac{\partial^2 u}{\partial \eta^2} + \frac{\partial^2 u}{\partial \zeta^2} + \frac{\partial u}{\eta} = 0.$$

c) The equation is parabolic on $\mathbf{R}^3$. It transforms to the canonical form by the change of variables:

$$\zeta = x + y, \quad \eta = -x + y, \quad \zeta = y + z.$$

d) The equation is elliptic on $\mathbf{R}^4$. It transforms to the canonical form by the change of variables:

$$\zeta = x, \quad \eta = -x + y, \quad \zeta = x - y + z, \quad \tau = 2x - 2y + z + t$$

Example 3.14 *Determine the canonical forms of the following equations:*

a) $\dfrac{\partial^2 u}{\partial x_1^2} + 2\displaystyle\sum_{k=2}^{n} \dfrac{\partial^2 u}{\partial x_k^2} - 2\displaystyle\sum_{k=1}^{n-1} \dfrac{\partial^2 u}{\partial x_k \partial x_{k+1}} = 0;$

b) $\dfrac{\partial^2 u}{\partial x_1^2} - 2\displaystyle\sum_{k=2}^{n} (-1)^k \dfrac{\partial^2 u}{\partial x_{k-1} \partial x_k} = 0;$

c) $\displaystyle\sum_{k=1}^{n} \dfrac{\partial^2 u}{\partial x_k^2} + 2\displaystyle\sum_{\ell=1}^{k} \ell \dfrac{\partial^2 u}{\partial x_\ell \partial x_k} = 0.$

Solutions.

a) The canonical from of the equation is $\displaystyle\sum_{k=1}^{n} \dfrac{\partial^2 u}{\partial \zeta_k^2} = 0$, where

$$\zeta_k = \sum_{\ell=1}^{k} x_\ell, \quad k = 1, 2, ..., n.$$

b) The canonical from of the equation is

$$\sum_{k=1}^{n} (-1)^{k+1} \frac{\partial^2 u}{\partial \zeta_k^2} = 0, \quad \zeta_k = \sum_{\ell=1}^{k} x_\ell, \ k = 1, 2, ..., n.$$

c) The canonical from of the equation is $\sum\limits_{k=1}^{n} \dfrac{\partial^2 u}{\partial \zeta_k^2} = 0$, where $\zeta_1 = x_1, \zeta_k = x_k - x_{k-1}$,

$k = 2, 3, ..., n$.

Example 3.15 *Show that the Laplace equation $\Delta u = 0$ is invariant under an orthogonal transformation:*

$$(x_1, ..., x_n) \mapsto (y_1, ..., y_n), \quad y_i = \sum_{j=1}^{n} a_{ij} x_j \quad (i = 1, 2, ..., n),$$

where $\sum\limits_{i=0}^{n} a_{ki} a_{\ell i} = \delta_{k\ell}$.

Exercise 3.16 *The Lorentz transform is given by*

$$(x_1, ..., x_n) \mapsto (y_1, ..., y_n), \quad y_i = \sum_{j=1}^{n} a_{ij} x_j \quad (i = 1, 2, ..., n),$$

where

$$a_{11}^2 - \sum_{i=2}^{n} a_{1i}^2 = 1, \quad a_{k1} a_{\ell 1} - \sum_{i=2}^{n} a_{ki} a_{\ell i} = \delta_{k\ell} \quad (k, \ell = 1, 2, ..., n, k + \ell > 2).$$

a) *Show that the wave equation is invariant under the Lorentz transformation*

$$\frac{\partial^2 u}{\partial x_1^2} - \left(\frac{\partial^2 u}{\partial x_2^2} + \cdots + \frac{\partial^2 u}{\partial x_n^2} \right) = 0.$$

b) *Show that in the case $n = 2$ the Lorentz transformation can be represented in the following from:*

$$(x_1, x_2) \mapsto (x_1 \cosh \theta + x_2 \cosh \theta, x_1 \sinh \theta + x_2 \sinh \theta),$$

where $\theta = -\ln_n |\alpha|$ and α is a nonzero parameter.

c) *Show that, in the case $n \geq 2$, the Lorentz transformation can be represented as a composition of the special Lorentz transformation (for the case $n = 2$) and orthogonal transformations.*

Example 3.17 *Determine the characteristic manifold of the partial differential equation:*

$$\frac{\partial^2 u}{\partial t^2} + \frac{c}{t} \frac{\partial u}{\partial t} - \left(\frac{\partial^2 u}{\partial x_1^2} + \cdots + \frac{\partial^2 u}{\partial x_n^2} \right) = 0,$$

where $u = u(x_1, \ldots, x_n, t)$.

Solution. The characteristic manifold of a quasi–linear equation is a manifolds M such that its unit normal $\eta = (\eta_t, \eta_{x_1}, \ldots, \eta_{x_n})$ satisfies for $(t, x) \in M$ the following relations:

$$\eta_t^2 + \sum_{i=1}^{n} \eta_{x_i}^2 = 1 \quad \text{and} \quad \eta_t^2 - \sum_{i=1}^{n} \eta_{x_i}^2 = 0.$$

The above is equivalent to $\eta_t = \pm 1/\sqrt{2}$, i.e., a manifold M is characteristic for the given equation if and only if the angle between M and $t-$ axis is $\pi/4$.

3.3 Wave, Potential and Heat Equation

The first of the three most important second order PDEs is the one–dimensional nonhomogeneous *wave equation*, given by

$$\frac{\partial^2 u}{\partial t^2} - a^2 \frac{\partial^2 u}{\partial x^2} = F(x, t) \quad (x \in [0, \ell], \ t \in [0, \infty)). \tag{3.35}$$

The wave equation describes, e.g., the displacement $u = u(x, t)$ of a taut string, fixed in its endpoints $x = 0$ and $x = \ell$, at a point $x \in [0, \ell]$ and at a moment $t > 0$. The positive constant $a > 0$ is depending on the density of the material the string was made from. The function F on the right–hand side is the external force exerted on the string. With (3.35) one usually imposes *initial conditions*

$$u(x, 0) = f(x), \quad \frac{\partial u}{\partial t}(x, 0) = g(x) \quad (x \in [0, \ell]), \tag{3.36}$$

and *boundary conditions*

$$u(0, t) = u(\ell, t) = 0 \quad (t \geq 0). \tag{3.37}$$

In (3.36), $f \in C^2[0, \ell]$ and $g \in C^1[0, \ell]$ are the initial deflection and the velocity of the string, while (3.37) merely expresses the fact that the string is fixed at the points $x = 0$ and $x = \ell$.

The next important second order PDE is the *Poisson's equation*, given by

$$\frac{\partial^2 u}{\partial x^2} + \frac{\partial^2 u}{\partial y^2} = G(x, y), \quad (x, y) \in Q, \tag{3.38}$$

where $u = u(x, y)$ is the potential at a point (x, y) from some region $Q \subset \mathbf{R}^2$, and G is a given function on Q. In particular, if $G = 0$ on Q, then (3.38) is called *Laplace's equation* (or *potential equation*). Let us denote by ∂Q the boundary of Q. Then one imposes on the function u either the *Dirichlet boundary condition*:

$$u\big|_{\partial Q} = f, \tag{3.39}$$

which specifies the value of u on ∂Q, or the *Neumann boundary condition*:

$$\left.\frac{\partial u}{\partial n}\right|_{\partial Q} = f, \tag{3.40}$$

which specifies the rate of change of u at points on ∂Q in a direction outwardly perpendicular to ∂Q.

The third important second order PDE is the *heat equation*, given by

$$\frac{\partial u}{\partial t} - a^2 \frac{\partial^2 u}{\partial x^2} = H(x,t), \tag{3.41}$$

where $u = u(x,t)$ is the temperature at a point $x \in (0, \ell)$ and at a moment $t > 0$ of a wire of length ℓ. Finally, H is the heat source affecting the wire and that it is made from a homogeneous material; $a > 0$ is a constant depending on this material. With (3.41) we impose the following initial condition

$$u(x,0) = f(x) \quad (0 \leq x \leq \ell), \tag{3.42}$$

i.e., f is the initial temperature of the wire, and, e.g.,

$$u(0,t) = u(\ell,t) = 0 \quad (t \geq 0), \tag{3.43}$$

i.e., the ends of the wire are being kept on the zero temperature.

Chapter 4

Hyperbolic Equations

4.1 Cauchy Problem for the One–dimensional Wave Equation

4.1.1 Preliminaries

The one–dimensional wave equation given on the set $\Omega_2 = \{(x,t) \mid x \in \mathbf{R}, t > 0\}$,

$$\frac{\partial^2 u}{\partial t^2} - a^2 \frac{\partial^2 u}{\partial x^2} = F(x,t), \tag{4.1}$$

for $a > 0$, with conditions

$$u(x,0) = f(x), \qquad \frac{\partial u(x,0)}{\partial t} = g(x) \quad (x \in \mathbf{R}) \tag{4.2}$$

where $f \in C^2(\mathbf{R})$, $g \in C^1(\mathbf{R})$ and $F \in C^2(\Omega_2)$ are given functions, is called the *Cauchy problem* for one dimensional wave equation. If $F = 0$, then we are dealing with a *homogeneous wave equation*, otherwise it is a *nonhomogeneous wave equation*.

The classical solution of problem (4.1), (4.2) is the function $u = u(x,t) \in C^2(\Omega_2)$ given by *D'Alambert's formula*

$$u(x,t) = \frac{f(x+at) + f(x-at)}{2} + \frac{1}{2a} \int_{x-at}^{x+at} g(s)ds$$

$$+ \frac{1}{2a} \int_0^t \int_{x-at+av}^{x+at-av} F(w,v)dwdv. \tag{4.3}$$

71

4.1.2   Examples and Exercises

Example 4.1 *Prove D'Alambert's formula (4.3) for homogeneous one dimensional wave equation given by*

$$\frac{\partial^2 u}{\partial t^2} - a^2 \frac{\partial^2 u}{\partial x^2} = 0 \qquad (x \in \mathbf{R}, \ t > 0),$$

with initial conditions

$$u(x,0) = f(x), \qquad \frac{\partial u(x,0)}{\partial t} = g(x) \qquad (x \in \mathbf{R}),$$

where a is a positive constant.

Solution. Let us change the variables $v = x + at$, $w = x - at$, and denote by

$$U(v,w) = u(x(v,w), t(v,w)),$$

then the considered wave equation can be written as

$$\frac{\partial^2 U}{\partial v \partial w} = 0.$$

The solution of the last equation is of the form

$$U(v,w) = F(v) + G(w),$$

where F and G are arbitrary continuous functions with continuous second derivatives. Therefore the solution of the considered equation is

$$u(x,t) = F(x + at) + G(x - at).$$

Using the initial conditions we get

$$u(x,0) = f(x) = F(x) + G(x) \qquad (x \in \mathbf{R}),$$

$$\frac{\partial u(x,0)}{\partial t} = g(x) = aF'(x) + aG'(x) \qquad (x \in \mathbf{R}).$$

From these two equations (after differentiation of the first one) we obtain

$$F'(x) = \frac{1}{2a}(af'(x) + g(x)),$$

wherefrom it follows that the functions F and G have the forms

$$F(x) = \frac{1}{2}f(x) + \frac{1}{2a}\int_0^x g(\xi)d\xi + C, \quad G(x) = \frac{1}{2}f(x) - \frac{1}{2a}\int_0^x g(\xi)d\xi - C$$

or

$$F(x+at) = \frac{1}{2}f(x+at) + \frac{1}{2a}\int_0^{x+at} g(\xi)d\xi + C,$$

$$G(x-at) = \frac{1}{2}f(x-at) - \frac{1}{2a}\int_0^{x-at} g(\xi)d\xi - C.$$

The solution of the considered problem is

$$u(x,t) = \frac{1}{2}\left(f(x+at) + f(x-at)\right) + \frac{1}{2a}\int_{x-at}^{x+at} g(\xi)d\xi. \tag{4.4}$$

Thus we obtain the D'Alambert's formula as in (4.3).

Example 4.2 *Check the validity of the formal solution given by relation (4.4) for the Cauchy problem for homogeneous one–dimensional wave equation.*

Solution. Let us check that the solution given by relation (4.4)

$$u(x,t) = \frac{1}{2}\left(f(x+at) + f(x-at)\right) + \frac{1}{2a}\int_{x-at}^{x+at} g(\xi)d\xi,$$

satisfies the wave equation with appropriate conditions. From

$$\frac{\partial u}{\partial t} = \frac{a}{2}\left(f'(x+at) - f'(x-at)\right) + \frac{1}{2}\left(g(x+at) + g(x-at)\right),$$

$$\frac{\partial^2 u}{\partial t^2} = \frac{a^2}{2}\left(f''(x+at) + f''(x-at)\right) + \frac{a}{2}\left(g'(x+at) - g'(x-at)\right),$$

$$\frac{\partial u}{\partial x} = \frac{1}{2}\left(f'(x+at) + f'(x-at)\right) + \frac{1}{2a}\left(g(x+at) - g(x-at)\right),$$

$$\frac{\partial^2 u}{\partial x^2} = \frac{1}{2}\left(f''(x+at) + f''(x-at)\right) + \frac{1}{2a}\left(g'(x+at) - g'(x-at)\right),$$

it follows that

$$\frac{\partial^2 u}{\partial t^2} = a^2 \frac{\partial^2 u}{\partial x^2}.$$

Also, we have

$$u(x,0) = f(x), \quad \frac{\partial u}{\partial t} = \frac{a}{2}\left(f'(x) - f'(x)\right) + \frac{1}{2}\left(g(x) + g(x)\right) = g(x).$$

Example 4.3 *Let the following Cauchy problem on the set $\{(x,t)|\, x \in \mathbf{R}, t > 0\}$ be given:*

$$\frac{\partial^2 u}{\partial t^2} = a^2 \frac{\partial^2 u}{\partial x^2},$$

$$u(x,0) = f(x), \qquad \frac{\partial u(x,0)}{\partial t} = g(x) \qquad (x \in \mathbf{R}),$$

where $a > 0$, $f \in C^2(\mathbf{R})$ and $g \in C^1(\mathbf{R})$. Show that the following properties hold.

a) *If the functions f and g are odd in x, for every fixed $t > 0$, meaning that $f(x,t) = -f(-x,t)$, then the function $u(0,t)$, for every fixed $t > 0$ is necessarily equal to 0.*

b) *If the functions f and g are even, for every fixed $t > 0$, then $\dfrac{\partial u(x,0)}{\partial x} = 0$.*

Solution. From D'Alambert's formula

$$u(x,t) = \frac{f(x+at) + f(x-at)}{2} + \frac{1}{2a} \int\limits_{x-at}^{x+at} g(\xi)d\xi,$$

one can conclude the following.

a) If the functions f and g are odd then

$$u(0,t) = \frac{f(at) + f(-at)}{2} + \frac{1}{2a} \int\limits_{x-at}^{x+at} g(\xi)d\xi = 0.$$

b) If the functions f and g are even then

$$\frac{\partial u(0,t)}{\partial x} = \frac{f'(at) + f'(-at)}{2} + \frac{g(at) + g(-at)}{2} = 0$$

(the first derivative of even function is the odd one).

Example 4.4 *Solve the following problems*

a) $\dfrac{\partial^2 u}{\partial t^2} = a^2 \dfrac{\partial^2 u}{\partial x^2} \qquad (x > 0, \quad t > 0),$

$$u(0,t) = 0, \qquad u(x,0) = f(x), \qquad \frac{\partial u(x,0)}{\partial t} = g(x) \qquad (x > 0),$$

b) $\dfrac{\partial^2 u}{\partial t^2} = a^2 \dfrac{\partial^2 u}{\partial x^2} \qquad (x > 0, \quad t > 0),$

$$\frac{\partial u(0,t)}{\partial x} = 0, \qquad u(x,0) = f(x), \qquad \frac{\partial u(x,0)}{\partial t} = g(x) \qquad (x > 0).$$

Solutions.

a) Let us introduce the functions

$$F(x) = \begin{cases} f(x), & x > 0 \\ -f(-x), & x < 0, \end{cases} \qquad G(x) = \begin{cases} g(x), & x > 0 \\ -g(-x), & x < 0. \end{cases}$$

Then the Cauchy problem

$$\frac{\partial^2 U}{\partial t^2} = a^2 \frac{\partial^2 U}{\partial x^2} \qquad (-\infty < x < \infty, \quad t > 0),$$

$$U(x,0) = F(x), \qquad \frac{\partial U(x,t)}{\partial t} = G(x) \qquad (-\infty < x < \infty),$$

(4.5)

has the solution given by

$$U(x,t) = \frac{F(x+at) + F(x-at)}{2} + \frac{1}{2a} \int_{x-at}^{x+at} G(\xi) d\xi.$$

The functions F and G are odd and therefore it holds

$$U(0,t) = 0$$
$$U(x,0) = F(x) = f(x), \qquad \frac{\partial U}{\partial t} = g(x) \qquad (x > 0).$$

This means that the solution of the problem (4.5), for $x > 0$, $t > 0$, is in fact the solution of the considered problem, i.e., $U(x,t) = u(x,t)$, for $x > 0$, $t > 0$, and it has the form

$$u(x,t) = \begin{cases} \dfrac{f(x+at) + f(x-at)}{2} + \dfrac{1}{2a} \displaystyle\int_{x-at}^{x+at} g(x) dx & \left(x > 0, \ t < \dfrac{x}{a}\right), \\[4mm] \dfrac{f(x+at) - f(at-x)}{2} + \dfrac{1}{2a} \displaystyle\int_{x-at}^{x+at} g(x) dx & \left(x > 0, \ t > \dfrac{x}{a}\right). \end{cases}$$

b) Introducing the even functions

$$F(x) = \begin{cases} f(x), & x > 0, \\ f(-x), & x < 0, \end{cases} \qquad G(x) = \begin{cases} g(x), & x > 0, \\ g(-x), & x < 0, \end{cases}$$

and solving the Cauchy problem

$$\frac{\partial^2 U}{\partial t^2} = a^2 \frac{\partial^2 U}{\partial x^2} \qquad (-\infty < x < \infty, \quad t > 0),$$

(4.6)

$$U(x,0) = F(x), \qquad \frac{\partial U(x,0)}{\partial t} = G(x) \qquad (-\infty < x < \infty),$$

we obtain

$$\frac{\partial U(x,0)}{\partial x} = 0$$

$$U(x,0) = F(x) = f(x), \qquad \frac{\partial U}{\partial t} = g(x) \qquad (x > 0).$$

In this case we also have, $U(x,t) = u(x,t)$ $(x > 0, t > 0)$. So we obtain the solution in the form

$$u(x,t) = \begin{cases} \dfrac{f(x+at) + f(x-at)}{2} + \dfrac{1}{2a} \displaystyle\int_{x-at}^{x+at} g(x)dx & \left(x > 0, \ t < \dfrac{x}{a} \right), \\[3em] \dfrac{f(x+at) - f(at-x)}{2} \\[2em] \quad + \dfrac{1}{2a} \left(\displaystyle\int_{0}^{x+at} g(x)dx - \displaystyle\int_{0}^{at-x} g(x)dx \right) & \left(x > 0, \ t > \dfrac{x}{a} \right). \end{cases}$$

Example 4.5 *Let us consider the nonhomogeneous equation*

$$\frac{\partial^2 u}{\partial t^2} = a^2 \frac{\partial^2 u}{\partial x^2} + f(x,t) \quad (-\infty < x < \infty, \quad t > 0),$$

with homogeneous conditions

$$u(x,0) = \frac{\partial u(x,0)}{\partial t} = 0 \qquad (-\infty < x < \infty).$$

Show the following states.

a) *If the function f is an odd one by x, for every fixed t, then $u(0,t) = 0$.*

b) *If the function f is an even function by x, for every fixed t, then*

$$\frac{\partial u(0,t)}{\partial x} = 0.$$

Solutions. Since the initial conditions are homogeneous the solution of the considered problem is

$$u(x,t) = \frac{1}{2a} \int_0^t \int_{x-a(t-\tau)}^{x+a(t-\tau)} f(z,\tau)\,dz\,d\tau.$$

a) Therefore, if f is an odd function by x for every fixed t, then it holds

$$u(0,t) = \frac{1}{2a} \int_0^t d\tau \int_{-a(t-\tau)}^{a(t-\tau)} f(z,\tau)\,dz = 0.$$

b) If f is an even function by x, for every fixed t, then it holds

$$\frac{\partial u(0,t)}{\partial x} = \frac{1}{2a} \int_0^t \left(f(a(t-\tau),\tau) - f(-a(t-\tau),\tau) \right) d\tau = 0.$$

Example 4.6 *Solve the following problems*

a) $\dfrac{\partial^2 u}{\partial t^2} = a^2 \dfrac{\partial^2 u}{\partial x^2} + f(x,t)$ \qquad $(x > 0, \quad t > 0),$

$u(x,0) = \dfrac{\partial u(x,0)}{\partial t} = 0$ \qquad $(x > 0),$ \qquad $u(0,t) = 0$ \quad $(t > 0).$

b) $\dfrac{\partial^2 u}{\partial t^2} = a^2 \dfrac{\partial^2 u}{\partial x^2} + f(x,t)$ \qquad $(x > 0, \quad t > 0),$

$u(x,0) = \dfrac{\partial u(x,0)}{\partial t} = 0$ \qquad $(x > 0),$ \qquad $\dfrac{\partial u(0,t)}{\partial x} = 0$ \quad $(t > 0).$

Solution.

a) Let us introduce the following odd function in x

$$F(x,t) = \begin{cases} f(x,t), & x > 0 \\ -f(-x,t), & x < 0. \end{cases}$$

Using the solution of of the Cauchy problem

$$\frac{\partial^2 U}{\partial t^2} = a^2 \frac{\partial^2 U}{\partial x^2} + F(x,t) \qquad (-\infty < x < \infty, \quad t > 0),$$

$$U(x,0) = \frac{\partial U(x,0)}{\partial t} = 0 \qquad (-\infty < x < \infty),$$

and Example 4.5, we obtain the solution of considered problem as

$$
u(x,t) = \begin{cases}
\dfrac{1}{2a} \displaystyle\int_0^t \int_{x-a(t-\tau)}^{x+a(t-\tau)} f(z,\tau)\,dz\,d\tau & \left(x>0,\ t<\dfrac{x}{a}\right), \\[2em]
\dfrac{1}{2a} \displaystyle\int_0^{t-x/a} \int_{a(t-\tau)-x}^{a(t-\tau)+x} f(z,\tau)\,dz\,d\tau & \\[2em]
\quad +\dfrac{1}{2a} \displaystyle\int_{t-x/a}^{t} \int_{x-a(t-\tau)}^{x+a(t-\tau)} f(z,\tau)\,dz\,d\tau & \left(x>0,\ t>\dfrac{x}{a}\right).
\end{cases}
$$

b) The solution of this problem is

$$
u(x,t) = \begin{cases}
\dfrac{1}{2a} \displaystyle\int_0^t \int_{x-a(t-\tau)}^{x+a(t-\tau)} f(z,\tau)\,dz\,d\tau & \left(x>0,\ t<\dfrac{x}{a}\right), \\[2em]
\dfrac{1}{2a} \displaystyle\int_0^{t-x/a} \left(\int_0^{a(t-\tau)-x} f(z,\tau)\,dz\,d\tau + \int_0^{x+a(t-\tau)} f(z,\tau)\,dz\,d\tau \right) & \\[2em]
\quad +\dfrac{1}{2a} \displaystyle\int_{t-x/a}^{t} \int_{x-a(t-\tau)}^{x+a(t-\tau)} f(z,\tau)\,dz\,d\tau & \left(x>0,\ t>\dfrac{x}{a}\right).
\end{cases}
$$

Example 4.7. *Let us consider the homogeneous wave equation (4.1), (4.2) ($F(x,t) = 0$), i.e.,*

$$
\frac{\partial^2 u}{\partial t^2} - c^2 \frac{\partial^2 u}{\partial x^2} = 0 \quad (x \in \mathbf{R}, \ t > 0),
$$

with initial conditions

$$
u(x,0) = f(x), \quad \frac{\partial u}{\partial t}(x,0) = g(x) \quad (x \in \mathbf{R}),
$$

where $c > 0$, $f \in C^2(\mathbf{R})$ and $g \in C^1(\mathbf{R})$ are given functions. Suppose there exist constants $A \neq 0$, $B \neq 0$ and $b > 0$, such that

$$
\lim_{|x|\to\infty} \frac{f(x)}{|x|^b} = A, \qquad \lim_{|x|\to\infty} \frac{g(x)}{|x|^{b-1}} = B.
$$

Show then the existence of a constant C with the property

$$\lim_{t\to\infty} \frac{u(x,t)}{t^b} = C.$$

Determine the constant C.

Solution. We know that $u(x,t)$ given by (4.3) is the solution of the problem (4.1), (4.2) with $F(x,t) = 0$. Introducing the variable s by $s = vt + x$, the first integral in (4.3) is transformed as

$$\int_{x-ct}^{x+ct} g(s)ds = t \int_{-c}^{c} g(vt + x)\,dv.$$

Fix $x \in \mathbf{R}$. Then we can write

$$\lim_{t\to\infty} \frac{u(x,t)}{t^b} = \lim_{t\to\infty} \frac{f(x+ct) + f(x-ct)}{2t^b} + \lim_{t\to\infty} \frac{1}{2c} \int_{-c}^{c} \frac{g(vt+x)}{t^{b-1}}\,dv$$

$$= Ac^b + \frac{1}{2c} \int_{-c}^{c} \left(\lim_{t\to\infty} \frac{\dfrac{g(vt+x)}{|vt+x|^{b-1}}}{\dfrac{t^{b-1}}{|vt+x|^{b-1}}} \right) dv = Ac^b + \frac{B}{c} \int_{0}^{c} |v|^{b-1}\,dv$$

$$= Ac^b + \frac{Bc^{b-1}}{b}.$$

Thus $C = Ac^b + \dfrac{Bc^{b-1}}{b}$.

The interchange of the limit as $t \to \infty$ and the integral was allowed in view of the assumptions on g.

Example 4.8. *Let the Cauchy problem for the homogeneous wave equation be given on the set $\{(x,t)|\, x \in \mathbf{R}, t > 0\}$, i.e.,*

$$\frac{\partial^2 u}{\partial t^2} - a^2 \frac{\partial^2 u}{\partial x^2} = F(x,t),$$

$$u(x,0) = f(x), \quad \frac{\partial u(x,0)}{\partial t} = g(x) \quad (x \in \mathbf{R}),$$

where $f \in C^2(\mathbf{R})$ and $g \in C^1(\mathbf{R})$ are given functions. Suppose the functions f and g satisfy the following inequalities:

$$m|x|^\alpha \le f(x) \le M|x|^\alpha, \quad m|x|^{\alpha-1} \le g(x) \le M|x|^{\alpha-1}, \tag{4.7}$$

for $|x| \geq \delta > 0$, $\alpha > 0$ and $0 < m < M$. Show then that for every $x_0 \in \mathbf{R}$ there exist constants t_0, C_1 and $C_2 > 0$, such that it holds

$$C_1 t^\alpha \leq u(x_0, t) \leq C_2 t^\alpha, \quad t \geq t_0. \tag{4.8}$$

Solution. Using formula (4.3) (for $F(x,t) = 0$), the solution of the given problem at the point (x_0, t) is

$$u(x_0, t) = \frac{1}{2} \left(f(x_0 + at) + f(x_0 - at) \right) + \frac{1}{2a} \int\limits_{x_0 - at}^{x_0 + at} g(\xi)\, d\xi.$$

We shall only prove the right–hand side inequality in (4.8). Suppose $x_0 > 0$ is given; the case $x_0 \leq 0$ is analogous and is left to the reader. Then for t sufficiently large from the first assumption in (4.7) it follows

$$\frac{f(x_0 + at) + f(x_0 - at)}{2} \leq \frac{M}{2} \left(|x_0 + at|^\alpha + |x_0 - at|^\alpha \right).$$

Next we have for $t > 0$ with the property $|x_0 - at| \geq \delta$:

$$\frac{1}{2a} \int\limits_{x_0 - at}^{x_0 + at} g(\xi) d\xi \leq \frac{M}{2a} \int\limits_{x_0 - at}^{x_0 + at} |\xi|^{\alpha - 1} d\xi = \frac{M}{2\alpha} \left(|x_0 + at|^\alpha - |x_0 - at|^\alpha \right).$$

Since it holds

$$\lim_{t \to \infty} \left(\frac{\frac{M}{2} \left(|x_0 + at|^\alpha + |x_0 - at|^\alpha \right)}{t^\alpha} \right) = M a^\alpha,$$

and

$$\lim_{t \to \infty} \left(\frac{\frac{M}{2} \left(|x_0 + at|^\alpha - |x_0 - at|^\alpha \right)}{t^\alpha} \right) = \frac{M}{2} a^{\alpha - 1} (1 - (-1)^\alpha),$$

there exist constants $C_2 = C_2(x_0, a, \alpha)$ and $t_0 = t_0(x_0, a, \alpha)$ such that for $t > t_0$ it holds

$$u(x_0, t) \leq C_2 |t|^\alpha.$$

4.2 Cauchy Problem for the n–dimensional Wave Equation

4.2.1 Preliminaries

We always suppose that the boundary ∂Q of a bounded region $Q \subset \mathbf{R}^n$ is sufficiently regular so that we can apply the Gauss-Ostrogradsky (divergence) theorem for $u \in C^1(\overline{Q})$ (this is satisfied if, e.g., Q is a locally quadratic region, see Chapter 8).

Theorem 4.1 (Gauss-Ostrogradsky) *Let $Q \subset \mathbf{R}^n$ be a bounded region with sufficiently regular boundary ∂Q and $u \in C^1(\overline{Q})$. Then*

$$\int_Q D_k u(x)\, dx = \int_{\partial Q} u(x) \mathbf{n}_k\, dS_x,$$

where $\mathbf{n} = (\mathbf{n}_1, \ldots, \mathbf{n}_n)$ is the exterior unit normal of ∂Q and dS_x is the surface element with integration on x.

The n–dimensional Cauchy problem $n \geq 1$, on $\Omega_{n+1} = \{(x, t) | x \in \mathbf{R}^n, t > 0\}$

$$\frac{\partial^2 u}{\partial t^2} - \Delta u = F(x, t), \tag{4.9}$$

with initial conditions

$$u(x, 0) = f(x), \qquad \frac{\partial u}{\partial t}(x, 0) = g(x) \qquad (x \in \mathbf{R}^n)$$

where $f \in C^3(\mathbf{R}^n)$, $g \in C^2(\mathbf{R}^n)$, $F \in C^2(\Omega_{n+1})$, are given functions has the classical solution $u \in C^2(\Omega_{n+1})$.

The solution for two–dimensional case ($n = 2$) is given by the *Poisson's formula*

$$
\begin{aligned}
u(x, t) \;=\; & \frac{1}{2\pi} \iint_{D(x,t)} \frac{g(y_1, y_2) dy_1 dy_2}{\sqrt{t^2 - (y_1 - x_1)^2 - (y_2 - x_2)^2}} \\[2mm]
& + \frac{\partial}{\partial t} \left[\frac{1}{2\pi} \iint_{D(x,t)} \frac{f(y_1, y_2) dy_1 dy_2}{\sqrt{t^2 - (y_1 - x_1)^2 - (y_2 - x_2)^2}} \right] \\[2mm]
& + \frac{1}{2\pi} \int_0^t \iint_{|y-x|<t-s} \frac{F(y_1, y_2, t) dy_1 dy_2 ds}{\sqrt{(t - s)^2 - (y_1 - x_1)^2 - (y_2 - x_2)^2}},
\end{aligned}
\tag{4.10}
$$

where $D(x, t)$ is a disc with the center in $x = (x_1, x_2)$ and diameter t.

The solution for the three–dimensional case ($n = 3$) is given by *Kirchoff's formula*

$$
\begin{aligned}
u(x, t) \;=\; & \frac{1}{4\pi} \int_{\partial S_1} g(x + tv) dS_V + \frac{1}{4\pi} \frac{\partial}{\partial t} \left[\frac{1}{t} \int_{\partial S_1} f(x + tv) dS_V \right] \\[2mm]
& + \frac{1}{4\pi} \int_{\partial S_t} \frac{1}{|y - x|} F(y_1, y_2, y_3, t - |y - x|) dy_1 dy_2 dy_3,
\end{aligned}
\tag{4.11}
$$

where $S_t = \{y \,||x - y| < 1\}$, and $\partial S_1 = \{y \,||x - y| = 1\}$.

4.2.2   Examples and Exercises

Example 4.9 *Let $Q \subset \mathbf{R}^n$ be a bounded region with sufficiently regular boundary ∂Q. Prove that for $u \in C^2(\overline{Q}), v \in C^1(\overline{Q})$ we have the first (antisymmetric) Green formula*

$$\int_Q v\Delta u\, dx = -\int_Q \nabla u \cdot \nabla v\, dx + \int_{\partial Q} v\frac{\partial u}{\partial \mathbf{n}}\, dS, \qquad (4.12)$$

where $\mathbf{n}$ is the normal on ∂Q from Q outside and $\dfrac{\partial u}{\partial \mathbf{n}}$ is the derivative of the function u in the direction of $\mathbf{n}$.

Solution. Apply on $\int_Q v\Delta u\, dx$ the product derivation formula and the Gauss-Ostrogradsky theorem.

Example 4.10 *Let $Q \subset \mathbf{R}^n$ be a bounded region with a sufficiently regular boundary ∂Q. Prove that for $u, v \in C^2(\overline{Q})$ we have the second (symmetric) Green formula*

$$\int_Q (v\Delta u - u\Delta v)\, dx = \int_{\partial Q} \left(v\frac{\partial u}{\partial \mathbf{n}} - u\frac{\partial v}{\partial \mathbf{n}} \right) dS. \qquad (4.13)$$

Solution. Changing the position of functions u and v in (4.12) and subtracting the obtained equality from (4.12) we obtain (4.13).

Example 4.11 *Show that the D'Alambert formula (4.3), giving the solution of the problem (4.1), (4.2) for $F = 0$, can be obtained from Poisson's formula (4.10) for $n = 1$, namely for the problem*

$$\frac{\partial^2 u}{\partial t^2} - \frac{\partial^2 u}{\partial x^2} = 0 \qquad (x \in \mathbf{R},\ t > 0),$$

$$u(x,0) = f(x), \quad \frac{u(x,0)}{\partial t} = g(x) \qquad (x \in \mathbf{R}),$$

where $f \in C^3(\mathbf{R}),\ g \in C^2(\mathbf{R})$.

Solution. Starting from (4.10) we obtain

$$u(x,t) = \frac{1}{2\pi}\int_{-t}^{t} f(x+\xi)d\xi \int_{-\sqrt{t^2-\xi^2}}^{\sqrt{t^2-\xi^2}} \frac{d\eta}{\sqrt{t^2-\xi^2-\eta^2}}$$

$$+\frac{1}{2\pi}\frac{\partial}{\partial t}\int_{-t}^{t} f(x+\xi)d\xi \int_{-\sqrt{t^2-\xi^2}}^{\sqrt{t^2-\xi^2}} \frac{d\eta}{\sqrt{t^2-\xi^2-\eta^2}}$$

$$= \frac{1}{2\pi}\int_{-t}^{t} g(x+\xi)d\xi + \frac{1}{2}\frac{\partial}{\partial t}\int_{-t}^{t} f(x+\xi)d\xi$$

$$= \frac{1}{2}\left(f(x+t)+g(x-t)\right) + \frac{1}{2}\int_{x-t}^{x+t} g(\tau)d\tau.$$

Exercise 4.12 *If f_1, $f_2 \in C^2(\mathbf{R})$ and g_1, $g_2 \in C^1(\mathbf{R})$ are given functions, then find the solutions two dimensional wave equation*

$$\frac{\partial^2 u}{\partial t^2} = a^2\left(\frac{\partial^2 u}{\partial x^2} + \frac{\partial^2 u}{\partial y^2}\right),$$

with the initial conditions

$$u(x,y,0) = f_1(x) + f_2(y), \qquad \frac{\partial u(x,y,0)}{\partial t} = g_1(x) + g_2(y).$$

Answer.

$$u(x,t) = \frac{f_1(x+at)+f_1(x-at)+f_2(y+at)+f_2(y-at)}{2}$$

$$+\frac{1}{2a}\int_{x-at}^{x+at} g_1(s)ds + \frac{1}{2a}\int_{y-at}^{y+at} g_1(s)ds.$$

Exercise 4.13 *Solve the following Cauchy problems for $u = u(x,y,t)$;*

a) $\dfrac{\partial^2 u}{\partial t^2} - \Delta u = x^3 - 3xy^2$, $u(x,y,0) = e^x\cos y$, $\dfrac{\partial u(x,y,0)}{\partial t} = e^y\sin x$;

b) $\dfrac{\partial^2 u}{\partial t^2} - \Delta u = t\sin y$, $u(x,y,0) = x^2$, $\dfrac{\partial u(x,y,0)}{\partial t} = \sin y$;

c) $\dfrac{\partial^2 u}{\partial t^2} - 2\Delta u = 0$, $u(x,y,0) = 2x^2 - y^2$, $\dfrac{\partial u(x,y,0)}{\partial t} = 2x^2 + y^2.$

Answers.

a) $\quad u(x,y,t) = \frac{1}{2}t^2(x^3 - 3xy^2) + e^x \cos y + te^y \sin x.$

b) $\quad u(x,y,t) = x^2 + t^2 + t\sin y.$

c) $\quad u(x,y,t) = 2x^2 - y^2 + (2x^2 + y^2)t + 2t^2 + 2t^3.$

Exercise 4.14 *Solve the following Cauchy problems for $u = u(x,y,z,t)$;*

a) $\quad \dfrac{\partial^2 u}{\partial t^2} - \Delta u = 2xyz, \quad u(x,y,z,0) = x^2 + y^2 - 2z^3, \quad \dfrac{\partial u(x,y,z,0)}{\partial t} = 1;$

b) $\quad \dfrac{\partial^2 u}{\partial t^2} - 8\Delta u = t^2 x^2, \quad u(x,y,z,0) = y^2, \quad \dfrac{\partial u(x,y,z,0)}{\partial t} = z^2.$

Answers.

a) $\quad u(x,y,t) = x^2 + y^2 - 2z^2 + t + t^2 xyz;$

b) $\quad u(x,y,t) = y^2 + 8t^2 + tz^2 + \dfrac{8}{3}t^3 + \dfrac{1}{12}t^4 x^2 + \dfrac{2}{45}t^6.$

Example 4.15 *The spheric mean value for a function $w \in C(\mathbf{R}^n)$ and $\partial B(0,r) = \{y|\ |x - y| = r\}$ is given by*

$$M_w(x,r) = \frac{1}{\sigma_n r^{n-1}} \int_{\partial B(0,r)} w(y)\, dS_y,$$

where $\sigma_n = \dfrac{2\pi^{n/2}}{\Gamma(n/2)}$. Prove that

a) $\quad M_w(x,r) = \dfrac{1}{\sigma_n} \displaystyle\int_{\partial B(0,1)} w(x + rv)\, dS_v;$ (4.14)

b) $\quad M_w$ *satisfies the Darboux equation*

$$\Delta_x u = \frac{\partial^2 u}{\partial r^2} + \frac{n-1}{r}\frac{\partial u}{\partial r}.$$ (4.15)

Solution.

a) Put $y = x + rv$ for $|v| = 1$. The equality (4.14) enables to extend M_w also for $r \leq 0$.

b) Differentiating (4.14) we obtain

$$\frac{\partial}{\partial r} M_w(x,r) = \frac{1}{\sigma_n} \int_{\partial B(0,1)} \sum_{i=1}^{n} w(x + rv)v_i \, dS_v.$$

Applying the divergence theorem on the integral at the right-hand side we obtain

$$\frac{\partial}{\partial r} M_w(x,r) = \frac{r}{\sigma_n} \int_{B(0,1)} \Delta_x w(x + rv)v_i \, dv$$

$$= \frac{r^{1-n}}{\sigma_n} \Delta_x \int_{B(x,r)} w(y) \, dy$$

$$= r^{1-n} \Delta_x \int_0^r s^{n-1} M_w(x,s) \, ds.$$

Multiplying both sides by r^{n-1} and applying the derivative with respect to r we obtain

$$\frac{\partial}{\partial r} \left(r^{n-1} \frac{\partial}{\partial r} M_w(x,r) \right) = \Delta_x r^{n-1} M_w(x,r),$$

which implies that M_w satisfies (4.15).

Example 4.16 *Prove that if a function $u = u(x,t) \in C^2(\Omega_{n+1})$ is a solution of the problem*

$$\frac{\partial^2 u}{\partial t^2} - \Delta u = F,$$

with initial conditions

$$u(x,0) = f(x), \quad \frac{\partial u}{\partial t}(x,0) = g(x) \qquad (x \in \mathbf{R}^n),$$

then the spheric mean value

$$M_u(x,r,t) = \frac{1}{\sigma_n} \int_{\partial B(0,1)} u(x + rv,t) \, dS_v \tag{4.16}$$

is a solution of the Cauchy (initial) problem for the Allure-Poisson-Darboux equation

$$\frac{\partial^2 M_u}{\partial t^2} = \frac{\partial^2 M_u}{\partial r^2} + \frac{n-1}{r} \frac{\partial M_u}{\partial r} \tag{4.17}$$

with initial conditions

$$M_u(x,y,0) = M_f(x,r), \quad \frac{\partial}{\partial t} M_u(x,r,0) = M_g(x,r). \tag{4.18}$$

Solution. By (4.15) we have for $M_u = M_u(x, r, t)$

$$\Delta_x M_u = \frac{\partial^2 M_u}{\partial r^2} + \frac{n-1}{r} \frac{\partial M_u}{\partial r}. \tag{4.19}$$

Since the function $u = u(x, t)$ is a solution of the wave equation we obtain by (4.16)

$$\Delta_x M_u = \frac{1}{\sigma_n} \int_{\partial B(0,1)} \Delta_x u(x + rv, t) \, dS_v$$

$$= \frac{\partial^2}{\partial t^2} \frac{1}{\sigma_n} \int_{\partial B(0,1)} u(x + rv, t) \, dS_v = \frac{\partial^2 M_u}{\partial t^2}.$$

Now substituting the last right-hand side in (4.19), we obtain (4.17).

Example 4.17 *Prove that the equation (4.17) for $n = 3$*

$$\frac{\partial^2 v}{\partial t^2} = \frac{\partial^2 v}{\partial r^2} + \frac{2}{r} \frac{\partial v}{\partial r} \qquad (0 \le r \le R, 0 < t \le T)$$

with the initial conditions

$$v(r, 0) = 0, \qquad \frac{\partial v(r, 0)}{\partial t} = f(r) \qquad (0 \le r \le R),$$

where $f \in C^1[0, R]$, has a unique solution

$$v(r, t) = \frac{1}{2r} \int_{r-t}^{r+t} g(s) \, ds,$$

where

$$g(s) = \frac{\partial}{\partial t}(sv(s, t))|_{t=0} = sf(s).$$

Solution. The function $w(r, t) = rv(r, t)$ satisfies the equation

$$\frac{\partial^2 w}{\partial t^2} - \frac{\partial^2 w}{\partial r^2} = 0$$

with the initial conditions

$$w(r, 0) = 0, \qquad \frac{\partial w(r, 0)}{\partial t} = rf(r)$$

and with the boundary condition $w(0, t) = 0$.
The last problem has a unique solution given by D'Alambert's formula

$$w(r, t) = \frac{1}{2} \int_{r-t}^{r+t} sf(s) \, ds = \frac{1}{2} \int_{r-t}^{r+t} g(s) \, ds.$$

Example 4.18 *Suppose that a function* $u = u(x_1, x_2, x_3, t)$ *is a solution of the Cauchy problem for the three-dimensional wave equation*

$$\frac{\partial^2 u}{\partial t^2} - \left(\frac{\partial^2 u}{\partial x_1^2} + \frac{\partial^2 u}{\partial x_2^2} + \frac{\partial^2 u}{\partial x_3^2} \right) = 0 \qquad (x = (x_1, x_2, x_3) \in \mathbf{R}^3, t > 0),$$

with the initial conditions

$$u(x, 0) = 0, \qquad \frac{\partial u(x, 0)}{\partial t} = g(x).$$

Prove that the function $v = \dfrac{\partial u}{\partial t}$ *also satisfies the same wave equation and the initial conditions*

$$v(x, 0) = 0, \qquad \frac{\partial v(x, 0)}{\partial t} = 0.$$

Solution. We have

$$\frac{\partial^2 v}{\partial t^2} = \frac{\partial}{\partial t} \left(\frac{\partial^2 u}{\partial t^2} \right) = \frac{\partial}{\partial t}(\Delta_x u) = \Delta_x \frac{\partial u}{\partial t} = \Delta_x v.$$

We have for the initial conditions

$$v(x, 0) = \frac{\partial u(x, 0)}{\partial t} = g(x),$$

$$\frac{\partial v(x, 0)}{\partial t} = \frac{\partial^2 u(x, 0)}{\partial t^2} = \Delta_x u(x, t)|_{t=0} = \Delta_x u(x, 0) = 0.$$

Example 4.19 (Kirchoff formula) *Prove that the Cauchy problem for three-dimensional wave equation*

$$\frac{\partial^2 u}{\partial t^2} - \Delta u = 0$$

with initial conditions

$$u(x, 0) = f(x), \qquad \frac{\partial u}{\partial t}(x, 0) = g(x) \quad (x \in \mathbf{R}^3),$$

where $f \in C^3(\mathbf{R}^3)$, $g \in C^2(\mathbf{R}^3)$ *are given functions has the solution*

$$
\begin{aligned}
u(x, t) &= \frac{1}{4\pi} \int_{\partial S_1} g(x + tv)\, dS_v + \frac{\partial}{\partial t} \left[\frac{1}{2\pi} \int_{\partial S_1} g(x + tv)\, dS_v \right] \\
&= t M_g(x, t) + \frac{\partial}{\partial t}[t M_f(x, t)].
\end{aligned}
$$

Solution. By Example 4.18 and linearity of the wave equation it is enough to prove that the function

$$v(x,t) = tM_f(x,t)$$

satisfies the equation

$$\frac{\partial^2 v}{\partial t^2} - \Delta_x v = 0$$

and the initial conditions

$$v(x,0) = 0, \quad \frac{\partial v(x,0)}{\partial t} = f(x) \qquad (x \in \mathbf{R}^3).$$

We have by Example 4.15 b)

$$\Delta_x v = t\Delta_x M_f$$

$$= t\left[\frac{\partial^2}{\partial r^2}(M_f) + \frac{2}{t}\frac{\partial M_f}{\partial r}\right] = t\frac{\partial^2 M_f}{\partial r^2} + 2\frac{\partial M_f}{\partial r}.$$

Therefore using the equality

$$\frac{\partial v}{\partial t} = M_f + t\frac{\partial M_f}{\partial r}$$

we obtain

$$\frac{\partial^2 v}{\partial t^2} = 2\frac{\partial M_f}{\partial r} + t\frac{\partial^2 M_f}{\partial r^2} = \Delta_x v.$$

We have for the initial conditions $v(x,0) = 0$ and

$$\frac{\partial v(x,0)}{\partial t} = M_f|_{t=0} = \frac{1}{4\pi}\int_{\partial B(0,1)} f(x_1,x_2,x_3)\,dS_v = f(x_1,x_2,x_3).$$

Example 4.20 (Poisson formula) *Prove that the Cauchy problem for two-dimensional wave equation*

$$\frac{\partial^2 u}{\partial t^2} - \Delta u = 0$$

with initial conditions

$$u(x,0) = f(x), \quad \frac{\partial u}{\partial t}(x,0) = g(x) \qquad (x \in \mathbf{R}^2),$$

where $f \in C^3(\mathbf{R}^2)$, $g \in C^2(\mathbf{R}^2)$ has the solution

$$u(x,t) = \frac{1}{2\pi}\int\int_{D(x,t)} \frac{g(y_1,y_2)dy_1 dy_2}{\sqrt{t^2 - (y_1-x_1)^2 - (y_2-x_2)^2}}$$

$$+ \frac{\partial}{\partial t}\left[\frac{1}{2\pi}\int\int_{D(x,t)} \frac{f(y_1,y_2)dy_1 dy_2}{\sqrt{t^2 - (y_1-x_1)^2 - (y_2-x_2)^2}}\right].$$

Solution. We suppose in Example 4.19 that the functions f and g are independent of x_3 as well as the solution $u = u(x_1, x_2, t)$. Let β be the angle $\angle ABB'$, where $A(x_1, x_2, 0)$, $B(x_1 + v_1 t, x_2 + v_2 t, x_3 + v_3 t)$ and $B'(y_1, y_2, 0)$. We have

$$dy_1 dy_2 = dS_v \cos \beta = \frac{(t^2 - (y_1 - x_1)^2 - (y_2 - x_2)^2)1/2}{t} dS_v.$$

Therefore

$$\frac{t}{4\pi} \int_{\partial B(0,1)} g(x + tv) \, dS_v = \frac{1}{4\pi t} \int_{\partial B(0,t)} g(x + tv) \, dS_v$$

$$= \frac{1}{4\pi} \int_{\partial B_t^+ \cup \partial B_t^-} g(x + tv) \frac{dS_t}{v}$$

$$= \frac{1}{2\pi} \int \int_{D(x,t)} \frac{g(y_1, y_2) dy_1 dy_2}{\sqrt{t^2 - (y_1 - x_1)^2 - (y_2 - x_2)^2}},$$

where ∂B_t^+ and ∂B_t^- are the half–spheres for $y_3 \geq 0$ and $y_3 \leq 0$, respectively. We can obtain in an analogous way the second summand in the Poisson formula.

4.3   The Fourier Method of Separation Variables

4.3.1   Preliminaries

Fourier Series

A function f is *piecewise continuous* on an interval $[a, b]$ if $[a, b]$ can be divided into a finite number of subintervals such that

(i) inside each of which f is continuous

(ii) the left-hand and right-hand limits exists at each point on the subintervals including their end points.

The left-hand and right-hand limits are defined, respectively, by

$$f(x^-) = \lim_{x \to 0^-} f(x), \quad f(x^+) = \lim_{x \to 0^+} f(x),$$

and the function f is continuous at the point x if $f(x^-) = f(x^+) = f(x)$.

Definition 4.2 *Let f be a $2\pi-$periodic piecewise continuous function on the interval $[-\pi, \pi]$. The trigonometric series*

$$A_0 + \sum_{n=1}^{\infty} (A_n \cos nx + B_n \sin nx), \tag{4.20}$$

is called the Fourier series *of function f, if the coefficients* A_n, $n = 0, 1, \ldots$, B_n, $n = 1, 2, \ldots$, *are given by*

$$A_0 = \frac{1}{2\pi} \int_{-\pi}^{\pi} f(x) dx,$$

$$A_n = \frac{1}{\pi} \int_{-\pi}^{\pi} f(x) \cos(nx) dx, \qquad n = 1, 2, \ldots, \qquad (4.21)$$

$$B_n = \frac{1}{\pi} \int_{-\pi}^{\pi} f(x) \sin(nx) dx, \qquad n = 1, 2, \ldots.$$

The coefficients A_n, $n = 0, 1, \ldots$, B_n, $n = 1, 2, \ldots$, *are called Fourier coefficients.*

Theorem 4.3 *The Fourier series of a periodic piecewise continuous function f on an interval* $[-\pi, \pi]$, *with piecewise continuous first derivative* f', *converges at any point* $x \in [\pi, \pi]$ *(converges pointwise for all values x). Then we have*

$$\frac{f(x^+) + f(x^-)}{2} = A_0 + \sum_{n=1}^{\infty} (A_n \cos nx + B_n \sin nx),$$

where A_n, $n = 0, 1, 2, \ldots$, B_n, $n = 1, 2, \ldots$, *are given by (4.21).*

Theorem 4.4 (Weierstrass test) *Let* $\sum_{n}^{\infty} u_n(x)$ *be a series of continuous functions* $u_n(x)$ *defined on* $[a, b]$. *Suppose there exists a sequence* $\{M_n\}_{n \in \mathbb{N}}$ *such that*

$$|u_n(x)| \leq M_n \qquad (x \in [a, b]),$$

and the series $\sum_{n=1}^{\infty} M_n$ *converges. Then the series* $\sum_{n=1}^{\infty} u_n(x)$ *converges uniformly on* $[a, b]$.

The following two theorems are giving two general criteria for uniform convergence of Fourier series.

Theorem 4.5 *Suppose the series* $\sum_{n=1}^{\infty} (|A_n| + |B_n|)$ *converges, where the Fourier coefficients* A_n *and* B_n *are given by (4.21). Then the Fourier series converges uniformly on every finite interval.*

Theorem 4.6 *The Fourier series of a continuous* $2\pi-$*periodic function f, with piecewise continuous first derivative* f', *converges uniformly on every finite interval.*

The following two theorems are giving the conditions for termwise differentiation and integration of Fourier series.

Theorem 4.7 *If f is a continuous $2\pi-$periodic function and both f' and f'' are piecewise continuous on the interval $[-\pi, \pi]$, then the Fourier series (4.20) can be differentiated termwise to the series*

$$f'(x) = \sum_{n=1}^{\infty} n \left(-A_n \sin nx + B_n \cos nx\right).$$

The last series converges pointwise to f' at the points where f'' exists.

Theorem 4.8 *If f is a piecewise continuous $2\pi-$periodic function of period 2π then its Fourier series (4.20) can be integrated termwise to the series*

$$\int_{-\pi}^{x} f(t)dt = A_0 \frac{x + \pi}{2} + \sum_{n=1}^{\infty} \frac{1}{n} \left(A_n \sin nx - B_n \cos nx + (-1)^n B_n\right).$$

The last series converges pointwise to the integral of f.

The change of variables $x = \pi t/\ell$, and $f(x) = f(\pi t/\ell) = g(t)$, imply the Fourier series for the function g as

$$A_0 + \sum_{n=1}^{\infty}(A_n \cos \frac{n\pi t}{\ell} + B_n \sin \frac{n\pi t}{\ell}), \qquad (4.22)$$

where the coefficients have the forms

$$A_0 = \frac{1}{2\ell} \int_{-\ell}^{\ell} g(x)dx,$$

$$A_n = \frac{1}{\ell} \int_{-\ell}^{\ell} g(x) \cos \frac{n\pi x}{\ell} dx, \qquad n = 1, 2, \ldots, \qquad (4.23)$$

$$B_n = \frac{1}{\ell} \int_{-\ell}^{\ell} g(x) \sin \frac{n\pi x}{\ell} dx, \qquad n = 1, 2, \ldots.$$

Note that the functions $\cos \frac{n\pi x}{\ell}$, $\sin \frac{n\pi x}{\ell}$ are the eigenfunctions of the Sturm-Liouville problem

$$f'' + \lambda f = 0,$$

with boundary conditions

$$f(-\ell) = f(\ell), \quad f'(-\ell) = f'(\ell)$$

(see Section 4.4).

The Fourier Method of Separation of Variables

The method of separation variables is the most often used techniques for obtaining the solution of partial differential equations. Let us consider the following linear differential homogeneous second order PDE

$$A(x)\frac{\partial^2 u}{\partial x^2} + B(x)\frac{\partial u}{\partial x} + C(x)u - D(y)\frac{\partial^2 u}{\partial y^2} - E(y)\frac{\partial u}{\partial y} - H(y)u = 0, \qquad (4.24)$$

where $0 < x < \ell$, $y > 0$, and either $D(y) > 0$, or $D(y) = 0$, $E(y) > 0$, for hyperbolic and parabolic equations.

We look for a particular solution of given equation which is a product of a function of x alone with a function of y alone, namely we assume that the solutions exist of the form

$$u(x, y) = X(x) \cdot Y(y),$$

and try to obtain ordinary differential equations for $X(x)$ and $Y(y)$.

Since each factor depends on only one variable we have

$$\frac{\partial^2 u}{\partial x^2} = X''(x)Y(y), \qquad \frac{\partial^2 u}{\partial y^2} = X(x)Y''(y),$$

$$\frac{\partial u}{\partial x} = X'(x)Y(y), \qquad \frac{\partial u}{\partial y} = X(x)Y'(y).$$

Substituting these expressions into (4.24) we obtain

$$\frac{1}{X(x)}\left(A(x)X''(x) + B(x)X'(x) + C(x)X(x)\right)$$

$$= \frac{1}{Y(y)}\left(D(y)Y''(y) + E(y)Y'(y) + H(y)Y(y)\right).$$

Let us remark that in previous equation the left-hand side contains only functions depending on x and the right-hand side contains only functions depending on y, meaning that left-hand side do not depend on y and the right-hand side do not depend on x. This can happen only if both sides are equal to a common constant $-\lambda$. So we obtain two ordinary differential equations

$$\frac{1}{X(x)}\left(A(x)X''(x) + B(x)X'(x) + C(x)X(x)\right) = -\lambda,$$

$$\frac{1}{Y(y)}\left(D(y)Y''(y) + E(y)Y'(y) + H(y)Y(y)\right) = -\lambda,$$

where λ is a separation constant. Thus our task is to solve these two ordinary differential equations.

The Mixed type Problem

The one–dimensional homogeneous wave equation given on the set
$\Omega_2 = \{(x,t)|\; x_2 \leq x \leq x_1,\; t \geq 0\}$, for $a > 0$

$$\frac{\partial^2 u}{\partial t^2} - a^2 \frac{\partial^2 u}{\partial x^2} = 0, \tag{4.25}$$

with initial conditions

$$u(x,0) = f_1(x), \quad \frac{\partial u(x,0)}{\partial t} = f_2(x) \quad (x_1 \leq x \leq x_2), \tag{4.26}$$

and boundary conditions

$$u(x_1,t) = g_1(t), \quad u(x_2,t) = g_2(t) \quad (t > 0), \tag{4.27}$$

where f_1, f_2 are functions defined on the interval (x_1, x_2), and g_1, g_2 are functions defined for $t > 0$, is called the *mixed type problem*.

Remark The solutions of one–dimensional wave equation (4.25) $u(x,t)$ are the transverse deflections of a string, which is stretched between fixed supports x_1 and x_2. The length of a string is ℓ. If $g_1 \equiv 0$ and $g_2 \equiv 0$, in boundary conditions (4.27) then we have zero displacement at the endpoints. The initial deflection of a string is given by the function $f_1(x)$ and the initial velocity is given $f_2(x)$ appearing in the initial conditions (4.26).

4.3.2   Examples and Exercises

Exercise 4.21 *Prove the following orthogonal relations:*

$$\textbf{a)} \quad \int_{-\ell}^{\ell} \cos \frac{k\pi x}{\ell} \cos \frac{n\pi x}{\ell}\, dx = \begin{cases} 0, & k \neq n \\ \ell, & k = n \neq 0, \\ 2\ell, & k = n = 0; \end{cases}$$

$$\textbf{b)} \quad \int_{-\ell}^{\ell} \sin \frac{k\pi x}{\ell} \sin \frac{n\pi x}{\ell}\, dx = \begin{cases} 0, & k \neq n \\ \ell, & k = n \neq 0, \\ 2\ell, & k = n = 0; \end{cases}$$

$$\textbf{c)} \quad \int_{-\ell}^{\ell} \cos \frac{k\pi x}{\ell} \sin \frac{n\pi x}{\ell}\, dx = \begin{cases} 0, & k \neq n \\ \ell, & k = n \neq 0, \\ 2\ell, & k = n = 0, \end{cases}$$

for all $k \in \mathbf{Z}$.

Hint: Use the trigonometric identities

$$\cos\alpha \cdot \cos\beta = \frac{1}{2}\left(\cos(\alpha-\beta)+\cos(\alpha+\beta)\right);$$

$$\sin\alpha \cdot \sin\beta = \frac{1}{2}\left(\cos(\alpha-\beta)-\cos(\alpha+\beta)\right);$$

$$\sin\alpha \cdot \cos\beta = \frac{1}{2}\left(\sin(\alpha-\beta)+\sin(\alpha+\beta)\right).$$

Example 4.22 *Determine the Fourier series for the following functions*

a) $f(x) = x^2$;

b) $f(x) = x$, *for* $x \in [-\ell, \ell]$, *and* $f(x+2\ell) = f(x)$.

Specially determine their Fourier series when $\ell = \pi$.

Solutions.

a) The function f is even and therefore it holds $B_n = 0, \quad n = 1, 2, \ldots$. The coefficients $A_n, n = 0, 1, \ldots$, can be written as

$$A_0 = \frac{1}{2\ell}\int_{-\ell}^{\ell} x^2 dx = \frac{2\ell^2}{3}$$

$$A_n = \frac{1}{\ell}\int_{\ell} x^2 \cos\frac{n\pi x}{\ell}dx = \frac{(-1)^n 4\ell^2}{n^2\pi^2}.$$

Thus, we obtain

$$f(x) = \frac{\ell^2}{3} + \frac{4\ell^2}{\pi^2}\sum_{n=1}^{\infty}\frac{(-1)^n}{n^2}\cos\frac{n\pi x}{\ell}. \tag{4.28}$$

If $\ell = \pi$, we have

$$f(x) = \frac{\pi^2}{3} + 4\sum_{n=1}^{\infty}\frac{(-1)^n}{n^2}\cos nx. \tag{4.29}$$

b) The function f is odd and the coefficients $A_n = 0, \quad n = 1, 2, \ldots$. In this case we have

$$B_n = \frac{1}{\ell}\int_{\ell} x\sin\frac{n\pi x}{\ell}dx$$

$$= \frac{2}{\ell}\left(-x\frac{\ell}{n\pi}\cos\frac{n\pi x}{\ell}\Big|_0^{\ell} + \frac{\ell}{n\pi}\int_0^{\ell}\cos\frac{n\pi x}{\ell}dx\right) \tag{4.30}$$

$$= -(-1)^{n+1}\frac{2\ell}{n\pi}.$$

So the Fourier series for the function f is

$$f(x) = \frac{2\ell}{\pi} \sum_{n=1}^{\infty} \frac{(-1)^{n+1}}{n} \sin \frac{n\pi x}{\ell}, \tag{4.31}$$

For $\ell = \pi$ we have

$$f(x) = 2 \sum_{n=1}^{\infty} \frac{(-1)^{n+1}}{n} \sin n\pi. \tag{4.32}$$

Example 4.23 *Find the formal solution of the problem*

$$a^2 \frac{\partial^2 u}{\partial x^2} - \frac{\partial^2 u}{\partial t^2} = 0 \qquad (0 < x < \ell, \ t > 0), \tag{4.33}$$

where a is a constant, with boundary conditions

$$u(0,t) = 0, \quad u(\ell,t) = 0 \qquad (t > 0), \tag{4.34}$$

and initial conditions

$$u(x,0) = f(x), \qquad \frac{\partial u(x,0)}{\partial t} = g(x) \qquad (0 < x < \ell). \tag{4.35}$$

Solution. The considered problem characterizes free oscillations of taut string with fixed ends with zero displacement.

Let us construct the solution of this problem by using the method of separation of variables. Taking

$$u(x,t) = X(x) \cdot T(t),$$

we obtain two differential equations

$$\frac{X''(x)}{X(x)} = -\lambda, \qquad \frac{T''(t)}{a^2 T(t)} = -\lambda,$$

or

$$X''(x) + \lambda X(x) = 0, \tag{4.36}$$

$$T''(t) + \lambda \cdot a^2 T(t) = 0. \tag{4.37}$$

From boundary conditions (4.34) it follows

$$X(0)T(t) = 0, \quad X(\ell)T(t) = 0, \qquad \text{i.e.,} \qquad X(0) = 0, \quad X(\ell) = 0. \tag{4.38}$$

Let us first solve the first equation (4.36) depending on variable X with the boundary conditions (4.38). This is Sturm-Liouville problem. First, we have to determine all values of parameter λ which allow nontrivial solutions of the given problem and then to find the solution. These special values λ are called *eigenvalues* and the solutions of the considered problem are called *eigenfunctions*. Therefore we need the following analyses.

(i) If $\lambda = -k^2 < 0$, then the separated solution is

$$X(x) = C_1 e^{kx} + C_2 e^{-kx}.$$

Using boundary conditions (4.38) we have

$$X(0) = C_1 + C_2 = 0, \quad C_1 e^{k\ell} + C_2 e^{-k\ell} = 0 \ ,$$

wherefrom it follows $C_1 = C_2 = 0$.

Clearly this gives us $X(x) = 0$, for all $x \in (-\ell, \ell)$, hence $u(x,t) = 0$, for $x \in (-\ell, \ell)$ $\quad (t > 0)$. But then the given problem has no solution if at least one of the functions f and g are nonzero.

(ii) If $\lambda = 0$, then from

$$X(x) = C_1 + C_2 x,$$

and from (4.38) we obtain the same conclusion.

(iii) If $\lambda = k^2 > 0$, then

$$X(x) = C_1 \cos kx + C_2 \sin kx,$$

and required boundary conditions (4.38) lead to

$$X(0) = C_1 = 0, \quad X(\ell) = C_2 \sin k\ell = 0,$$

wherefrom, it follows that $C_1 = 0$. If $C_2 = 0$, we obtain the trivial solution of X again. Therefore let us take $C_2 \neq 0$. Then the second equation is equal zero if $\sin k\ell = 0$, which is true for $k = \dfrac{n\pi}{\ell}$, $n = 1, 2, \ldots$. (We do not have to take $n = -1, -2, \ldots$, because they do not give any new solutions.)

So the eigenvalues of the considered problem are

$$\lambda = \lambda_n = \frac{n^2 \pi^2}{\ell^2} \quad (n \in \mathbf{N}), \tag{4.39}$$

and the corresponding eigenfunctions have the forms

$$X_n(x) = \sin \frac{n\pi x}{\ell} \quad (n \in \mathbf{N}), \tag{4.40}$$

where we took $C_2 = 1$.

For λ given by (4.39) the solution of the ordinary differential equation (4.37) has the form

$$T(t) = A_n \cos \frac{na\pi t}{\ell} + B_n \sin \frac{na\pi t}{\ell} \qquad (n \in \mathbf{N}), \qquad (4.41)$$

where A_n and B_n are arbitrary constants.

Multiplying (4.40) and (4.41) we obtain the solution of the considered problem

$$u_n(x,y) = X(x) \cdot T(t) = \left(A_n \cos \frac{na\pi t}{\ell} + B_n \sin \frac{na\pi t}{\ell} \right) \sin \frac{n\pi x}{\ell} \qquad (n \in \mathbf{N}), \quad (4.42)$$

Using the superposition principle on the set of solutions we take the solution of the given problem as an infinite series

$$u(x,t) = \sum_{n=1}^{\infty} \left(A_n \cos \frac{na\pi t}{\ell} + B_n \sin \frac{na\pi t}{\ell} \right) \sin \frac{n\pi x}{\ell}. \qquad (4.43)$$

The series (4.43) converges uniformly if the series $\sum_{n=1}^{\infty}(|A_n|+|B_n|)$, where A_n and B_n are Fourier coefficients, converges.

Remark 4.23.1 It is known that the series $\sum_{n=1}^{\infty} n^k |\phi_n|$, where

$$\phi_n = \frac{2}{\ell} \int_0^{\ell} \phi(x) \sin \frac{n\pi}{\ell} dx,$$

for $k = 0,1,2$ converges, if the function ϕ has continuous second derivative and the third derivative is piecewise continuous and it holds

$$\phi(0) = \phi(\ell) = 0, \qquad \phi''(0) = \phi''(\ell) = 0.$$

The series $\sum_{n=1}^{\infty} n^k |\psi_n|$, where

$$\psi_n = \frac{2}{\ell} \int_0^{\ell} \psi(x) \sin \frac{n\pi}{\ell} dx,$$

for $k = -1,0,1$, converges if the function ψ has continuous first derivative and the second derivative is piecewise continuous and it holds

$$\psi(0) = \psi(\ell) = 0, \qquad \psi''(0) = \psi''(\ell) = 0.$$

The solution u given by (4.43) has to satisfy the initial conditions, and therefore for $t = 0$ we obtain two Fourier series

$$u(x,0) = f(x) = \sum_{n=1}^{\infty} A_n \sin \frac{n\pi x}{\ell};$$

$$\frac{\partial u(x,0)}{\partial t} = g(x) = \sum_{n=1}^{\infty} \frac{na\pi}{\ell} B_n \sin \frac{n\pi x}{\ell}.$$

The coefficients can be determined from

$$A_n = \frac{2}{\ell} \int_0^{\ell} f(x) \sin \frac{n\pi x}{\ell} dx, \qquad n = 1, 2, \ldots \qquad (4.44)$$

$$\frac{nc\pi}{\ell} B_n = \frac{2}{\ell} \int_0^{\ell} g(x) \sin \frac{n\pi x}{\ell} dx, \qquad n = 1, 2, \ldots \qquad (4.45)$$

If the functions $f(x)$ and $g(x)$, $0 < x < \ell$, given by the initial condition for wave equation i.e.,

$$u(x,0) = f(x), \quad \frac{\partial u(x,0)}{\partial t} = g(x) \quad (0 < x < \ell).$$

are continuous and their first derivatives are piecewise continuous and if

$$f(0) = f''(0) = f(\ell) = f''(\ell) = 0, \qquad g(0) = g(\ell) = 0,$$

then the Fourier series with coefficients

$$A_n = \frac{2}{\ell} \int_0^{\ell} f(x) \sin \frac{n\pi x}{\ell} dx, \qquad \frac{nc\pi}{\ell} B_n = \frac{2}{\ell} \int_0^{\ell} g(x) \sin \frac{n\pi x}{\ell} dx \qquad (n \in \mathbf{N}).$$

converge uniformly for all periodic extensions of these functions.

Example 4.24

a) *Prove D'Alambert's formula (4.3) for homogeneous wave equation given by*

$$\frac{\partial^2 u}{\partial t^2} - a^2 \frac{\partial^2 u}{\partial x^2} = 0 \quad (0 < x < \ell, \ t > 0),$$

with initial conditions

$$u(x,0) = f(x), \quad \frac{\partial u(x,0)}{\partial t} = g(x) \quad (0 < x < \ell),$$

where a is a constant.

b) *Explain the boundary conditions*

$$u(0,t) = 0, \quad u(\ell,t) = 0 \quad (t > 0).$$

Solution.

a) Similarly as in Example 4.1.

b) Let us now apply in relation (4.4) the boundary conditions. From the first one it follows that

$$0 = \frac{1}{2}\left(f(at) + f(-at)\right) + \frac{1}{2a}\int_{-at}^{at} g(\xi)d\xi,$$

implies

$$0 = \frac{1}{2}\left(f(at) + f(-at)\right) \quad \text{and} \quad 0 = \frac{1}{2a}\int_{-at}^{at} g(\xi)d\xi.$$

This means that the function f and g must be extended (from the domain $0 < x < \ell$) as odd functions.

The second boundary conditions gives us the following

$$0 = \frac{1}{2}\left(f(\ell + at) + f(\ell - at)\right) \quad \text{and} \quad 0 = \frac{1}{2a}\int_{\ell-at}^{\ell+at} g(\xi)d\xi.$$

This implies that the odd extension of the functions f and g must be periodic for all real arguments, with basic period 2ℓ. These extensions of functions f and g are denoted usually as f_0 and g_0 respectively.

Example 4.25 *Show that the solution obtained by using the method of separation variables, given by (4.43), can be written in the form of D'Alambert's formula, (4.4), for $f \in C^2$, $g \in C^1$.*

Solution. The solution of the form (4.43) obtained by separation of variables can be transformed as

$$u(x,t) = \sum_{n=1}^{\infty} A_n \cos\frac{na\pi t}{\ell} \cdot \sin\frac{n\pi x}{\ell} + B_n \sin\frac{na\pi t}{\ell} \sin\frac{n\pi x}{\ell}.$$

Since $f \in C^2$, $g \in C^1$, this Fourier series converges uniformly. We can transform it as follows

$$u(x,t) = \frac{1}{2}\sum_{n=1}^{\infty} A_n \left(\sin\frac{na\pi(x + at)}{\ell} + \sin\frac{na\pi(x - at)}{\ell}\right)$$

$$-\frac{1}{2}\sum_{n=1}^{\infty} B_n \left(\cos\frac{na\pi(x + at)}{\ell} - \cos\frac{na\pi(x - at)}{\ell}\right).$$

From relations (4.44) and (4.45) it follows

$$
u(x,t) \;=\; \frac{f(x+at)+f(x-at)}{2} + \frac{1}{2}\sum_{n=1}^{\infty} B_n \frac{n\pi}{\ell} \int_{x-at}^{x+at} \sin\frac{n\pi x}{\ell}\,dx
$$

$$
=\; \frac{f(x+at)+f(x-at)}{2} + \frac{1}{2}\int_{x-at}^{x+at} \sum_{n=1}^{\infty} B_n \frac{n\pi}{\ell} \sin\frac{n\pi x}{\ell}\,dx
$$

$$
=\; \frac{f(x+at)+f(x-at)}{2} + \frac{1}{2a}\int_{x-at}^{x+at} g(\xi)\,d\xi.
$$

Example 4.26 *Find the formal solution of the problem*

$$
a^2\frac{\partial^2 u}{\partial x^2} - \frac{\partial^2 u}{\partial t^2} = 0 \qquad (0 < x < \ell,\ t > 0),
$$

$$
u(0,t) = 0, \qquad u(\ell,t) = 0 \qquad (t > 0),
$$

$$
u(x,0) = f(x), \quad \frac{\partial u(x,0)}{\partial t} = 0 \qquad (0 < x < \ell),
$$

where

a) $\quad f(x) = \begin{cases} x, & 0 < x < \dfrac{\ell}{2} \\[2mm] \ell - x, & \dfrac{\ell}{2} < x < \ell; \end{cases}$

b) $\quad f(x) = x(\ell - x) \qquad (0 < x < \ell).$

Solution.

a) Using the method of separation variables we obtain the solution of the form (4.43) with $B_n = 0$, because the initial condition is zero. So the solution has the form

$$
u(x,t) = \sum_{n=1}^{\infty} A_n \cos\frac{n a \pi t}{\ell} \sin\frac{n\pi x}{\ell},
$$

where in our case we have

$$A_n = \frac{2}{\ell} \int_0^\ell f(x) \sin \frac{n\pi x}{\ell} dx$$

$$= \frac{2}{\ell} \int_0^{\ell/2} x \sin \frac{n\pi x}{\ell} dx + \frac{2}{\ell} \int_{\ell/2}^\ell (\ell - x) \sin \frac{n\pi x}{\ell} dx$$

$$A_n = \frac{4}{\ell} \int_0^{\ell/2} x \sin \frac{n\pi x}{\ell} dx = \frac{4\ell}{n^2\pi^2} \cos \frac{n\pi x}{\ell} \Big|_{\ell/2}^\ell + \frac{4}{n\pi} \int_0^{\ell/2} x \cos \frac{n\pi x}{\ell} dx$$

$$= \frac{4\ell}{n^2\pi^2} \sin \frac{n\pi}{2} = \frac{4\ell}{(2n-1)^2\pi^2}(-1)^{n+1} \quad (n \in \mathbf{N}).$$

So the solution can be written as

$$u(x,t) = \frac{4}{\pi^2} \sum_{n=1}^{\infty} \frac{(-1)^{n+1}}{(2m+1)^2} (\cos \frac{(2n-1)a\pi t}{\ell} \sin \frac{(2n-1)\pi x}{\ell}.$$

b) The solution of this problem has the form

$$u(x,t) = \sum_{n=1}^{\infty} A_n \cos \frac{na\pi t}{\ell} \sin \frac{n\pi x}{\ell}.$$

The initial condition $u(x,0) = x(\ell - x)$, lead us to Fourier series

$$x(\ell - x) = \sum_{n=1}^{\infty} A_n \sin \frac{n\pi x}{\ell} \quad (0 < x < \ell),$$

whose coefficients can be obtained from

$$A_n = \frac{2}{\ell} \int_0^\ell x(\ell - x) \sin \frac{n\pi x}{\ell} dx.$$

Integrating by parts we get

$$A_n = \frac{4\ell^2(1 + (-1)^{n+1})}{n^3\pi^3} \quad (n \in \mathbf{N}),$$

and therefore the solution of the considered problem has the form

$$u(x,t) = \frac{8\ell^2}{\pi^3} \sum_{n=1}^{\infty} \frac{1}{(2n-1)^3} \cos \frac{(2n-1)a\pi t}{\ell} \sin \frac{(2n-1)\pi x}{\ell}.$$

Example 4.27 *A horizontal bar of length ℓ is originally at rest and unstraint along the $x-$axes. The end point $x = 0$ is fixed while the right end is subjected to a constant elongating force F per unit area. Determine the longitudinal displacement of the cross section.*

Solution. The problem

$$a^2 \frac{\partial^2 u}{\partial x^2} = \frac{\partial^2 u}{\partial t^2} \qquad (0 < x < \ell, \quad t > 0)$$

$$u(0, t) = 0, \qquad E \frac{\partial u(\ell, t)}{\partial x} = F \qquad (t > 0),$$

$$u(x, 0) = 0, \qquad \frac{\partial u(x, 0)}{\partial t} = 0 \qquad (0 < x < \ell),$$

describe this displacement of the cross section (E is Young's modulus of elasticity of the material in tension and compression and $a > 0$ is a constant depending on the material the bar was made from).

Let us remark that in this problem the boundary conditions are not homogeneous. Thus we introduce the functions S, (depending on x) and v, (depending on x, t,) such that

$$u(x, t) = S(x) + v(x, t). \tag{4.46}$$

Then the considered problem can be written as

$$S''(x) + \frac{\partial^2 v(x, t)}{\partial x^2} = a^{-2} \frac{\partial v(x, t)}{\partial t} \qquad (0 < x < \ell, \quad t > 0),$$

$$S(0) + v(0, t) = 0, \qquad ES'(\ell) + E \frac{\partial v(\ell, t)}{\partial x} = F \qquad (t > 0)$$

$$S(x) + v(x, 0) = 0, \qquad \frac{\partial u(x, 0)}{\partial t} = 0 \qquad (0 < x < \ell).$$

This problem can be treated as two problems

$$S''(x) = 0, \qquad S(0) = 0, \qquad S'(\ell) = \frac{F}{E}, \tag{4.47}$$

and well known problem with homogeneous boundary conditions

$$\frac{\partial^2 v(x, t)}{\partial x^2} = a^{-2} \frac{\partial^2 v(x, t)}{\partial t^2} \qquad (0 < x < \ell, \quad t > 0),$$

$$v(0, t) = 0, \qquad \frac{\partial v(\ell, t)}{\partial x} = 0 \qquad (t > 0), \tag{4.48}$$

$$v(x, 0) = -S(x), \qquad \frac{\partial u(x, 0)}{\partial t} = 0 \qquad (0 < x < \ell).$$

The solution of problem (4.47) is

$$S(x) = \frac{F}{E}x.$$

The method of separation variables leads the problem (4.48) to the systems of ODEs

$$X''(x) - \lambda X(x) = 0, \quad X(0) = 0, \quad X'(\ell) = 0,$$

$$T''(t) - \lambda a^2 T(t) = 0, \quad T'(0) = 0,$$

wherefrom, we obtain the solution of problem (4.48) as

$$v(x,t) = \sum_{n=1}^{\infty} C_n \cos\left(\frac{(2k-1)t}{2\ell}\pi a\right) \sin\frac{2k-1}{2\ell}\pi x \quad (0 < x < \ell).$$

The values C_n can be obtained by using the first initial conditions as the coefficients of Fourier series

$$-\frac{F}{E}x = \sum_{n=1}^{\infty} C_n \sin\frac{2n-1}{2\ell}\pi x.$$

They are

$$C_n = -\frac{2}{\ell}\int_0^\ell \frac{F}{E}x \sin\left(\frac{2n-1}{2\ell}\pi x\right) dx = \frac{8\ell F}{E\pi^2} \cdot \frac{(-1)^n}{(2n-1)^2}.$$

Therefore the solution of the considered problem is

$$u(x,t) = \frac{F}{E}\left(x + \frac{8\ell}{\pi^2}\sum_{n=1}^{\infty}\frac{(-1)^n}{(2n-1)^2}\sin\left(\frac{(2n-1)\pi x}{2\ell}\right)\cos\left(\frac{(2n-1)\pi at}{2\ell}\right)\right).$$

Example 4.28 *Solve the nonhomogeneous problem*

$$\frac{\partial^2 u}{\partial x^2} = \frac{\partial^2 u}{\partial t^2} + 2e^{-3x} \quad (0 < x < 1, \quad t > 0),$$

$$u(0,t) = 0, \quad u(1,t) = 0 \quad (t > 0), \tag{4.49}$$

$$u(x,0) = \frac{2}{9}\left(e^{-3x} - 1\right), \quad \frac{\partial u(x,0)}{\partial t} = 0 \quad (0 < x < 1).$$

Solution. Taking $u(x,t) = S(x) + v(x,t)$ the problem (4.49) can be written as

$$S''(x) + \frac{\partial^2 v}{\partial x^2} = \frac{\partial^2 v}{\partial t^2} + 2e^{-3x} \quad (0 < x < 1, \quad t > 0),$$

$$S(0) + v(0, t) = 0, \qquad S(1) + v(1, t) = 0 \qquad (t > 0),$$

$$S(x) + v(x, 0) = \frac{2}{9} \left(e^{-3x} - 1 \right), \qquad \frac{\partial v(x, 0)}{\partial t} = 0 \qquad (0 < x < 1).$$

The solution of the problem

$$S'' = 2e^{-3x}, \quad S(0) = 0, \quad S(1) = 0,$$

is

$$S(x) = \frac{2}{9} \left(e^{-3x} - 1 \right) + \frac{2}{9} \left(1 - e^{-3} \right) x.$$

The solution of second part of the considered problem, i.e., of the problem

$$\frac{\partial^2 v}{\partial x^2} = \frac{\partial^2 v}{\partial t^2} \qquad (0 < x < 1, \quad t > 0),$$

$$v(0, t) = 0, \qquad v(1, 0) = 0 \qquad (t > 0), \tag{4.50}$$

$$v(x, 0) = -\frac{2}{9} \left(1 - e^{-3} \right) x, \qquad \frac{\partial v(0, t)}{\partial t} = 0 \qquad (0 < x < 1),$$

is

$$v(x, t) = \sum_{n=1}^{\infty} C_n \sin(\pi n x) \cos(t n \pi).$$

Using the initial conditions

$$v(x, 0) = -\frac{2}{9} (1 - e^{-3}) = \sum_{n=1}^{\infty} C_n \sin n \pi x,$$

we obtain

$$C_n = -\frac{4}{9} \left(1 - e^{-3} \right) \int_0^1 x \sin(n \pi x) dx = \frac{4(-1)^n}{9 n \pi} (1 - e^{-3}) \qquad (n \in \mathbf{N}),$$

and the final solution of our problem has the form

$$u(x, t) = \frac{2}{9} \left(e^{-3x} - 1 \right) + \frac{2}{9} \left(1 - e^{-3} \right) x$$

$$+ (1 - e^{-3}) \frac{4}{9 \pi} \sum_{n=1}^{\infty} \frac{(-1)^n \sin(\pi n x)}{n} \cos(-t n \pi).$$

Example 4.29 *Let us consider the following boundary problem for the wave equation*

$$\frac{\partial^2 u}{\partial x^2} = \frac{\partial^2 u}{\partial t^2} \quad (0 < x < \ell_1, \quad 0 < t < \ell_2),$$

$$u(0,t) = 0, \quad u(\ell_1, t) = 0 \quad (0 < t < \ell_2)$$

$$u(x,0) = 0, \quad u(x, \ell_2) = 0 \quad (0 < x < \ell_1).$$

Show that the solution of this problem is not unique on the given rectangular.

Solution. The method of separation variables leads us to the systems

$$X''(x) + \lambda X = 0, \quad X(0) = 0, \quad X(\ell_1) = 0,$$

$$T''(x) + \lambda T = 0, \quad T(0) = 0, \quad T(\ell_2) = 0.$$

The solution of the first system is

$$X(x) = A \sin \lambda x + B \cos \lambda x.$$

From the condition $X(0) = 0$ it follows $B = 0$ and from $X(\ell_1) = 0$ it follows $A \sin \lambda \ell_1 = 0$. If $A \neq 0$, then we have

$$\lambda \ell_1 = n\pi, \qquad \lambda = \frac{n\pi}{\ell_1}, \qquad n = 1, 2 \dots.$$

Also, from the condition $T(0) = 0$ it follows $D = 0$ and from $T(\ell_2) = 0$ it follows $C \sin \lambda \ell_2 = 0$. If $C \neq 0$, then we have

$$\lambda \ell_2 = k\pi, \qquad \lambda = \frac{k\pi}{\ell_1}, \qquad k = 1, 2 \dots.$$

From above it follows

$$\lambda = \frac{n\pi}{\ell_1} = \frac{n\pi}{\ell_2} \quad \text{and} \quad \frac{\ell_2}{\ell_1} = \frac{k}{n}.$$

So we get that the solution of this problem is not unique on a rectangle, if $\frac{\ell_2}{\ell_1}$ is rational. If $\frac{\ell_2}{\ell_1}$ is irrational then there is no nontrivial separable solution.

Remark 4.29.1 This is the Dirichlet problem for the wave equation.

4.4   The Sturm–Liouville Problem

4.4.1   Preliminaries

Let us consider the homogeneous second order differential equation

$$\frac{d}{dx}(p(x)y') + p'(x)y' + (q(x) + \lambda r(x))\,y = 0 \qquad (a < x < b), \qquad (4.51)$$

or $p(x)y'' + p'(x)y' + (q(x) + \lambda r(x))\,y = 0$, where $p(x) > 0$, $r(x) > 0$, $x \in (a, b)$ and p', q, and r are continuous functions on $[a, b]$. The function r is called the *weight function*. The equation (4.51) with homogeneous boundary conditions

$$B_1[u] = \alpha_{11}y(a) + \alpha_{12}y'(a) = 0 \qquad (\alpha_{11}^2 + \alpha_{12}^2 \neq 0),$$

$$B_2[u] = \alpha_{21}y(b) + \alpha_{22}y'(b) = 0 \qquad (\alpha_{21}^2 + \alpha_{22}^2 \neq 0),$$

$$(4.52)$$

is called regular *Sturm-Liouville problem*. In (4.52), $\alpha_{11}, \alpha_{12}, \alpha_{21}, \alpha_{22}$ are arbitrary constants.

The series

$$f(x) = \sum_{n=1}^{\infty} c_n \phi_n(x) \qquad (a < x < b), \qquad (4.53)$$

where the set of functions $\phi_n(x)$ are orthogonal with respect to the given weighting function $r(x) > 0$, satisfying

$$\int_a^b r(x)\phi_k(x)\phi_n(x)dx = 0, \qquad k \neq n,$$

with the coefficients c_n given as

$$c_n = ||\phi_n||^{-2} \int_a^b r(x)f(x)\phi_n(x)dx \qquad n \in \mathbf{N}, \qquad (4.54)$$

where

$$||\phi_n|| = \sqrt{\int_a^b r(x)\,(\phi_n(x))^2\,dx}, \qquad (4.55)$$

are called *Sturm-Liouville series*.

Theorem 4.9 *If a function f and its first derivative f' are piecewise continuous on $[a, b]$ then the series*

$$f(x) = \sum_{n=1}^{\infty} c_n \phi_n(x) \qquad (x \in (a, b)), \qquad (4.56)$$

where $\phi_n(x)$ are eigenfunctions of the regular Sturm-Liouville system and c_n are of the form (4.54), converges pointwise to $f(x)$ at each point $x \in [a, b]$ of continuity of the function f.

Theorem 4.10 *If the function f and its derivative f' are piecewise continuous on $[a, b]$ and f satisfies the boundary conditions of the Sturm-Liouville problem on $[a, b]$, then generalized Fourier series (4.53) converges uniformly to f on $[a, b]$.*

Theorem 4.11 (Fourier Integral Theorem) *Suppose the function f and its first derivative f' are piecewise continuous on every finite interval and that f is an absolutely integrable function on $\mathbf{R}$ (i.e.,*

$$\int_{-\infty}^{\infty} |f(x)|dx$$

converges on the interval $-\infty < x < \infty$). Then it holds

$$\frac{f(x^-) + f(x^+)}{2} = \frac{1}{\pi} \int_0^{\infty} (A(s)\cos(sx) + B(s)\sin(sx))\, ds, \qquad (4.57)$$

where

$$A(s) = \frac{1}{\pi} \int_{-\infty}^{\infty} f(t)\cos(st)dt, \qquad (4.58)$$

$$B(s) = \frac{1}{\pi} \int_{-\infty}^{\infty} f(t)\sin(st)dt. \qquad (4.59)$$

Special Functions

The Legendre equation is given by

$$((1 - x^2)P'(x))' + \lambda P(x) = 0$$

on the interval $[-1, 1]$ with the eigenvalues

$$\lambda = n(n+1) \qquad (n \in \mathbf{Z}_+),$$

and with eigenfunctions - *Legendre polynomials*

$$P_n(x) = \frac{1}{2^n n!} \frac{d^n}{dx^n} (x^2 - 1)^n \qquad (n \in \mathbf{Z}_+).$$

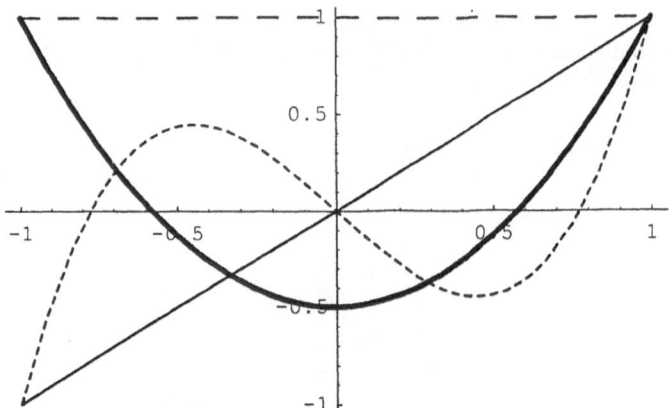

Figure 4.1 Legendre polynomials for $n = 0, 1, 2, 3$ with long dashed, thin, thick and short dashed line, respectively.

We consider the more general equation (for $m = 0$ we obtain the previous Legendre equation)

$$((1 - x^2)y'(x))' + (\lambda - \frac{m}{1 - x^2})y(x) = 0$$

for $m \in \mathbf{N}$. The eigenvalues are

$$\lambda_n = n(n + 1) \text{ for } n = m, m + 1, \dots$$

and the corresponding eigenfunctions are the Legendre functions P_n^m given by

$$P_n^m(x) = (1 - x^2)^{\frac{m}{2}} P_n^{(m)}(x) \text{ for } n = m, m + 1, \dots,$$

where P_n is the Legendre polynomial.

The functions

$$Y_n^m(\theta, \varphi) = P_n^m(\cos \theta) \cos m\varphi,$$
$$Y_n^{-m}(\theta, \varphi) = P_n^m(\cos \theta) \sin m\varphi,$$

for $m = 0, 1, \dots, n = m, m + 1, \dots$, are the *spheric functions*. Using them we introduce the ball functions

$$r^n Y_n^m(\theta, \varphi), \ r^{-(n+1)} Y_n^m(\theta, \varphi) \text{ for } n \in \mathbf{Z}_+, -n \le m \le n.$$

Specially for $m = 0$ the ball functions are axial symmetric, i.e., independent of φ.

The Bessel equation is given by

$$x^2 y''(x) x y' + (x^2 - n^2)y(x) = 0$$

'with the regular solution *Bessel function* of order n

$$J_n(x) = \sum_{k=0}^{\infty} \frac{(-1)^k}{\Gamma(n+k+1)\Gamma(k+1)} \left(\frac{x}{2}\right)^{n+2k} \qquad (n \in \mathbf{Z}_+).$$

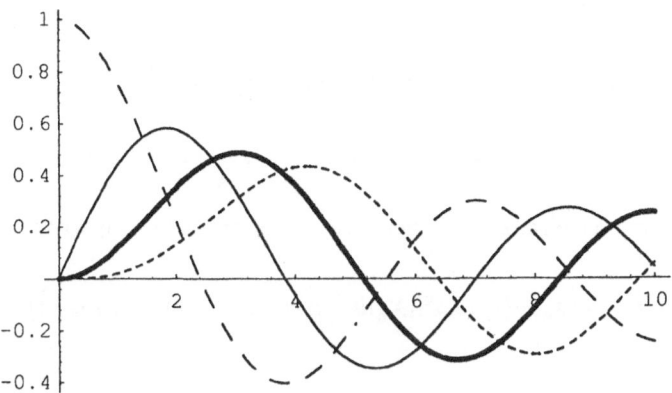

Figure 4.2 Bessel functions for $n = 0, 1, 2, 3$ with long dashed, thin, thick and short dashed line, respectively.

4.4.2   Examples and Exercises

Example 4.30 *Solve the nonhomogeneous problem*

$$\frac{\partial^2 u}{\partial x^2} = a^{-2}\frac{\partial^2 u}{\partial t^2} - r(x,t) \qquad (0 < x < \ell,\ t > 0),$$

$$B_1[u] = 0, \qquad B_2[u] = 0 \qquad (t > 0), \tag{4.60}$$

$$u(x,0) = f(x), \qquad \frac{\partial u(x,0)}{\partial t} = g(x) \qquad (0 < x < \ell).$$

Solution. Let us suppose that the solution of the problem (4.60) has the form of generalized Fourier series

$$u(x,t) = \sum_{n=1}^{\infty} T_n(t)X_n(x), \tag{4.61}$$

where X_n are eigenfunctions of the Sturm-Liouville problem

$$X'' + \lambda X = 0, \qquad B_1(X) = 0, \quad B_2(X) = 0.$$

If the termwise differentiation is allowed then we have

$$\frac{\partial^2 u(x,t)}{\partial t^2} = \sum_{n=1}^{\infty} T_n''(t) X_n(x),$$

$$\frac{\partial^2 u(x,t)}{\partial x^2} = \sum_{n=1}^{\infty} T_n(t) X_n''(x) = \lambda_n \sum_{n=1}^{\infty} T_n(t) X_n(x),$$

where λ_n are eigenfunctions.

The partial differential equation (4.60) can be written as

$$a^2 r(x,t) = \sum_{n=1}^{\infty} \left(T_n''(t) + a^2 \lambda_n T_n(t) \right) X_n(x).$$

For fixed t this is a generalized Fourier series wherefrom we can write

$$T_n''(t) + a^2 \lambda_n T_n(t) = a^2 ||X_n||^{-2} \int_0^{\ell} r(x,t) X_n(x) \qquad (n \in \mathbf{N}).$$

where $||X_n||^{-2}$ is given by (4.55) i.e.,

$$||X_n||^2 = \int_0^{\ell} (X_n(x))^2 dx \qquad (n \in \mathbf{N}).$$

Denoting by

$$v_n(t) = ||X_n(x)||^{-2} \int_0^{\ell} r(x,t) X_n(x) dx,$$

we obtain the ordinary differential equation

$$T_n''(t) + a^2 \lambda_n T_n(t) = a^2 v_n(t),$$

with the solution

$$T_n(t) = \frac{a}{\sqrt{\lambda_n}} \int_0^t v_n(\tau) \sin\left(\sqrt{\lambda_n} a(t - \tau) \right) d\tau + A_n \cos(at\sqrt{\lambda_n}) + B_n \sin(at\sqrt{\lambda_n}),$$

where A_n and B_n are arbitrary constants for $n \in \mathbf{N}$. The solution of the given equation with boundary conditions is

$$u(x,t) = \sum_{n=1}^{\infty} \left(\frac{a}{\sqrt{\lambda_n}} \int_0^t v_n(\tau) \sin\left(\sqrt{\lambda_n} a(t - \tau) \right) d\tau \right.$$

$$\left. + A_n \cos(at\sqrt{\lambda_n}) + B_n \sin(at\sqrt{\lambda_n}) \right) X_n(x).$$

Using the initial conditions we obtain

$$u(x,0) = f(x) = \sum_{n=1}^{\infty} A_n X_n(x),$$

$$\frac{u(x,0)}{\partial t} = g(x) = \sum_{n=1}^{\infty} \sqrt{\lambda_n} a B_n X_n(x),$$

wherefrom we get

$$A_n = ||X_n||^{-2} \int_0^{\ell} f(x) X_n(x) dx,$$

$$\sqrt{\lambda_n} B_n = ||X_n||^{-2} \int_0^{\ell} f(x) X_n(x) dx.$$

Example 4.31 *Determine the solution of the problem*

$$\frac{\partial^2 u}{\partial x^2} = a^{-2} \frac{\partial^2 u}{\partial t^2} - R(x) \sin \omega t \qquad (0 < x < \ell, \quad t > 0),$$

$$u(0,t) = 0, \qquad u(\ell,t) = 0 \qquad (t > 0), \tag{4.62}$$

$$u(x,0) = f(x), \qquad \frac{\partial u(0,t)}{\partial t} = g(x) \qquad (0 < x < \ell).$$

Solution. This is nonhomogeneous problem and it describes the motion of an elastic string with a time-dependent, sinusoidal, external force. The solution is the sum of the homogeneous solution $u_h(x,t)$, which is the solution of of the boundary value problem

$$\frac{\partial^2 u}{\partial x^2} = a^{-2} \frac{\partial^2 u}{\partial t^2} \qquad (0 < x < \ell, \quad t > 0),$$

$$u(0,t) = 0, \qquad u(\ell,t) = 0 \qquad (t > 0),$$

and the particular one $u_p(x,t)$, which is the solution of the problem

$$\frac{\partial^2 u}{\partial x^2} = a^{-2} \frac{\partial^2 u}{\partial t^2} - R(x) \sin \omega t \qquad (0 < x < \ell, \quad t > 0),$$

$$u(0,t) = 0, \qquad u(\ell,t) = 0 \qquad (t > 0),$$

$$\tag{4.63}$$

i.e.,

$$u(x,t) = u_h(x,t) + u_p(x,t).$$

We have already obtained the homogeneous solution as

$$u_h(x,t) = \sum_{n=1}^{\infty} \left(A_n \cos \frac{n a \pi t}{\ell} + B_n \sin \frac{n a \pi t}{\ell} \right) \sin \frac{n \pi x}{\ell}. \tag{4.64}$$

The particular solution can be found by using a modification of the method of undetermined coefficient for ordinary differential equations. Let us suppose that the particular solution can be written as

$$u_p(x,t) = C(x)\sin\omega t + D(x)\cos\omega t, \qquad (4.65)$$

where $C(x)$ and $D(x)$ are the undetermined coefficients. Replacing

$$\frac{\partial u_p}{\partial t} = \omega C(x)\cos\omega t - \omega D(x)\sin\omega t,$$

$$\frac{\partial^2 u_p}{\partial t^2} = -\omega^2 C(x)\sin\omega t - \omega^2 D(x)\cos\omega t,$$

in equation (4.63) we get

$$C''(x)\sin\omega t + D''(x)\cos\omega t = \left(-\left(\frac{\omega}{a}\right)^2 C(x) - R(x)\right)\sin\omega t - \left(\frac{\omega}{a}\right)^2 D(x)\cos\omega t,$$

wherefrom we obtain two equations

$$C''(x) + \left(\frac{\omega}{a}\right)^2 C(x) = -R(x), \qquad D''(x) + \left(\frac{\omega}{a}\right)^2 D(x) = 0.$$

The solution of the first of those two equations is $C(x)$, which satisfy the conditions $C(0) = C(\ell) = 0$. The solution of second equation we can take to be zero i.e., $D(x) = 0$. Therefore the solution of the consider problem is

$$u(x,t) = u_h(x,t) + u_p(x,t) = C(x)\sin\omega t + \sum_{n=1}^{\infty}\left(A_n\cos\frac{na\pi t}{\ell} + B_n\sin\frac{na\pi t}{\ell}\right)\sin\frac{n\pi x}{\ell}.$$

Using the initial conditions $u(x,0) = f(x)$, $\dfrac{\partial u(x,0)}{\partial t} = g(x)$ $(0 < x < \ell)$, we obtain

$$f(x) = \sum_{n=1}^{\infty} A_n(t)\sin\frac{n\pi x}{\ell},$$

$$g(x) - \omega C(x) = \sum_{n=1}^{\infty}\frac{na\pi}{\ell}B_n\sin\frac{n\pi x}{\ell},$$

wherefrom

$$A_n = \frac{2}{\pi}\int_0^{\ell} f(x)\sin\frac{n\pi x}{\ell}dx, \qquad \frac{na\pi}{\ell}B_n = \frac{2}{\pi}\int_0^{\ell}(g(x) - \omega C(x))\sin\frac{n\pi x}{\ell}dx.$$

Example 4.32 *Determine the solution of the problem*

$$\frac{\partial^2 u}{\partial x^2} = a^{-2}\frac{\partial^2 u}{\partial t^2} - R\sin\omega t \qquad (0 < x < \ell, \quad t > 0),$$

$$u(0,t) = 0, \qquad u(\ell,t) = 0 \qquad (t > 0), \tag{4.66}$$

$$u(x,0) = 0, \qquad \frac{\partial u(0,t)}{\partial t} = 0 \qquad (0 < x < \ell),$$

where ω, R *are constants.*

Solution. (Let us remark that the force $R\sin\omega t$ is proportional to the wind force.) This is a special form of the previous example. The homogeneous solution has the form given in relation (4.64). The particular solution given by (4.65) can be determined from the problem

$$C''(x) + \left(\frac{\omega}{a}\right)^2 C(x) = -R, \qquad C(0) = 0, \quad C(\ell) = 0.$$

It has the form

$$C(x) = R\left(\frac{\omega}{a}\right)^2\left(\frac{\left(1 - \cos\dfrac{\omega\ell}{a}\right)\sin\left(\dfrac{\omega x}{a}\right)}{\sin\left(\dfrac{\omega\ell}{a}\right)} + \cos\left(\frac{\omega x}{a}\right) - 1\right).$$

The solution of the considered problem has the form

$$u(x,t) = R\left(\frac{\omega}{a}\right)^2\left(\frac{\left(1 - \cos\dfrac{\omega\ell}{a}\right)\sin\left(\dfrac{\omega x}{a}\right)}{\sin\left(\dfrac{\omega\ell}{a}\right)} + \cos\left(\frac{\omega x}{a}\right) - 1\right)\sin\omega t$$

$$+ \sum_{n=1}^{\infty}\left(A_n\cos\frac{n a\pi t}{\ell} + B_n\sin\frac{n a\pi t}{\ell}\right)\sin\frac{n\pi x}{\ell}.$$

Using the initial conditions we obtain two Fourier series

$$u(x,0) = 0 = \sum_{n=1}^{\infty} A_n\sin\left(\frac{n\pi x}{\ell}\right),$$

wherefrom the coefficients $A_n = 0$ and from

$$\frac{\partial u(x,0)}{\partial t} = 0 \; = \; \omega R \left(\frac{\omega}{a}\right)^2 \left(\frac{\left(1 - \cos \frac{\omega\ell}{a}\right) \sin \left(\frac{\omega x}{a}\right)}{\sin \left(\frac{\omega\ell}{a}\right)} + \cos \left(\frac{\omega x}{a}\right) - 1 \right)$$

$$+ \sum_{n=1}^{\infty} \frac{n\pi a}{\ell} B_n \sin \left(\frac{n\pi x}{\ell}\right),$$

we get

$$\frac{n\pi a}{\ell} B_n = -\frac{2R\omega^3}{a^2\ell} \int_0^\ell \left(\frac{\left(1 - \cos \frac{\omega\ell}{a}\right) \sin \left(\frac{\omega x}{a}\right)}{\sin \left(\frac{\omega\ell}{a}\right)} + \cos \left(\frac{\omega x}{a}\right) - 1 \right) \sin \frac{n\pi x}{\ell} dx$$

$$= \begin{cases} -\dfrac{4\omega R}{n\ell \left(\dfrac{n^2\pi^2}{\ell^2} - \dfrac{\omega^2}{a^2}\right)} & n = 1, 3, 5, \ldots, \\[4mm] 0, & n = 0, 2, 4, \ldots. \end{cases}$$

The solution of considered problem is

$$u(x,t) \; = \; \frac{4R a\omega}{\ell} \sum_{n=1}^{\infty} \frac{\sin \left(\dfrac{(2n-1)at}{\ell}\right) \sin \left(\dfrac{(2n-1)x}{\ell}\right)}{(2n-1)^2\pi^2 - \dfrac{\omega^2}{a^2}}$$

$$+ R \left(\frac{a}{\omega}\right)^2 \left(\frac{\left(1 - \cos \frac{\omega\ell}{a}\right) \sin \left(\frac{\omega x}{a}\right)}{\sin \left(\frac{\omega\ell}{a}\right)} + \cos \left(\frac{\omega x}{a}\right) - 1 \right).$$

$$(4.67)$$

This solution can be obtained by using the other method. Namely let us consider solution in a form

$$u(x,t) = \sum_{n=1}^{\infty} T_n(t) \sin n\pi x.$$

After termwise differentiation we obtain the generalized Fourier series

$$a^2 R \sin \omega t = \sum_{x=0}^{\infty} \left(T_n''(t) + \frac{n^2\pi^2}{a^2\ell^2} T_n(t) \right) \sin \left(\frac{n\pi x}{\ell}\right).$$

The coefficients of this series can be obtained as

$$T_n''(t) + \frac{n^2\pi^2}{a^2\ell^2}T_n(t) = a^2 || \sin\left(\frac{n\pi x}{\ell}\right) ||^{-2} R \sin\omega t \int_0^\ell \sin\left(\frac{n\pi x}{\ell}\right) dx$$

$$= a^2 \frac{2}{n\pi} R \sin\omega t (1 - (-1)^n) = a^2 \begin{cases} \dfrac{4\ell}{n\pi} R \sin\omega t & n = 1,2,5,\ldots, \\[2mm] 0 & n = 2,4,6,\ldots. \end{cases}$$

The solution of previous differential equation has the form

$$T_n(t) = \begin{cases} \dfrac{4Ra}{\ell n^2} \displaystyle\int_0^t \sin\left(\frac{n a\pi(t-\tau)}{\ell}\right) \sin\omega\tau d\tau \\[4mm] \qquad + A_n \cos\left(\dfrac{n a\pi t}{\ell}\right) + B_n \sin\left(\dfrac{n a\pi t}{\ell}\right) & n = 1,3,5,\ldots, \\[4mm] A_n \cos\left(\dfrac{n a\pi t}{\ell}\right) + B_n \sin\left(\dfrac{n a\pi t}{\ell}\right) & n = 2,4,6,\ldots. \end{cases}$$

Example 4.33 *Determine the solution of the problem*

$$\frac{\partial^2 u}{\partial x^2} = \frac{\partial^2 u}{\partial t^2} - 1 \quad (0 < x < 1,\ t > 0),$$

$$u(0,t) = 0, \quad u(1,t) = 1 + t \quad (t > 0),$$

$$u(x,0) = x, \quad \frac{\partial u(x,0)}{\partial t} = 0 \quad (0 < x < \ell).$$

Solution. Taking $u(x,t) = V(x,t) + W(x,t)$, we obtain the following problem

$$\frac{\partial^2 V}{\partial x^2} + \frac{\partial^2 W}{\partial x^2} = \frac{\partial V^2}{\partial t^2} + \frac{\partial^2 W}{\partial t^2} - 1 \quad (0 < x < 1,\ t > 0),$$

$$V(0,t) + W(0,t) = 0, \quad V(1,t) + W(1,t) = 1 + t \quad (t > 0),$$

$$V(x,0) + W(x,0) = x, \quad \frac{\partial V(x,0)}{\partial t} + \frac{\partial W(x,0)}{\partial t} = 0 \quad (0 < x < 1).$$

Let us first introduce the function

$$V(x,t) = (1 + t)x,$$

which satisfies the conditions $V(0,t) = 0,\quad V(1,t) = 1+t,\quad V(x,0) = x$. So we have to solve the nonhomogeneous problem with homogeneous conditions

$$\frac{\partial^2 W}{\partial x^2} = \frac{\partial^2 W}{\partial t^2} - 1 \quad (0 < x < 1,\ t > 0),$$

$$W(0,t) = 0, \quad W(1,t) = 0 \quad (t > 0),$$

$$W(x,0) = 0, \quad \frac{\partial W(x,0)}{\partial t} = 0 \quad (0 < x < 1).$$

The function $W(x,t)$ can be found in the following form $W(x,t) = S(x)+v(x,t)$, where $S(x)$ is the solution of problem

$$S''(x) = -1, \quad S(0) = 0, \quad S(1) = 0.$$

The solution of this problem is $S(x) = -\dfrac{x^2 - x}{2}$.

We still have to find the function v satisfying the equation

$$\frac{\partial^2 v}{\partial x^2} - \frac{\partial^2 v}{\partial t^2} = 0 \quad (0 < x < 1,\quad t > 0),$$

and the conditions

$$v(0,t) = 0 \quad v(1,t) = 0 \quad (t > 0),$$

$$v(x,0) = \frac{x^2 - x}{2}, \quad \frac{\partial v(x,0)}{\partial t} = 0 \quad (0 < x < 1).$$

The solution of this problem has the form

$$v(x,t) = \sum_{n=1}^{\infty} A_n \sin(n\pi x)\cos(n\pi t),$$

Using the initial condition $v(x,0) = \dfrac{x^2 - x}{2}$, we obtain the Fourier series

$$\frac{x^2 - x}{2} = \sum_{n=1}^{\infty} A_n \sin(n\pi x),$$

where

$$A_n = 2\int_0^1 \frac{x^2 - x}{2}\sin(n\pi x)\,dx = \frac{(-1)^n - 1}{n^3 \pi^3}.$$

The solution of the considered problem is

$$u(x,t) = (1+t)x - \frac{x^2 - x}{2} - \frac{2}{\pi^3}\sum_{n=1}^{\infty} \frac{\sin((2n-1)\pi x)\cos((2n-1)\pi t)}{(2n-1)^3}.$$

Example 4.34 *Determine the solution of the problem*

$$\frac{\partial^2 u}{\partial x^2} = \frac{\partial^2 u}{\partial t^2} - 1 \quad (0 < x < 1,\ t > 0),$$

$$u(0,t) = \frac{t^2}{2}, \qquad u(1,t) = -\cos t \quad (t > 0),$$

$$u(x,0) = -x, \qquad \frac{\partial u(x,0)}{\partial t} = 0 \quad (0 < x < 1).$$

Solution. Introducing the function

$$V(x,t) = \frac{t^2}{2} - \left(\cos t + \frac{t^2}{2}\right) x,$$

which satisfies the conditions $V(0,t) = \frac{t^2}{2}$, $V(1,t) = -\cos t$, $V(x,0) = -x$. If we take $u(x,t) = V(x,t) + W(x,t)$ we get the problem

$$\frac{\partial^2 W}{\partial x^2} = \frac{\partial^2 W}{\partial t^2} + x(\cos t - 1) \quad (0 < x < 1,\ t > 0),$$

$$W(0,t) = 0, \qquad W(1,t) = 0 \quad (t > 0),$$

$$W(x,0) = 0, \qquad \frac{\partial W(x,0)}{\partial t} = 0 \quad (0 < x < 1).$$

The function $W(x,t)$ can be written as $W(x,t) = S(x) + v(x,t)$, where $S(x)$ is the solution of the following problem

$$S''(x) = -x, \quad S(0) = 0, \quad S(1) = 0. \tag{4.68}$$

The solution of this problem is

$$S(x) = \frac{x^3 - x}{6}.$$

We still have to find the function v satisfying the equation

$$\frac{\partial^2 v}{\partial x^2} - \frac{\partial^2 v}{\partial t^2} = x \cos t \quad (0 < x < 1,\ t > 0), \tag{4.69}$$

and the conditions

$$v(0,t) = 0 \quad v(1,t) = 0 \quad (t > 0),$$

$$v(x,0) = -\frac{x^3 - x}{6}, \qquad \frac{\partial v(x,0)}{\partial t} = 0 \quad (0 < x < 1).$$

The solution of this problem is $v(x,t) = v_h(x,t) + v_p(x,t)$. The homogeneous part of the solution is

$$v_h(x,t) = \sum_{n=1}^{\infty} A_n \sin(n\pi x)\cos(n\pi t).$$

The particular solution can be taken as

$$u_p(x,t) = C(x)\sin t + D(x)\cos t,$$

where $C(x)$ and $D(x)$ are undetermined coefficients. The equation (4.69) implies

$$C''(x)\sin t + D(x)\cos t = -C(x)\sin t - D(x)\cos t,$$

wherefrom we obtain two differential equations

$$C''(x) + C(x) = -x, \qquad D''(x) + D(x) = 0.$$

The solution of second equation we can take to be zero i.e., $D(x) = 0$. The solution of the first of those two equations and conditions $C(0) = 0$, $C(1) = 0$ is $C(x) = \dfrac{\sin x}{\sin 1} - x$. So the solution of the problem (4.69) has the form

$$v(x,y) = \frac{\sin x}{\sin 1} - x + \sum_{n=1}^{\infty} A_n \sin(n\pi x)\cos(n^2\pi^2 t).$$

Using the initial condition $v(x,0) = -\dfrac{x^3 - x}{6}$, we obtain the Fourier series

$$-\frac{x^3 - x}{6} - \frac{\sin x}{\sin 1} + x = \sum_{n=1}^{\infty} A_n \sin(n\pi x),$$

where

$$A_n = -2\int_0^1 \left(-\frac{x^3 - x}{6} - \frac{\sin x}{\sin 1} + x\right)\sin(n\pi x)\,dx.$$

The solution of the considered problem is

$$u(x,t) = \frac{t^2(1 - x)}{2} + x\cos t$$

$$+\frac{2}{\pi}\sum_{n=1}^{\infty}\left(\frac{1 - \cos n\pi t}{n^2\pi^2} + \frac{2(\cos t - \cos n\pi t)}{n^2\pi^2 - 1}\right)\frac{(-1)^{n-1}}{n}\sin n\pi x.$$

Example 4.35 *Determine the transverse vibration of a uniform beam with simply supported ends described by*

$$\frac{\partial^2 u}{\partial t^2} + a^2 \frac{\partial^4 u}{\partial x^4} = 0 \qquad (0 < x < \ell, \ t > 0),$$

$$u(0, t) = 0, \quad u(\ell, t) = 0 \qquad (t > 0),$$

$$\frac{\partial^2 u(0, t)}{\partial t^2} = 0, \quad \frac{\partial^2 u(0, t)}{\partial t^2} = 0 \qquad (t > 0),$$

$$u(x, 0) = x \frac{\pi x}{\ell}, \quad \frac{\partial u(x, 0)}{\partial t} = 0 \qquad (0 < x < \ell).$$

Solution. The method of separate variables lead us to the following equations

$$X^{(4)}(x) - \lambda X(x) = 0, \qquad T''(t) + \lambda a^2 T(t) = 0.$$

The first differential equation with boundary conditions in this case also, implies trivial solutions for $\lambda < 0$ and $\lambda = 0$.

Taking $\lambda = k^4$, we obtain the general solutions of the differential equation

$$X'''' - k^4 X = 0,$$

in the form

$$X(x) = A \cos kx + B \sin kx + C e^{kx} + D e^{-kx}.$$

From boundary conditions

$$X(0) = X(\ell) = X''(0) = X''(\ell) = 0,$$

we obtain that $A = C = D = 0$ and $B \neq 0$ for $k = \dfrac{n\pi}{\ell}$. The solution of equation

$$T'' + a^2 \frac{n^4 \pi^4}{\ell^4} T = 0 \quad \text{with condition} \quad T'(0) = 0,$$

is

$$T(t) = A \cos \frac{n^2 \pi^2 a t}{\ell^2}.$$

Using the superposition principle the solution of the considered problem is

$$u(x, t) = \sum_{n=1}^{\infty} A_n \sin \frac{n\pi x}{\ell} \cos \frac{n^2 \pi^2 a t}{\ell^2}.$$

From the second initial condition we obtain the Fourier series

$$x \sin \frac{\pi x}{\ell} = \sum_{n=1}^{\infty} A_n \sin \frac{n\pi x}{\ell},$$

whose coefficients are

$$A_n = \frac{2}{\ell} \int\limits_0^\infty x \sin\frac{\pi x}{\ell} \sin\frac{n\pi x}{\ell} dx = \begin{cases} \dfrac{\ell}{2}, & n = 1 \\[3mm] \dfrac{-4n\ell(1+(-1)^n)}{(n^2-1)^2\pi^2}, & n > 1. \end{cases}$$

So we obtain the solution of the considered problem

$$u(x,t) = \frac{\ell}{2}\sin\frac{\pi x}{\ell}\cos\frac{\pi^2 at}{\ell^2} - \frac{16\ell}{\pi^2}\sum_{n=1}^\infty \frac{n}{(4n^2-1)^2}\sin\frac{2n\pi x}{\ell}\cos\frac{4n^2\pi^2 at}{\ell^2}.$$

Exercise 4.36 *Solve the problem*

$$\frac{\partial^2 u}{\partial t^2} = a^2\frac{\partial^2 u}{\partial x^2} + F(x,t) \qquad (0 \le x \le \ell,\ t > 0),$$

$$a_1\frac{\partial u(0,t)}{\partial x} + b_1 u(0,t) = p(t), \qquad a_2\frac{\partial u(\ell,t)}{\partial x} + b_2 u(\ell,t) = q(t) \qquad (t > 0),$$

$$u(x,0) = f(x), \qquad \frac{\partial u(x,0)}{\partial t} = g(x) \qquad (0 \le x \le \ell),$$

where $a_1^2 + b_1^2 > 0$, $a_2^2 + b_2^2 > 0$ and f, p, q, f, g and F are given functions.

Example 4.37 *Determine the solution $u = u(x,t)$ of the problem:*

$$\frac{\partial^2 u}{\partial t^2} = a^2\frac{\partial^2 u}{\partial x^2}, \qquad u(0,t) = 0 \qquad (0 < x < \infty, 0 < t < \infty),$$

$$u(x,0) = f(x), \qquad \frac{\partial u(x,0)}{\partial t} = g(x) \qquad (0 < x < \infty),$$

such that u and $\dfrac{\partial u}{\partial x}$ are bounded when $x \to \infty$. Let the functions f, f', g and g' be piecewise continuous functions and absolutely integrable.

Solution. We determine the solution by the Fourier method of separation of variables in the form $u(x,t) = X(x)T(t)$. The solutions of the problem

$$X'' + \lambda X = 0, \qquad X(0) = 0,$$

where $X(x)$ and $X'(x)$ are bounded when $x \to \infty$, are

$$X(x) = \sin(sx), \qquad s > 0 \tag{4.70}$$

($s < 0$ does not produce a new solution). We have here a continuum of eigenvalues generated by $\lambda = s^2$. The solutions of the problem

$$T_n'' + \lambda a^2 T_n = 0,$$

for these eigenvalues are

$$T(t) = A(s)\cos(ast) + B(s)\sin(ast) \quad (s > 0), \qquad (4.71)$$

where $A(s)$ and $B(s)$ are arbitrary functions. Using the superposition principle, from (4.70) and (4.71) we obtain

$$u(x,t) = \int_0^\infty (A(s)\cos(sat) + B(s)\sin(sat))\sin(sx)ds. \qquad (4.72)$$

The functions A and B can be found from the initial condition, as in Fourier series method. From

$$u(x,0) = f(x) = \int_0^\infty A(s)\sin(sx)ds,$$

it follows

$$A(s) = \frac{2}{\pi}\int_0^\infty f(x)\sin(sx)dx.$$

From

$$\frac{u(x,0)}{\partial t} = g(x) = \int_0^\infty saB(s)\sin(sx)ds,$$

it follows

$$saB(s) = \frac{2}{\pi}\int_0^\infty g(x)\sin(sx)dx.$$

The conditions that the functions f, f', g and g' are piecewise continuous functions and absolutely integrable, enable the convergence of integral (4.72).

Exercise 4.38 *Determine the solution $u = u(x,t)$ of the problem*

$$\frac{\partial^2 u}{\partial t^2} = a^2\frac{\partial^2 u}{\partial x^2}, \quad u(0,t) = 0 \quad (0 < x < \infty, 0 < t < \infty),$$

$$u(x,0) = 0, \quad \frac{\partial u(x,0)}{\partial t} = e^{-x} \quad (0 < x < \infty),$$

such that u and $\dfrac{\partial u}{\partial x}$ are bounded when $x \to \infty$.

Answer. This is a special case of the previous example. The solution of this problem is

$$u(x,t) = \int\limits_0^\infty (A(s)\cos(sat) + B(s)\sin(sat))\sin(sx)ds.$$

From

$$u(x,0) = 0 = \int\limits_0^\infty A(s)\sin(sx)dx,$$

it follows $A(s) = 0$. From

$$\frac{u(x,0)}{\partial t} = e^{-x} = \int\limits_0^\infty saB(s)\sin(sx)dx,$$

it follows

$$saB(s) = \frac{2}{\pi}\cdot\int\limits_0^\infty e^{-x}\sin(sx)dx,$$

and

$$B(s) = \frac{2}{a\pi}\frac{1}{1+s^2}.$$

Therefore the solution has the form

$$u(x,t) = \frac{2}{a\pi}\int\limits_0^\infty \frac{\sin(sat)\sin(sx)}{1+s^2}ds.$$

Example 4.39. *Determine the transversal oscillation of a rectangular membrane, whose length is a and width b, if the edges of the membrane are fixed, its initial shape is given with a continuous function f and its initial velocity is zero.*

Solution. The physical assumptions lead us to the following problem over the set $\{(x,y,t)|\, 0 \leq x \leq a,\ 0 \leq y \leq b,\ t \geq 0\}$:

$$\frac{\partial^2 u}{\partial^2 t} = c^2\left(\frac{\partial^2 u}{\partial x^2} + \frac{\partial^2 u}{\partial y^2}\right), \tag{4.73}$$

$$\begin{cases} u(0,y,t) = u(a,y,t) = 0 & (0 \leq y \leq b,\ t \geq 0), \\[2mm] u(x,0,t) = u(x,b,t) = 0 & (0 \leq x \leq a,\ t \geq 0), \end{cases} \tag{4.74}$$

$$u(x,y,0) = f(x,y),\quad \frac{\partial u(x,y,0)}{\partial t} = 0. \tag{4.75}$$

Let us find the solution in the form

$$u(x,y,t) = V(x,y)\cdot T(t), \tag{4.76}$$

which, in view of (4.73) takes us to the equation

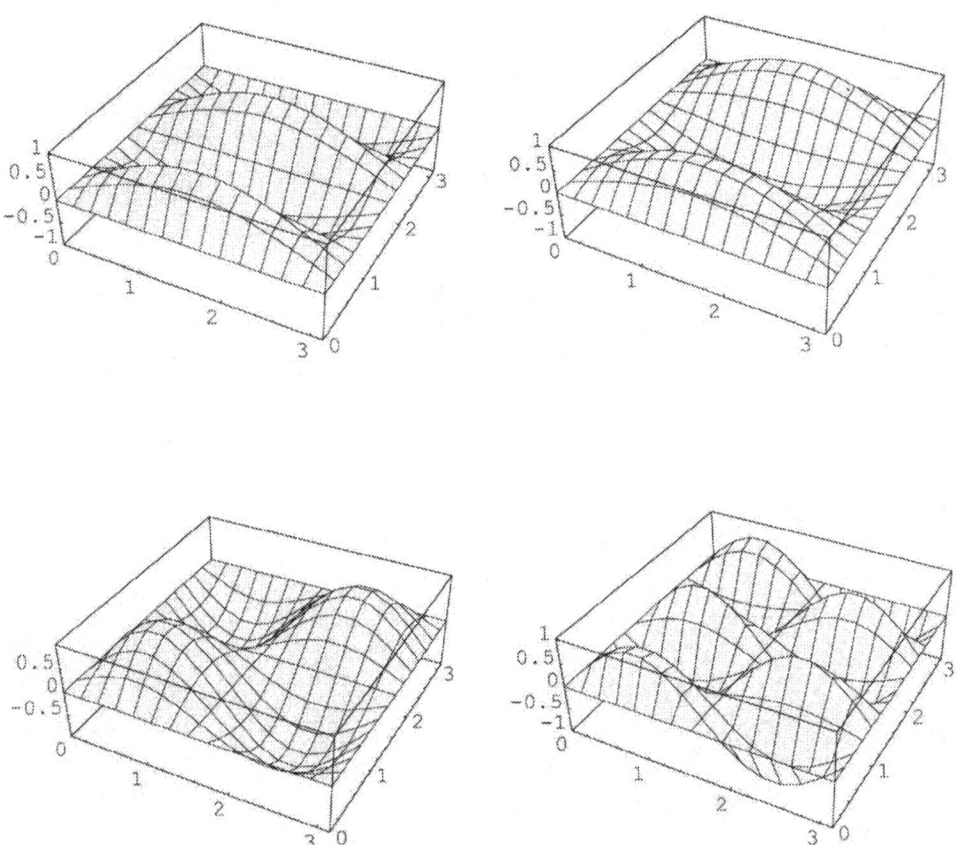

Figure 4.3 $u(x, y, 0) = \sin(nx)\sin(my)$ for $(n, m) = (1, 4), (1, 3), (2, 2), (2, 5)$.

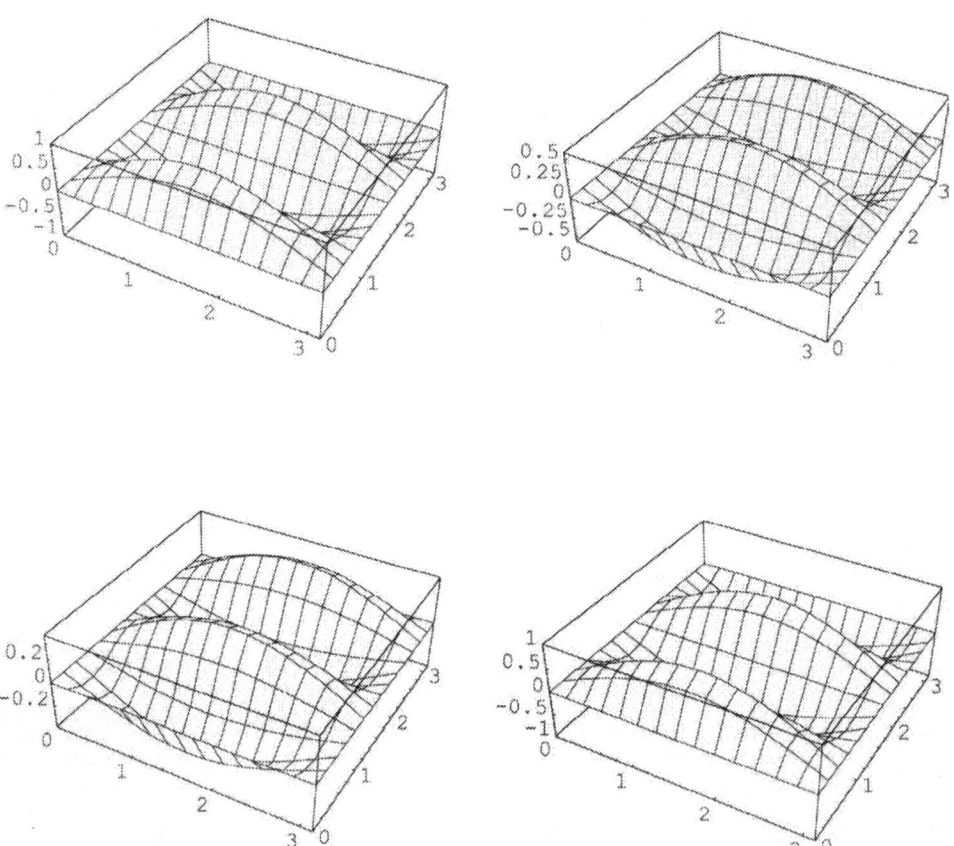

Figure 4.4 $u(x, y, t) = \sin x \sin 4y \cos(\sqrt{17}\, t)$ for $t = 0, 1, 2, 3$.

$$\frac{\dfrac{\partial^2 V}{\partial^2 x} + \dfrac{\partial^2 V}{\partial^2 y}}{V} = \frac{\dfrac{\partial^2 T}{\partial^2 t}}{c^2\, T} = -\lambda. \tag{4.77}$$

Hence λ must be a constant, which implies

$$\frac{\partial^2 V}{\partial^2 x} + \frac{\partial^2 V}{\partial^2 y} + \lambda V = 0. \tag{4.78}$$

The conditions (4.74) give

$$\begin{cases} V(0,y) = V(a,y) = 0, & \text{for } 0 \le y \le b, \quad \text{and} \\[2mm] V(x,0) = V(x,b) = 0, & \text{for } 0 \le x \le b. \end{cases} \tag{4.79}$$

The PDE (4.78) is elliptic. In order to find the solution of the problem (4.78), (4.79), we shall apply the Fourier method of separation of variables, i.e., let us put

$$V(x,y) = X(x) \cdot Y(y). \tag{4.80}$$

Equation (4.78) with conditions (4.79) gives

$$\frac{X''(x)}{X} = -\left(\frac{Y''(y)}{Y} + \lambda\right) = -\mu,$$

and $X(0) = X(a) = 0$, $Y(0) = X(b) = 0$. One easily checks that nontrivial solutions exist only if $\lambda > \mu > 0$. Then we have

$$X_n(x) = \sin\frac{n\pi x}{a}, \quad Y_m(y) = \sin\frac{m\pi y}{b},$$

$$\mu_n = \frac{n^2 \pi^2}{a^2} \quad \text{and} \quad \lambda - \mu_n = \frac{m^2 \pi^2}{b^2}.$$

The boundary conditions (4.74) give now the eigenvalues

$$\lambda_{m,n} = \left(\frac{n^2}{a^2} + \frac{m^2}{b^2}\right)\pi^2$$

and the eigenfunctions

$$F_{m,n} = X_n(x) \cdot Y_m(y) = \sin\frac{n\pi x}{a} \cdot \sin\frac{m\pi y}{b} \qquad (m,n \in \mathbf{N}).$$

The second ODE from (4.77) is $T''(t) - \lambda a^2 T = 0$, and from the last condition in (4.75) we get $T'(0) = 0$. We know that we must take $\lambda = \lambda_{m,n}$, hence the corresponding $T_{m,n}$ are of the form

$$T_{m,n}(t) = A_{m,n} \cdot \cos\left(a\sqrt{\lambda_{m,n}}\right) \qquad (m,n \in \mathbf{N}).$$

We shall determine the constants $A_{m,n}$ from (4.75). From (4.76) we get

$$u(x,y,t) = \sum_{m=1}^{\infty} \sum_{n=1}^{\infty} A_{m,n} \sin\frac{n\pi x}{a} \cdot \sin\frac{m\pi y}{b} \cdot \cos\left(c\sqrt{\frac{n^2}{a^2} + \frac{m^2}{b^2}}\,\pi t\right). \qquad (4.81)$$

The first initial condition in (4.75) gives

$$f(x,y) = \sum_{m=1}^{\infty} \sum_{n=1}^{\infty} A_{m,n} \sin\frac{n\pi x}{a} \sin\frac{m\pi y}{b} \qquad (m,n \in \mathbf{N}),$$

i.e., $A_{m,n}$ are the Fourier coefficients of the double Fourier series of the function f. As in the case of the Fourier series, in our case we have (check!)

$$A_{m,n} = \frac{4}{ab} \int_0^a \left(\int_0^b f(x,y) \sin\frac{n\pi x}{a} \sin\frac{m\pi y}{b}\, dy \right) dx, \qquad (m,n \in \mathbf{N}).$$

Presumably the reader suspects that analysis of the convergence of the series (4.81) and its formal derivatives is rather complicated. Of course, as in the one–dimensional case, additional suppositions on the smoothness of the initial function f might imply the convergence of the solution given by (4.81). In particular, if $f(x,y) = x$ on $(x,y) \in [0,a] \times [0,b]$, then the solution of the problem (4.73), (4.74) and (4.75) is

$$u(x,y,t) = \frac{8a}{\pi^2} \sum_{m=1}^{\infty} \sum_{n=1}^{\infty} \frac{(-1)^{n-1}}{mn} \sin\frac{n\pi x}{a} \cdot \sin\frac{m\pi y}{b} \cdot \cos\left(c\sqrt{\frac{n^2}{a^2} + \frac{m^2}{b^2}}\,\pi t\right).$$

Figure 4.3 represents some eigenfunctions $F_{m,n}$ for $a = \pi$ and $b = \pi$.

Figure 4.4 represents the solution u for special initial condition $f(x,y) = \sin x \sin 4y$ for $t = 0,1,2,3$, for $a = \pi$ and $b = \pi$.

Example 4.40 *Show that the Laplacian* $\Delta u = \dfrac{\partial^2 u}{\partial x^2} + \dfrac{\partial^2 u}{\partial y^2}$ *in polar coordinates* $y(x,y) = U(r,\varphi)$ *can be written as*

$$\Delta U = \frac{\partial^2 U}{\partial r^2} + \frac{1}{r}\frac{\partial U}{\partial r} + \frac{1}{r^2}\frac{\partial U}{\partial \varphi}. \qquad (4.82)$$

Solution. The connection between the rectangular and polar coordinates is

$$x = r\cos\varphi, \qquad y = r\sin\varphi.$$

From

$$\frac{\partial u}{\partial x} = \frac{\partial U}{\partial r} \cdot \frac{\partial r}{\partial x} + \frac{\partial U}{\partial \varphi} \cdot \frac{\partial \varphi}{\partial x} = \cos\varphi\frac{\partial U}{\partial r} - \frac{\sin\varphi}{r}\frac{\partial U}{\partial \varphi},$$

$$\frac{\partial u}{\partial y} = \frac{\partial U}{\partial r}\cdot\frac{\partial r}{\partial y} + \frac{\partial U}{\partial \varphi}\cdot\frac{\partial \varphi}{\partial y} = \sin\varphi\frac{\partial U}{\partial r} + \frac{\cos\varphi}{r}\frac{\partial U}{\partial \varphi},$$

it follows

$$\frac{\partial^2 u}{\partial x^2} = \cos^2\varphi\,\frac{\partial^2 U}{\partial r^2} + \frac{2\cos\varphi\sin\varphi}{r^2}\frac{\partial U}{\partial \varphi} - \frac{2\sin\varphi\cos\varphi}{r}\frac{\partial^2 U}{\partial r\partial\varphi} + \frac{\sin^2\varphi}{r}\frac{\partial U}{\partial r} + \frac{\sin^2\varphi}{r^2}\frac{\partial^2 U}{\partial\varphi^2},$$

$$\frac{\partial^2 u}{\partial y^2} = \sin^2\varphi\,\frac{\partial^2 U}{\partial r^2} - \frac{2\cos\varphi\sin\varphi}{r^2}\frac{\partial U}{\partial \varphi} + \frac{2\sin\varphi\cos\varphi}{r}\frac{\partial^2 U}{\partial r\partial\varphi} + \frac{\cos^2\varphi}{r}\frac{\partial U}{\partial r} + \frac{\cos^2\varphi}{r^2}\frac{\partial^2 U}{\partial\varphi^2},$$

wherefrom the relation (4.82) follows.

Example 4.41 The Oscillation of the Round Membrane (Drum)
Solve the following mixed problem for the two–dimensional wave equation on the disc
$D = D(0,1)$

$$\frac{\partial^2 u}{\partial t^2} - c^2\Delta u = 0 \quad \text{on } D\times(0,\infty) \tag{4.83}$$

$$u|_{\partial D} = 0, \tag{4.84}$$

$$u(x,y,0) = m_1(x,y), \quad \frac{\partial u(x,y,0)}{\partial t} = m_2(x,y) \quad ((x,y)\in D). \tag{4.85}$$

Solution. First we apply the Fourier method of the separation of variables and we are looking for the solution in the form

$$X(x,y)\cdot T(t) \quad ((x,y)\in D).$$

Putting this in the equation (4.83) we obtain

$$\frac{1}{c^2}T''X - T\Delta X = 0$$

and separating the variables

$$\frac{T''}{c^2 T} = \frac{\Delta X}{X} = -\lambda,$$

where λ is a constant. Therefore we have

$$T'' + \lambda c^2 T = 0 \quad (t\in(0,\infty)), \tag{4.86}$$

$$\Delta X + \lambda X = 0 \text{ on } D. \tag{4.87}$$

The boundary condition (4.84) implies

$$X|_{\partial D} = 0. \tag{4.88}$$

We shall solve the problem (4.87), (4.88) in the polar coordinate system (r, φ) and using again the Fourier method. We take

$$X(x, y) = R(r) \cdot \Phi(\varphi),$$

where $x = r \cos \varphi$, $y = r \sin \varphi$. Then separating the variables and introducing a constant γ we obtain the equation

$$\Phi'' + \gamma \Phi = 0, \tag{4.89}$$

where γ is a constant, and the Bessel equation

$$r(rR')' + (\lambda r^2 - \gamma) = 0. \tag{4.90}$$

The equation (4.89) has its eigenvalues $\gamma_k = k^2$ $(k \in \mathbf{Z}_+)$. The corresponding eigenfunctions are $\cos k\varphi, \sin k\varphi$ $(k \in \mathbf{Z}_+)$. Taking $\gamma = k^2$ in the equation (4.90) we obtain the Bessel equation (for $x = r\sqrt{\lambda}$ for $\lambda \geq 0$) with solutions $J_n(r\sqrt{\lambda})$. The boundary condition $R(1) = 0$ gives $J_n(\sqrt{\lambda}) = 0$, which implies that the eigenvalues are given by

$$\lambda_{nj}(x_{nj})^2 \qquad (j \in \mathbf{N}),$$

where $x_{n1} < x_{n2} < \ldots$ are the positive zeros of the function J_n.

The corresponding eigenfunctions are $R_{kj}(r) = J_k(rx_{kj})$ $(j \in \mathbf{N})$. The boundary condition (4.88) implies $R(1) = 0$.

Therefore the (formal) solution of the problem (4.83)- (4.85) is given by

$$
\begin{aligned}
u(r, \varphi, t) = & \sum_{k=0}^{\infty} \sum_{j=1}^{\infty} ((a_{kj} \cos cx_{kj}t + b_{kj} \sin cx_{kj}t) \cos k\varphi \\
& + (c_{kj} \cos cx_{kj}t + d_{kj} \sin cx_{kj}t) \sin k\varphi) J_k(rx_{kj}),
\end{aligned}
$$

where the coefficients $a_{kj}, b_{kj}, c_{kj}, d_{kj}$ are given by

$$a_{kj} = \frac{2}{\pi (J_k'(x_{kj}))^2} \int_0^1 \int_0^{2\pi} m_1(r, \varphi) r J_k(rx_{kj}) \cos k\varphi \, d\varphi dr,$$

$$b_{kj} = \frac{2}{c\pi x_{kj}(J_k'(x_{kj}))^2} \int_0^1 \int_0^{2\pi} m_2(r, \varphi) r J_k(rx_{kj}) \cos k\varphi \, d\varphi dr,$$

$$c_{kj} = \frac{2}{\pi (J_k'(x_{kj}))^2} \int_0^1 \int_0^{2\pi} m_1(r, \varphi) r J_k(rx_{kj}) \sin k\varphi \, d\varphi dr,$$

$$d_{kj} = \frac{2}{c\pi x_{kj}(J_k'(x_{kj}))^2} \int_0^1 \int_0^{2\pi} m_2(r, \varphi) r J_k(rx_{kj}) \sin k\varphi \, d\varphi dr.$$

4.5 Miscellaneous Problems

Example 4.42. *Let us consider the hyperbolic equation*

$$L(u) = \frac{\partial^2 u}{\partial x \partial y} + \frac{1}{2}u = 0 \quad ((x,y) \in \mathbf{R}^2), \tag{4.91}$$

with the initial conditions

$$u(x,x) = f(x), \quad \frac{\partial u(x,x)}{\partial y} = g(x) \quad (x \in \mathbf{R}) \tag{4.92}$$

where f and g are given functions from the class $C^2(\mathbf{R})$.

Let $M(x_0, y_0)$ be an arbitrary point in the first quadrant ($x_0 > 0$, $y_0 > 0$), and let us denote by A and B the points (x_0, x_0) and (y_0, y_0) respectively (see Figure 4.5).

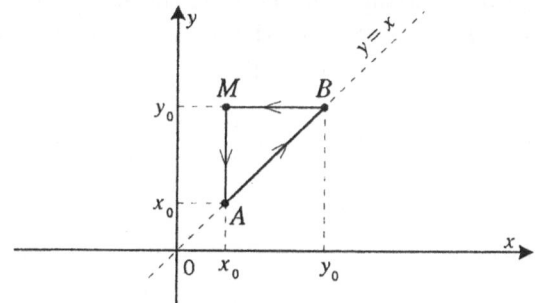

Figure 4.5

a) *Prove that for any u and v from* $C^2(\mathbf{R}^2)$ *it holds*

$$\int\int_T (v \cdot Lu - u \cdot Lv)\, dx\, dy = \oint_{\triangle ABM} \left(u\frac{\partial v}{\partial x}\, dx + v\frac{\partial u}{\partial y}\, dy \right),$$

where T denotes the interior of the triangle ABM.

b) *Prove the equality*

$$u(M)\cdot v(M) = u(B)\cdot v(B) - \int_{\overline{AM}} u\frac{\partial v}{\partial x}\, dy + \int_{\overline{BM}} u\frac{\partial v}{\partial y}\, dx + \int_{\overline{AB}} \left(u\frac{\partial v}{\partial x}\, dx + v\frac{\partial u}{\partial y}\, dy \right),$$

provided that u and v are the solutions of (4.42).

(We used the usual notation $u(P) = u(x,y)$, *when the point* $P = (x,y)$.

c) *Let* $v = v(x,y)$ *be a known particular solution of (4.42) such that* $v(x_0, y_0) \neq 0$,

$$\frac{\partial v}{\partial x} = 0 \quad for \ (x,y) \in \overline{BM}$$

$$\frac{\partial v}{\partial y} = 0 \quad for \ (x,y) \in \overline{AM}. \tag{4.93}$$

Determine then the solution $u = u(x,y)$ *of the problem (4.91), (4.92) at the point* (x_0, y_0).

d) *Show that the function* $v(x,y) = w((x - x_0)(y - y_0))$ *is a particular solution of (4.91) and satisfies the conditions (4.93), provided that the function* $w = w(z)$ *is a solution of the ODE*

$$z\,w''(z) + w'(z) + \frac{1}{2}w(z) = 0.$$

Solution. We shall construct the solution using the *Riemann's method*. For simplicity, we shall assume $0 < y_0 < x_0$; the other cases can be handled similarly.

a) By the definition of the operator L, we have

$$\int\int_T (v \cdot Lu - u \cdot Lv) \, dx \, dy = \int\int_T \left(v \cdot \frac{\partial^2 u}{\partial x \partial y} - u \cdot \frac{\partial^2 v}{\partial x \partial y} \right) dx \, dy. \tag{4.94}$$

Further on, Green's formula gives

$$\oint_{\triangle ABM} \left(u \frac{\partial v}{\partial x} \, dx + v \frac{\partial u}{\partial y} \, dy \right) = \int\int_T \left(v \cdot \frac{\partial^2 u}{\partial y \partial x} - u \cdot \frac{\partial^2 v}{\partial x \partial y} \right) dx \, dy. \tag{4.95}$$

The assumptions $u, v \in C^2(\mathbf{R}^2)$ imply that the mixed second order partial derivatives are equal, hence the right–hand sides of (4.94) and (4.95) are equal. This gives us the equality in a).

b) If u and v are two solutions of (4.91), then a) gives us

$$0 = \oint_{\triangle ABM} \left(u \frac{\partial v}{\partial x} \, dx + v \frac{\partial u}{\partial y} \, dy \right) = \int_{\overline{MA}} + \int_{\overline{BM}} + \int_{\overline{AB}}$$

$$= \int_{\overline{BM}} u \frac{\partial v}{\partial x} \, dx + \int_{\overline{MA}} v \frac{\partial u}{\partial y} \, dy + \int_{\overline{AB}} \left(u \frac{\partial v}{\partial x} \, dx + v \frac{\partial u}{\partial y} \, dy \right).$$

Now using the equality

$$\int_{\overline{BM}} u \frac{\partial v}{\partial x} \, dx = \int_{\overline{BM}} \left(\frac{\partial}{\partial x}(uv) - v \frac{\partial u}{\partial x} \right) dx = u(M)v(M) - u(B)v(B) - \int_{\overline{BM}} v \frac{\partial u}{\partial x} \, dx,$$

we obtain b).

c) From b), using the conditions (4.92), we get the solution u as

$$u(x_0, y_0) \;=\; \frac{1}{v(x_0, y_0)} \Big(f(x_0)\, v(x_0, y_0) $$

$$+ \int_{x_0}^{y_0} f(x) \frac{\partial v(x, x)}{\partial x}\, dx + \int_{x_0}^{y_0} g(y) v(x, x)\, dy \Big).$$

Note that all the quantities on the right–hand side are known.

d) Left to the reader.

Remark 4.42.1 It is clear that Riemann's method can be applied also when the segment $\overline{AB}$ from the line $y = x$ from Figure 4.5 is replaced with a smooth increasing curve ℓ, see Figure 4.6.

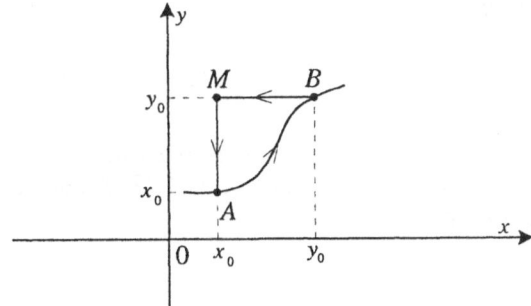

Figure 4.6

Example 4.43. *Find the function* $u = u(x, t)$, $0 \le x \le \ell$, $t \ge 0$, *such that*

$$\frac{\partial^2 u}{\partial t^2} = a^2 \frac{\partial^2 u}{\partial x^2}, \qquad \frac{\partial u(0, t)}{\partial x} = 0, \qquad \frac{\partial u(\ell, t)}{\partial x} = A \cdot e^{-t},$$

$$u(x, 0) = A \cdot \frac{a \cdot \cosh(x/a)}{\sinh(\ell/a)}, \qquad \frac{\partial u(x, 0)}{\partial t} = -A \cdot \frac{a \cdot \cosh(x/a)}{\sinh(\ell/a)}.$$

Solution. Let us put $u(x, t) = v(x, t) + e^{-t} w(x)$, where the function v satisfies the homogeneous wave equation and the conditions

$$\frac{\partial v(0, t)}{\partial x} = \frac{\partial v(\ell, t)}{\partial x} = 0 \quad \text{for} \quad t \ge 0,$$

$$v(x, 0) = \frac{\partial v(x, 0)}{\partial t} = 0 \quad \text{for} \quad 0 \le x \le \ell.$$

Then the function w should satisfy the conditions

$$w(x) = a^2 w''(x), \qquad w'(0) = 0 \quad \text{and} \quad w'(\ell) = A.$$

Clearly the function $v = v(x,t)$ is identically equal to 0, while the function w is easily found to be $w(x) = A \cdot \dfrac{a \cdot \cosh(x/a)}{\sinh(\ell/a)}$. Thus the sought after solution is

$$u(x,t) = A\, e^{-t} \cdot \frac{a \cdot \cosh(x/a)}{\sinh(\ell/a)}.$$

Example 4.44. *Prove that the following Cauchy's problem on the set* $\{(x,y)\mid |x| < 1,\ 0 < y < 1\}$:

$$\frac{\partial^2 u}{\partial x \partial y} = 0,$$

$$u(x, x^2) = 0, \qquad \frac{\partial u(x, x^2)}{\partial y} = g(x) \qquad (|x| < 1)$$

has a solution only if g is an even continuous function on $[-1,1]$ such that the function $xg(x)$ is continuously differentiable on $[-1,1]$. Prove then that the unique solution of the given problem is

$$u(x,y) = 2 \int_x^{\sqrt{y}} z\, g(z)\, dz. \tag{4.96}$$

Solution. Since the solution of the given PDE is of the form $u(x,y) = A(x) + B(y)$, for some functions $A \in C^1([-1,1])$ and $B \in C^1([0,1])$, the initial conditions give that

$$A(x) + B(x^2) = 0 \quad \text{and} \quad B'(x^2) = g(x) \qquad (x \in [-1,1]),$$

hence g must be an even function. Multiplying the last equality with $2x$ gives

$$B(x^2) = 2 \int_0^x z\, g(z)\, dz,$$

hence

$$B(y) = 2 \int_0^{\sqrt{y}} z\, g(z)\, dz \quad \text{if} \quad 0 \le y \le 1.$$

Thus we obtain formula (4.96).

Example 4.45. *Prove that the nonhomogeneous wave equation on* $\mathbf{R}^2$

$$\frac{\partial^2 u}{\partial x^2} - \frac{\partial^2 u}{\partial t^2} = 6(x+t),$$

which satisfies the initial conditions

$$u(x,x) = 0, \quad \frac{\partial u(x,x)}{\partial x} = g(x)$$

has a solution only if $g(x) = 3x^2 + C$, *for some constant* C. *Prove then that the solution is* not *unique, but all the solutions are of the form*

$$u(x,t) = x^3 - t^3 + f(x-t) - f(0) + (x-t)(g(0) - f'(0)), \tag{4.97}$$

where f *is an arbitrary function from* $C^2(\mathbf{R})$.

Solution. Using the change of variables $\xi = x+t$, $\eta = x-t$, $v(\xi,\eta) = u(x,t)$, we obtain

$$\frac{\partial^2 v}{\partial \xi \partial \eta} = \frac{3\xi}{2}.$$

Thus

$$v(\xi,\eta) = \frac{3}{4}\xi^2\eta + \phi(\xi) + \psi(\eta), \quad \phi,\psi \in C^2(\mathbf{R}^1),$$

or

$$u(x,y) = \frac{3}{4}(x+t)^2(x-t) + \phi(x+t) + \psi(x-t).$$

The first initial condition gives us

$$0 = u(x,x) = \phi(2x) + \psi(0), \quad \text{hence} \quad \phi(x) = -\psi(0), \quad x \in \mathbf{R},$$

which implies

$$u(x,t) = \frac{3}{4}(x+t)^2(x-t) - \psi(0) + \psi(x-t). \tag{4.98}$$

The second initial condition gives

$$g(x) = \frac{\partial u(x,x)}{\partial x} = \frac{3}{4}(x+t)^2 + \frac{3}{4}2(x+t)(x-t) + \psi'(x-t)\Big|_{x=t} = 3x^2 + \psi'(0),$$

which is, in fact, the sought after condition for the function g with $C = \psi'(0)$.
  So we obtain from (4.98) the solution u as

$$u(x,t) = x^3 - t^3 + \psi_1(x-t) - \psi_1(0),$$

for some $\psi_1 \in C^2(\mathbf{R})$. Applying Taylor's formula we get

$$u(x,t) = x^3 - t^3 + (x-t)(g(0) - \psi_1'(0)) + \psi_2(x-t) - \psi_2(0),$$

for some $\psi_2 \in C^2(\mathbf{R})$. The obtained form of the solution u is equivalent to (4.97).

Example 4.46. *Let the following two second order PDEs be given:*

$$\frac{\partial^2 u}{\partial x_1^2} + y_1 \frac{\partial^2 u}{\partial y_1^2} + A \frac{\partial u}{\partial y_1} = 0, \quad \frac{1}{2} < A < 1, \qquad (4.99)$$

and

$$T(A,B)(v) = \frac{\partial^2 v}{\partial x \partial y} - \frac{B \frac{\partial v}{\partial x}}{x - y} + \frac{A \frac{\partial v}{\partial y}}{x - y} = 0, \qquad (4.100)$$

for $0 < A < 1$, $0 < B < 1$, $A + B \neq 1$, where $u = u(x_1, y_1)$ and $v = v(x,y)$ are the unknown functions, respectively.

a) *Prove that in the half plane $\{(x_1, y_1)|\, y_1 < 0\}$ equation (4.99) is hyperbolic. Then transform (4.99) with the change of variables $x = x_1 - 2\sqrt{-y_1}$, $y = x_1 + 2\sqrt{-y_1}$.*

b) *Check that both functions f and g, where*

$$f_{A,B}(x, y, t) = (t - x)^{-A}(y - t)^{-B}, \qquad t \ \text{a real parameter,}$$

and

$$g(x,y) = \int_x^y \Phi(t)\, f_{A,B}(x, y, t)\, dt, \qquad \Phi \ \text{an arbitrary continuous function on } \mathbf{R},$$

are solutions of equation (4.100).

c) *Prove that the function $v^{A,B}(x,y) = (y - x)^{1-A-B}v^{1-A,1-B}(x,y)$ is a solution of (4.100), whenever $v^{1-A,1-B}$ is.*

d) *Using b) and c), find all solutions of equation (4.100), and then solve also equation (4.99).*

Hints. We leave to the reader to check that the given change of variables transforms equation (4.99) into

$$T(A - 1/2, B - 1/2)(v) = 0, \qquad (4.101)$$

where $v(x,y) = u(x_1, y_1)$. Also, parts b) and c) are omitted. However, they give us the form of the general solution of (4.100), namely it is

$$v(x,y) = \int_x^y \Phi_1(t)\, f_{A,B}(x, y, t)\, dt + (y - x)^{1-A-B} \int_x^y \Phi_2(t)\, f_{1-B,1-A}(x, y, t)\, dt,$$

where Φ_1 and Φ_2 are arbitrary continuous functions on $\mathbf{R}$. Thus the general solution of equation (4.99) is

$$u(x_1, y_1) = \int_{x_1 - 2\sqrt{-y_1}}^{x_1 + 2\sqrt{-y_1}} \Phi_1(t) \left(t - x_1 + 2\sqrt{-y_1}\right)^{1/2-A} \left(x_1 + 2\sqrt{-y_1} - t\right)^{1/2-A} dt$$

$$+ \left(4\sqrt{-y_1}\right)^{-2A} \int_{x_1 - 2\sqrt{-y_1}}^{x_1 + 2\sqrt{-y_1}} \Phi_2(t) \left(t - x_1 + 2\sqrt{-y_1}\right)^{1/2+A} \left(x_1 + 2\sqrt{-y_1} - t\right)^{1/2+A} dt.$$

Example 4.47 (Poincare's inequality on n–dimensional parallelepiped)
Prove that on n–dimensional parallelepiped

a) $P_1 = (0,1) \times \cdots \times (0,1)$ *and* $f \in C^1(P_1)$ *we have the following inequality*

$$\int_{P_1} |f(x)|^2 \, dx \leq \left| \int_{P_1} f(x) \, dx \right|^2 + \frac{n}{2} \int_{P_1} \sum_{i=1}^{n} |D_i f(x)|^2 \, dx, \qquad (4.102)$$

b) $P = (a_1, b_1) \times \cdots \times (a_n, b_n)$ *and* $f \in C^1(P)$ *we have the following inequality*

$$\int_{P} |f(x)|^2 \, dx \quad \leq \quad \frac{1}{(b_1 - a_1) \cdots (b_n - a_n)} \left| \int_{P} f(x) \, dx \right|^2$$

$$+ \frac{n}{2} \int_{P} \sum_{i=1}^{n} (b_i - a_i)^2 |D_i f(x)|^2 \, dx.$$

(4.103)

Solution.

a) The basic integral formula gives us

$$f(y) - f(x) = \sum_{i=1}^{n} \int_{x_i}^{y_i} D_i f(x_1, \ldots, x_{i-1}, s_i, y_{i+1}, \ldots, y_n) \, ds_i \qquad (4.104)$$

for $x, y \in P_1$. Multiplying the equality (4.104) by the conjugate equality (4.104) we obtain

$$(f(y) - f(x))(\overline{f(y)} - \overline{f(x)}) \;=\; \left(\sum_{i=1}^{n} \int_{x_i}^{y_i} D_i f(x_1, \ldots, x_{i-1}, s_i, y_{i+1}, \ldots, y_n) \, ds_i \right)$$

$$\overline{\left(\sum_{i=1}^{n} \int_{x_i}^{y_i} D_i f(x_1, \ldots, x_{i-1}, s_i, y_{i+1}, \ldots, y_n) \, ds_i \right)}.$$

Hence by the triangle inequality, the inequality

$$\left(\sum_{i=1}^{n} |z_i|\right)^2 \leq n\left(\sum_{i=1}^{n} |z_i|^2\right)$$

and inequality $|\int \cdot| \leq \int |\cdot|$

$$|f(y)|^2 + |f(x)|^2 - f(y)\overline{f(x)} - f(x)\overline{f(y)}$$

$$\leq n \left(\sum_{i=1}^{n} \int_0^1 |D_i f(x_1, \ldots, x_{i-1}, s_i, y_{i+1}, \ldots, y_n)|^2 \, ds_i\right).$$

Applying the integral with respect to x and y on parallelepiped P_1 we obtain (using the iterated integrals - see also the Fubini theorem from 8.1

$$\int_{P_1} |f(y)|^2 \, dy + \int_{P_1} |f(x)|^2 \, dx - \left(\int_{P_1} f(y) \, dy\right)\left(\overline{\int_{P_1} f(x) \, dx}\right)$$

$$- \left(\int_{P_1} f(x) \, dx\right)\left(\overline{\int_{P_1} f(y) \, dy}\right) \leq n \sum_{i=1}^{n} \int_{P_1} |D_i f(x)|^2 \, dx,$$

which gives the desired inequality (4.102).

b) The inequality (4.103) reduces on (4.102) by changing the variables

$$\overline{x_i} = \frac{x_i - a_i}{b_i - a_i} \qquad (i = 1, \ldots, n).$$

Exercise 4.48 *Prove that the solution $u = u(x,t)$ of the mixed problem*

$$\frac{\partial^2 u}{\partial t^2} - \frac{\partial\left(p(x)\dfrac{\partial u}{\partial x}\right)}{\partial x} + q(x)u = F(x,t) \text{ for}$$

$$p \in C^1[0,b], \quad q \in C[0,b] \qquad (p \geq p_0 > 0, q \geq 0),$$

with initial conditions

$$u(x,0) = f(x), \quad \frac{\partial u(x,0)}{\partial t} = g(x) \qquad (0 \leq x \leq a),$$

and boundary conditions

$$u(0,t) = r_1(t), \ u(b,t) = r_2(t) \qquad (0 \leq t \leq T,)$$

where $f(0) = r_1(0), f(b) = r_2(0), g(0) = r_1'(0), g(b) = r_2'(0)$, depends continuously from the initial conditions f and g.

Hints. Using the energy integral

$$E(t) = \int\limits_0^b \left((\frac{\partial u}{\partial t})^2 + p(\frac{\partial u}{\partial x})^2 + qu^2 \right) dx$$

show that for two solutions u_1 and u_2 with the same boundary conditions and initial conditions f, g and f_1, g_1, respectively, for every $\varepsilon > 0$ there exists $\delta > 0$ such that

$$\max(\max_{x\in[a,b]} |f_1(x) - f_2(x)|, \max_{x\in[a,b]} |f_1'(x) - f_2'(x)|, \max_{x\in[a,b]} |g_1(x) - g_2(x)|) < \delta$$

implies

$$|u_1(x,t) - u_2(x,t)| < \varepsilon \text{ for } x \in [0,b] \text{ and } t \in [0,T].$$

Exercise 4.49 (Volterra's method) *The equation*

$$\frac{\partial^2 u}{\partial x \partial y} + a(x,y)\frac{\partial u}{\partial x} + b(x,y)\frac{\partial u}{\partial y} + c(x,y)u = F(x,y), \qquad (4.105)$$

where $a, b, c, F \in C(P), P = [x_0, x_1] \times [y_0, y_1]$, *with the boundary conditions*

$$u(x_0, y) = r_1(y) \text{ for } y_0 \le y \le y_1, \quad \frac{\partial u(x, y_0)}{\partial y} = r_2(x) \text{ for } x_0 \le x \le x_1,$$

where $r_1 \in C^1[y_0, y_1], r_2 \in C^1[x_0, x_1]$ *and* $r_1(y_0) = r_2(x_0)$, *has a unique solution on* P.

Hints. Let $\dfrac{\partial u}{\partial x} = v$ and $\dfrac{\partial u}{\partial y} = w$. Then the equation (4.105) reduces on the equation

$$\frac{\partial v}{\partial y} = \frac{\partial w}{\partial x} = F - av - bw - cu.$$

Therefore we have

$$v(x,y) = v(x,y_0) + \int\limits_{y_0}^y (F - av - bw - cu)\, dy = r_2'(x) + \int\limits_{y_0}^y (F - av - bw - cu)\, dy$$

and

$$w(x,y) = w(x_0,y) + \int\limits_{x_0}^x (F - av - bw - cu)\, dx = r_1'(y) + \int\limits_{x_0}^x (F - av - bw - cu)\, dx.$$

The third integral equation is (integrating $\dfrac{\partial u}{\partial y} = w$)

$$u(x,y) = r_2(x) + \int\limits_{y_0}^y w\, dy.$$

The system of three integral equations with three unknown functions u, v and w is equivalent to the starting boundary problem. This system can be solved by the method of iterated approximations.

Exercise 4.50 *Solve the following quasi-linear equations*

a) the Riccati equation $\dfrac{\partial v}{\partial x} - \dfrac{1}{2}v^2 = F(x)$

where F is an arbitrary continuous function and $v = v(x, y)$;

b) the Liouville equation $\dfrac{\partial^2 u}{\partial x \partial y} = ce^u$,

where c is an arbitrary constant different from zero and $u = u(x, y)$;

c) $\dfrac{\partial^2 U}{\partial x^2} + \dfrac{\partial^2 U}{\partial y^2} = 4ce^U$,

where c is an arbitrary constant different from zero and $U = U(x, y)$.

Hints.

a) Taking (the Schwartz derivative)

$$F(x) = \frac{p'''(x)}{p'(x)} - \frac{3}{2}\frac{p''^2(x)}{p'^2(x)}$$

for $p \in C^3$, we can easily see that

$$v(x, y) = \frac{p''(x)}{p'(x)} - \frac{2p'(x)}{p(x) + q(y)},$$

for an arbitrary function q of the class C^1, is the solution of Riccati equation.

b) Differentiating the Liouville equation with respect to x we obtain

$$\frac{\partial^2 u}{\partial y \partial x} = ce^u \frac{\partial u}{\partial x}.$$

Eliminating ce^u from the preceding equation and Liouville equation and introducing a new function v by $v = \dfrac{\partial u}{\partial x}$ we obtain the Riccati equation. Hence

$$e^u = \frac{2}{c}\frac{p'(x)q'(x)}{(p(x) + q(y))^2}.$$

c) Introduce new variables $z = x + iy, \bar{z} = x - iy$ and

$$u(z, \bar{z}) = U(\frac{z + \bar{z}}{2}, \frac{z - \bar{z}}{2i})$$

and use b) to prove that the desired solution is given by

$$e^U = e^u = \frac{2}{c}\frac{p'(z)\overline{p'(z)}}{(p(z) + \overline{p(z)})^2},$$

where p is an arbitrary analytic function.

Exercise 4.51 *Find a solution of the sine–Gordon equation*

$$\frac{\partial^2 u}{\partial x \partial y} = \sin u$$

in the form $u(x, y) = p(ax + by)$, *where a and b are constants of the same sign.*

Hints. The problem reduces to the ordinary differential equation

$$p''(z) = \frac{1}{ab} \sin p,$$

where $z = ax + by$. Multiplying it by $2p'(x)$ and integrating we reduce it to the ordinary differential equation of the first order

$$p'^2 = -\frac{2}{ab} \cos p + \frac{2C_1}{ab},$$

where C_1 is an arbitrary constant. We obtain a real solution (for $C_1 \geq 1$) from

$$\int_0^{p(x)} \frac{ds}{\sqrt{C_1 - \cos s}} = \pm(\frac{2}{ab})^{1/2} z + C_2,$$

where C_2 is an arbitrary constant.

Example 4.52 *Prove that the Cauchy problem*

$$\frac{\partial^2 u}{\partial x \partial y} = 0 \ on \ K = \{(x, y)|\ |x| < 1, |y| < 1\},$$

with initial conditions

$$u(x, x^3) = |x|^\alpha, \qquad \frac{\partial u(x, x^3)}{\partial x} = 0$$

has a solution only for $\alpha = 0$ or $\alpha \geq 6$. Prove that for these cases the unique solution is given by $u(x, y) = |y|^{\alpha/3}$.

Solution. Since the general solution is of the following form $u(x, y) = p(x) + q(y)$ for arbitrary functions $p, q \in C^1(K)$ we obtain by the initial conditions $p(x) + q(x^3) = |x|^\alpha$ and $p'(x) = 0$. Hence $p(x) = C$ and $q(x^3) = |x|^\alpha - C$, where C is an arbitrary constant. Therefore $q(x) = |x|^{\alpha/3} - C$, which implies

$$u(x, y) = C + |y|^{\alpha/3} - C = |y|^{\alpha/3}.$$

This function u is the solution of the considered initial problem if $u \in C^2(K)$, what is satisfied only for $\alpha = 0$ or $\frac{\alpha}{3} \geq 2$, i.e., $\alpha \geq 6$.

Example 4.53

a) *Consider the following functional equation*

$$f(x) + kf(a(x)) = b(x), \qquad (4.106)$$

where f is an unknown function, k a real constant and a and b continuous functions defined on $\mathbf{R}$. We denote the composition of the function a with itself in the following way:

$$a^{o(0)} = Id, \quad a^{o(1)} = a, \quad a^{o(m)} = a(a^{o(m-1)}).$$

If there exists $0 < M < 1$ such that

$$|k^m b(a^{o(m)}(x))| \leq M^m \qquad (m \in \mathbf{N}, x \in \mathbf{R}), \qquad (4.107)$$

then the solution of the equation (4.106) is given by

$$f(x) = \sum_{m=0}^{\infty} (-1)^m k^m b(a^{o(m)}(x)) \qquad (x \in \mathbf{R}). \qquad (4.108)$$

b) *Using the result from a) find the solution of the problem*

$$\frac{\partial^2 u}{\partial x^2} - \frac{\partial^2 u}{\partial y^2} = 0, \quad u(x,0) = \sin x, \quad u\left(x, \frac{x}{4}\right) = x$$

on the angle $\left\{(x,y)\vert\ x > 0, 0 < y < \frac{x}{4}\right\}$.

Solution.

a) The condition (4.107) implies the uniform convergence of the series (4.108) on $\mathbf{R}$. Hence (4.108) defines a continuous function on $\mathbf{R}$. Taking it in the equation (4.106) we verify that it is a solution of this equation.

b) The general solution of the given PDE is given by

$$u(x,y) = p(x + y) + q(x - y),$$

where $p, q \in C^1(\mathbf{R})$. The given conditions imply

$$p(x) + q(x) = \sin x, \quad p\left(\frac{5x}{4}\right) + q\left(\frac{3x}{4}\right) = x.$$

Taking $y = \dfrac{5x}{4}$ we obtain a functional equation of the form (4.106)

$$q(x) - q\left(\frac{3}{5}x\right) = \sin x - \frac{4x}{5}, \qquad (4.109)$$

where $k = -1, a(x) = \frac{3}{5}x, b(x) = \sin x - \frac{4x}{5}$. It is easy to prove that $a^{o(m)}(x) = (\frac{3}{5})^m x$ for $m = 0, 1, \ldots$, and

$$b(a^{o(m)}(x)) = \sin((\frac{3}{5})^m x) - \frac{4}{5} \cdot (\frac{3}{5})^m x.$$

If x belongs to some compact subset K of $\mathbf{R}$, then we have

$$|(-1)^m (\sin((\frac{3}{5})^m x) - \frac{4}{5} \cdot (\frac{3}{5})^m x)| \leq \frac{9}{5} \cdot |x| \cdot (\frac{3}{5})^m \leq M^m,$$

where $M = M(K) \in (0, 1)$. Hence the condition (4.107) is satisfied for $m \geq m_0(K)$. Therefore by a) the solution of the equation (4.109) is given by

$$q(x) = \sum_{m=0}^{\infty} \left(\sin((\frac{3}{5})^m x) - \frac{4}{5} \cdot (\frac{3}{5})^m x \right).$$

Then the solution u of the considered problem is given by

$$
\begin{aligned}
u(x, y) = \ & \sin(x + y) - \sum_{m=0}^{\infty} \left(\sin((\frac{3}{5})^m (x + y)) - \frac{4}{5} \cdot (\frac{3}{5})^m (x + y) \right) \\
& + \sum_{m=0}^{\infty} \left(\sin((\frac{3}{5})^m (x - y)) - \frac{4}{5} \cdot (\frac{3}{5})^m (x - y) \right).
\end{aligned}
$$

4.6 The Vibrating String

Let us consider a perfectly elastic string (for example a violin string) which is stretch with tension T between two fixed points $x = 0$ and $x = \ell$ on x axis. We assume that the string is uniformly covered with constant density ρ.

If the string is somehow set into motion in the vertical plane, such that the deflections $u(x, t)$ are very small, then we shall analyze the subsequent motion of the string. We assume that the gravity force is ignored.

Let x and $x + \Delta x$ be two points on the string ($\Delta x \approx 0$) and assume that only vertical components of the tension T are acting at these two points. Let α_1 and α_2 be the angles between T and the x−axis at the points x and $x + \Delta x$, respectively. Since by assumption these angles are small, it holds $\sin \alpha \approx \tan \alpha$. So we have

$$-T \sin \alpha_1 \approx -T \tan \alpha_1 = -\frac{\partial u}{\partial x}, \qquad \text{and}$$

$$T \sin \alpha_2 \approx -T \tan \alpha_2 = -\frac{\partial u(x, x + \Delta x)}{\partial x}.$$

(4.110)

From the Newton's second low of motion it follows that the sum of the forces (4.110) is equal to the inertial force $\rho \Delta x \dfrac{\partial^2 u}{\partial t^2}$:

$$T \left(\frac{\partial u(x, x + \Delta x)}{\partial x} - \frac{\partial u}{\partial x} \right) = \rho \frac{\partial^2 u}{\partial t^2} \Delta x.$$

Let us divide the previous relation with $T \Delta x$ and take the limit as $\Delta x \to 0$. Then we have

$$\lim_{\Delta x \to 0} \frac{\dfrac{\partial u(x, x + \Delta x)}{\partial x} - \dfrac{\partial u}{\partial x}}{\Delta x} = \frac{\rho}{T} \frac{\partial^2 u}{\partial t^2},$$

which leads us to the *one dimensional homogeneous wave equation*

$$\frac{\partial^2 u}{\partial x^2} = a^{-2} \frac{\partial^2 u}{\partial t^2}.$$

If an external force F is acting to the string, then we obtain the *one dimensional nonhomogeneous wave equation*

$$\frac{\partial^2 u}{\partial x^2} = a^{-2} \frac{\partial^2 u}{\partial t^2} + F(x, t).$$

Chapter 5

Elliptic Equations

5.1 Dirichlet Problem

5.1.1 Preliminaries

We consider the equation given by

$$L(u) = \sum_{i=1}^{n}\sum_{j=1}^{n} a_{ij}(x)\frac{\partial^2 u}{\partial x_i \partial x_j} + \sum_{i=1}^{n} b_i(x)\frac{\partial u}{\partial x_i} + c(x)u = F,$$

where $a_{ij} = a_{ji}$ $i,j = 1,2,\ldots n$, $a_{ij}, b_{ij}, c \in C(Q)$ and L is an elliptic operator, i.e.,

$$\sum_{i=1}^{n}\sum_{j=1}^{n} a_{ij}(x)p_i p_j > 0,$$

for all $(p_1, p_2, \ldots, p_n) \in \mathbf{R}^n \setminus \{(0,0,\ldots,0)\}$ and for every $x \in Q$, where Q is a bounded region of $\mathbf{R}^n$ with a boundary ∂Q, which has piecewise continuous normal at each point.

Let us introduce the following problems where $f \in C(\partial Q)$.

1) *Dirichlet problem* (the first boundary value problem) Find $u \in C^2(Q) \cap C(\overline{Q})$ such that

$$L(u) = F \quad \text{on } Q \qquad u|_{\partial Q} = f.$$

2) *von Neumann problem* (the second boundary value problem) Find $u \in C^2(Q) \cap C^1(\overline{Q})$ such that

$$L(u) = F \quad \text{on } Q, \qquad \frac{\partial u}{\partial \mathbf{n}}\bigg|_{\partial Q} = f,$$

where $\dfrac{\partial u}{\partial \mathbf{n}}$ is the directional derivative of u along the outward normal.

143

3) (The third boundary problem) Find $u \in C^2(Q) \cap C^1(\overline{Q})$ such that

$$L(u) = F \quad \text{on } Q, \qquad \frac{\partial u}{\partial n} + a = f \quad (x \in \partial Q).$$

where $a \in C(\partial Q)$.

A function $u \in C^2(Q)$ which satisfies the Laplace equation

$$\Delta u = 0 \quad (x \in Q),$$

is called *harmonic function* on Q.

Let an arbitrary continuous function defined on the unit disc f be given in polar coordinates $f(x,y) = A(\varphi)$, for $x = \cos\varphi$, $y = \sin\varphi$ on interval $[-\pi, \pi]$, such that $A(-\pi) = A(\pi)$.

The function u given by (the Poisson integral formula)

$$u(r,\varphi) = \begin{cases} A(\varphi), & r = 1 \\ \dfrac{1}{2\pi} \displaystyle\int_{-\pi}^{\pi} \dfrac{A(t)(1-r^2)}{1 - 2r\sin(t-\varphi) + r^2} dt, & r^2 \leq 1, \end{cases}$$

represent the solution of Dirichlet problem for $\Delta u = 0$ on the unit circle

$$\{(r,\varphi)|0 \leq r \leq 1, \ -\pi \leq \varphi \leq \pi\}.$$

5.1.2   Examples and Exercises

Example 5.1 *Determine the formal solution of the Dirichlet problem*

$$\frac{\partial^2 u}{\partial x^2} + \frac{\partial^2 u}{\partial y^2} = 0 \quad (0 < x < \ell, \ 0 < y < \ell_1),$$

$$u(0,y) = 0, \quad u(\ell,y) = 0 \quad (0 < y < \ell_1), \tag{5.1}$$

$$u(x,0) = 0, \quad u(x,\ell_1) = f(x) \quad (0 < x < \ell).$$

Solution. We use the method of separation of variables. Taking

$$u(x,y) = X(x) \cdot Y(y),$$

we obtain two differential equations

$$\frac{X''(x)}{X(x)} = -\lambda, \qquad \frac{Y''(y)}{Y(y)} = \lambda.$$

Using the conditions from (5.1) we get two problems

$$X''(x) + \lambda X(x) = 0 \qquad (0 < x < \ell) \qquad X(0) = 0, \quad X(\ell) = 0, \qquad (5.2)$$

$$Y''(y) - \lambda Y(y) = 0 \qquad (0 < y < \ell_1), \qquad Y(\ell_1) = 0. \qquad (5.3)$$

Let us first solve the equation (5.2) depending on variable x. The eigenvalues of the considered problem are

$$\lambda = \lambda_n = \frac{n^2 \pi^2}{\ell^2} \qquad (n \in \mathbf{N}), \qquad (5.4)$$

and the corresponding eigenfunctions have the forms

$$X_n(x) = \sin \frac{n\pi x}{\ell} \qquad (n \in \mathbf{N}).$$

For λ given by (5.4) the solution of ordinary differential equation in problem (5.3) has the form

$$Y_n(y) = A_n \cosh \frac{n\pi y}{\ell} + B_n \sinh \frac{n\pi y}{\ell} \qquad (n \in \mathbf{N}),$$

where A_n and B_n are arbitrary constants. From the boundary condition $Y(0) = 0$, we obtain that $A_n = 0$.

The formal solution of the considered problem is

$$u(x, y) = \sum_{n=1}^{\infty} B_n \sin \frac{n\pi x}{\ell} \cdot \sinh \frac{n\pi y}{\ell},$$

for $0 < x < \ell, \quad 0 < y < \ell_1$.

This solution has to satisfy the condition $u(x, \ell_1) = f(x)$ and therefore for $y = \ell_1$ we obtain the Fourier series

$$f(x) = \sum_{n=1}^{\infty} B_n \sin \frac{n\pi x}{\ell} \sinh \frac{n\pi \ell_1}{\ell},$$

where

$$B_n = \frac{2}{\ell \sinh \dfrac{n\pi \ell_1}{\ell}} \int_0^{\ell} f(x) \sin \frac{n\pi x}{\ell} dx \qquad (n \in \mathbf{N}).$$

Example 5.2 *Find the formal solution of the problem*

$$\frac{\partial^2 u}{\partial x^2} + \frac{\partial^2 u}{\partial y^2} = 0 \qquad (0 < \ \cdot \ < \ell, \ 0 < y < \ell_1),$$

with the conditions

$$u(0, y) = 0, \quad u(\ell, y) = 0 \qquad (0 < y < \ell_1),$$
$$u(x, 0) = 2, \quad u(x, \ell_1) = 0 \qquad (0 < x < \ell).$$

Solution. Using the method of separation of variables we obtain the problem (5.2) with the eigenvalues

$$\lambda = \lambda_n = \frac{n^2 \pi^2}{\ell^2} \qquad (n \in \mathbf{N}),$$

and the corresponding eigenfunctions

$$X_n(x) = \sin \frac{n \pi x}{\ell} \qquad (n \in \mathbf{N}).$$

Also, we get the problem

$$Y''(y) - \lambda Y(y) = 0 \quad 0 < y < \ell_1, \qquad Y(\ell_1) = 0.$$

The solution of the equation has the form

$$Y_n(y) = A_n \cosh \frac{n \pi y}{\ell} + B_n \sinh \frac{n \pi y}{\ell} \qquad (n \in \mathbf{N}),$$

where A_n and B_n are arbitrary constants. From the boundary condition $Y(\ell_1) = 0$, we obtain

$$0 = A_n \cosh \frac{n \pi \ell_1}{\ell} + B_n \sinh \frac{n \pi \ell_1}{\ell},$$

wherefrom it follows that

$$B_n = -A_n \frac{\cosh \dfrac{n \pi \ell_1}{\ell}}{\sinh \dfrac{n \pi \ell_1}{\ell}},$$

the solution $Y_n(y)$ can be written as

$$
\begin{aligned}
Y_n(y) &= A_n \cosh \frac{n \pi y}{\ell} - A_n \frac{\cosh \dfrac{n \pi \ell_1}{\ell}}{\sinh \dfrac{n \pi \ell_1}{\ell}} \cdot \sinh \frac{n \pi y}{\ell} \\[2ex]
&= A_n \frac{1}{\sinh \dfrac{n \pi \ell_1}{\ell}} \left(\cosh \frac{n \pi y}{\ell} \cdot \sinh \frac{n \pi \ell_1}{\ell} - \cosh \frac{n \pi \ell_1}{\ell} \cdot \sinh \frac{n \pi y}{\ell} \right) \\[2ex]
&= A_n \frac{1}{\sinh \dfrac{n \pi \ell_1}{\ell}} \cdot \sinh \frac{n \pi (\ell_1 - y)}{\ell}.
\end{aligned}
$$

The formal solution of the considered problem is

$$u(x, y) = \sum_{n=1}^{\infty} A_n \frac{1}{\sinh \dfrac{n \pi \ell_1}{\ell}} \sin \frac{n \pi x}{\ell} \cdot \sinh \frac{n \pi (\ell_1 - y)}{\ell},$$

for $0 < x < \ell$, $0 < y < \ell_1$.

This solution has to satisfy the condition $u(x,0) = 2$ and therefore for $y = 0$ we obtain the Fourier series

$$u(x,0) = 2 = \sum_{n=1}^{\infty} A_n \sin \frac{n\pi x}{\ell},$$

where

$$A_n = \frac{2}{\ell} \int_0^\ell 2 \sin \frac{n\pi x}{\ell} dx = 4 \frac{1 + (-1)^{n+1}}{n\pi} \qquad (n \in \mathbf{N}).$$

The formal solution of given elliptic equation has the form

$$u(x,t) = \frac{8}{\pi} \sum_{n=1}^{\infty} \frac{1}{(2n-1) \sinh \dfrac{(2n-1)\pi \ell_1}{\ell}} \sin \frac{(2n-1)\pi x}{\ell} \sinh \frac{(2n-1)\pi (\ell_1 - y)}{\ell}.$$

Example 5.3 *Find the solution of the problem*

$$\frac{\partial^2 u}{\partial x^2} + \frac{\partial^2 u}{\partial y^2} = 0 \qquad (0 < x < 1, \ 0 < y < 2),$$

with the conditions

$$\frac{\partial u(0,y)}{\partial x} = 0, \qquad \frac{\partial u(1,y)}{\partial x} = 0 \qquad (0 < y < 2)$$

$$u(x,0) = x^2, \qquad \frac{\partial u(x,2)}{\partial y} = 0 \qquad (0 < x < 1).$$

Solution. Taking $u(x,y) = X(x) \cdot Y(y)$, we obtain the problem

$$X''(x) + \lambda X(x) = 0 \qquad (0 < x < 1), \qquad X'(0) = X'(\ell) = 0,$$

which eigenfunctions are

$$X_n(x) = \cos(n\pi x) \qquad (n \in \mathbf{N}).$$

Now, the solution of the equation

$$Y''(y) - \lambda Y(y) = 0 \qquad (0 < y < \ell_1),$$

is given by

$$Y_n(y) = \begin{cases} Cy + D, & n = 0 \\ A_n \cosh(n\pi y) + B_n \sinh(n\pi y), & n \geq 1. \end{cases}$$

The solution of the considered problem is

$$u(x,y) = Cy + D + \sum_{n=1}^{\infty} \left(A_n \cosh(n\pi y) + B_n \sinh(n\pi y) \right) \cdot \cos(n\pi)x.$$

Using the condition $u(x,0) = x^2$ we obtain

$$u(x,0) = x^2 = D + \sum_{n=1}^{\infty} A_n \cos(n\pi x),$$

where

$$D = A_0 = \int_0^1 x^2 dx = \frac{1}{3},$$

$$A_n = 2 \int_0^1 x^2 \cos(n\pi x)dx = 4\frac{(-1)^n}{n^2\pi^2} \qquad (n \geq 1).$$

From the conditions $\dfrac{\partial u(x,2)}{\partial y} = 0$, we get

$$\frac{\partial u(x,2)}{\partial y} = 0 = C + \sum_{n=1}^{\infty} \frac{1}{n\pi} \left(A_n \sinh(n\pi 2) + B_n \sin\cosh(n\pi 2) \right) \cdot \cos(n\pi)x,$$

wherefrom we have

$$A_n \sinh(n\pi 2) + B_n \sin\cosh(n\pi 2) = 0.$$

Hence

$$B_n = -4\frac{(-1)^n \sinh(2n\pi)}{n^2\pi^2 \cosh 2n\pi}.$$

The solution of given elliptic equation has the form

$$u(x,y) = \frac{4}{\pi^2} \sum_{n=1}^{\infty} \frac{(-1)^n \left(\cosh(n\pi y)\cosh(2n\pi) - \sinh(n\pi y) - \sinh(2n\pi) \right) \cos(n\pi x)}{n^2 \cosh n\pi}$$

$$+\frac{1}{3} = \frac{1}{3} + \frac{4}{\pi^2} \sum_{n=1}^{\infty} \frac{(-1)^n}{n^2 \cosh n\pi} \cdot \cosh(n\pi(2 - y)) \cos(n\pi x).$$

Example 5.4 *Determine the harmonic function* $u = u(x,y)$*, on rectangle* $P(a,b) = \{(x,y)|\, 0 < x < a, 0 < y < b\}$ *satisfying the conditions*

$$u(0,y) = u(a,y) = 0 \qquad\qquad (0 < y < b),$$

$$u(x,0) = 0, \qquad u(x,b) = x(1 - x) \qquad (0 < x < a).$$

Solution. The function u on P is harmonic if it satisfy

$$u \in C^2(R), \quad \frac{\partial^2 u}{\partial x^2} + \frac{\partial^2 u}{\partial y^2} = 0 \quad (x, y) \in P.$$

The eigenvalues of the problem

$$X''(x) + \lambda X(x) = 0 \quad (0 < x < a), \quad X(0) = 0, \quad X(a) = 0,$$

are of the form

$$\lambda_n = \frac{n^2 \pi^2}{a^2} \quad (n \in \mathbf{N}),$$

and eigenfunctions are

$$X_n(x) = \sin \frac{n\pi x}{a} \quad (n \in \mathbf{N}).$$

For these values λ_n the solutions of the problem

$$Y''(y) - \lambda Y(y) = 0 \quad (0 < y < b), \quad Y(b) = 0,$$

are given by

$$Y_n(y) = A_n \cosh \frac{n\pi y}{a} + B_n \sinh \frac{n\pi y}{a} \quad (n \in \mathbf{N}), \tag{5.5}$$

where A_n and B_n are arbitrary constants. From the boundary condition $Y(0) = 0$, we obtain $A_n = 0$ and therefore the solution of the considered problem is

$$u(x, y) = \sum_{n=1}^{\infty} B_n \sinh \frac{n\pi y}{a} \sin \frac{n\pi x}{a}.$$

This solution has to satisfy the condition $u(x, b) = x(1 - x)$ and therefore for $y = b$, we obtain two Fourier series

$$u(x, b) = x(1 - x) = \sum_{n=1}^{\infty} B_n \sinh \frac{n\pi b}{a} \sin \frac{n\pi x}{a} \quad (0 < x < a),$$

where

$$B_n \cdot \sinh \frac{n\pi b}{a} = \frac{2}{a} \int_0^a x(1 - x) \sin \frac{n\pi x}{a} dx$$

$$= 2a^2 \frac{1 + (-1)^n}{n^2 \pi^2} = \begin{cases} 0, & n = 2k + 1 \\ \dfrac{4a^2}{n^2 \pi^2}, & n = 2k \end{cases} \quad (k \in \mathbf{N}).$$

So the solution u can be written as

$$u(x, y) = \frac{4a^2}{\pi^2} \sum_{n=1}^{\infty} \frac{1}{n^2 \sinh \dfrac{n\pi b}{a}} \cdot \sinh \frac{2n\pi y}{a} \cdot \sin \frac{2n\pi x}{a}.$$

Example 5.5 *Find the solution of the problem*

$$\frac{\partial^2 u}{\partial x^2} + \frac{\partial^2 u}{\partial y^2} = u \qquad (0 < x < \ell, \ 0 < y < \ell_1),$$

with the conditions

$$u(0, y) = 0, \quad u(\ell, y) = 0 \qquad (0 < y < \ell_1),$$
$$u(x, 0) = f(x), \quad u(x, \ell_1) = g(x) \qquad (0 < x < \ell).$$

Solution. Taking $u(x, y) = X(x) \cdot Y(y)$, we obtain the problem

$$X''(x) + \lambda X(x) = 0 \qquad (0 < x < \ell), \qquad X(0) = 0, \quad X(\ell) = 0,$$

with the solutions

$$X_n(x) = \sin \frac{n\pi x}{\ell} \qquad (n \in \mathbf{N}),$$

for $\lambda_n = \dfrac{n^2 \pi^2}{\ell^2}$. For these eigenvalues the solution of equation

$$Y''(y) - (1 - \lambda)Y(y) = 0 \qquad (0 < y < \ell_1),$$

is

$$Y_n(y) = A_n \cosh \sqrt{1 + \frac{n\pi y}{\ell}} + B_n \sinh \sqrt{1 + \frac{n\pi y}{\ell}} \qquad (n \in \mathbf{N}),$$

where A_n and B_n are arbitrary constants.

The solution of the considered problem is

$$u(x, y) = \sum_{n=1}^{\infty} \left(A_n \cosh \sqrt{1 + \frac{n\pi y}{\ell}} + B_n \sinh \sqrt{1 + \frac{n\pi y}{\ell}} \right) \sin \frac{n\pi x}{\ell}.$$

Using the conditions $u(x, 0) = f(x)$, $u(x, \ell_1) = g(x)$ we obtain

$$f(x) = \sum_{n=1}^{\infty} (A_n \cosh 1 + B_n \sinh 1) \sin \frac{n\pi x}{\ell} \qquad (0 < x < \ell)$$

and

$$g(x) = \sum_{n=1}^{\infty} \left(A_n \cosh \sqrt{1 + \frac{n\pi \ell_1}{\ell}} + \sinh \sqrt{1 + \frac{n\pi \ell_1}{\ell}} \right) \sin \frac{n\pi x}{\ell} \qquad (0 < x < \ell).$$

Therefore the coefficients A_n and B_n are given by

$$A_n \cosh 1 + B_n \sinh 1 = \frac{2}{\ell} \int_0^{\ell} f(x) \sin \frac{n\pi x}{\ell} dx,$$

$$B_n = \frac{1}{\sinh \sqrt{1 + \dfrac{n\pi \ell_1}{\ell}}} \left(\frac{2}{\ell} \int_0^{\ell} g(x) \sin \left(\frac{n\pi x}{\ell} \right) dx - A_n \cosh \sqrt{1 + \frac{n\pi \ell_1}{\ell}} \right).$$

Example 5.6 *Find the solution of the problem*

$$\frac{\partial^2 u}{\partial x^2} + \frac{\partial^2 u}{\partial y^2} - ku = 0 \qquad (0 < x < \ell, \ 0 < y < \ell_1),$$

for $k > 0$ with the conditions

$$u(0,y) = \varphi(y), \qquad \frac{\partial u(\ell, y)}{\partial x} = \psi(y) \qquad (0 < y < \ell_1),$$

$$u(x,0) = f(x), \quad u(x,\ell_1) = g(x) \qquad (0 < x < \ell).$$

Solution. The equations appearing in this problem and in previous one are called Helmholtz equations and for $k = 0$ it becomes Laplace's equation. Let us introduce the functions v and w such that $u(x,y) = v(x,y) + w(x,y)$, and the function v satisfies the problem

$$\frac{\partial^2 v}{\partial x^2} + \frac{\partial^2 v}{\partial y^2} - kv = 0 \qquad (0 < x < \ell, \ 0 < y < \ell_1),$$

$$v(0,y) = 0, \qquad \frac{\partial v(\ell, y)}{\partial x} = 0 \qquad (0 < y < \ell_1), \tag{5.6}$$

$$v(x,0) = f(x), \qquad v(x,\ell_1) = g(x) \qquad (0 < x < \ell),$$

while the function w satisfies the following problem

$$\frac{\partial^2 w}{\partial x^2} + \frac{\partial^2 w}{\partial y^2} - kw = 0 \qquad (0 < x < \ell, \ 0 < y < \ell_1),$$

$$w(0,y) = \varphi(y), \qquad \frac{\partial w(\ell, y)}{\partial x} = \psi(y) \qquad (0 < y < \ell_1), \tag{5.7}$$

$$w(x,0) = 0 \qquad w(x,\ell_1) = 0 \qquad (0 < x < \ell).$$

Let us first solve the problem (5.6). Taking $v(x,y) = X(x) \cdot Y(y)$, we obtain the problem

$$X''(x) + \lambda X(x) = 0 \qquad (0 < x < \ell), \qquad X(0) = X'(\ell) = 0.$$

The eigenfunctions of this problem are $\lambda_n = \dfrac{(2n-1)^2 \pi^2}{(2\ell)^2}$, and the corresponding eigenfunctions are

$$X_n(x) = \sin \frac{(2n-1)\pi x}{2\ell} \qquad (n \in \mathbf{N}).$$

The solution of equation

$$Y''(y) - (k - \lambda) \qquad (0 < y < \ell_1), \qquad Y(y) = 0$$

is

$$Y_n(y) = A_n \cosh \sqrt{k + \frac{(2n-1)\pi y}{2\ell}} + B_n \sinh \sqrt{k + \frac{(2n-1)\pi y}{2\ell}} \qquad (n \in \mathbf{N}),$$

where A_n and B_n are arbitrary constants.

The solution of the problem (5.6) is

$$v(x,y) = \sum_{n=1}^{\infty} \left(A_n \cosh \sqrt{k + \frac{(2n-1)\pi y}{2\ell}} + B_n \sinh \sqrt{k + \frac{(2n-1)\pi y}{2\ell}} \right)$$

$$\cdot \sin \frac{(2n-1)\pi x}{2\ell}.$$

Using the conditions $v(x,0) = f(x)$, $u(x,\ell_1) = g(x)$ we obtain

$$A_n \cosh \sqrt{k} + B_n \sinh \sqrt{h} k = \frac{2}{\ell} \int_0^\ell f(x) \sin \frac{(2n-1)n\pi x}{2\ell} dx, \qquad (5.8)$$

and

$$B_n = \frac{1}{\sinh \sqrt{k + \frac{(2n-1)\pi \ell_1}{2\ell}}} \frac{2}{\ell} \int_0^\ell g(x) \sin \frac{(2n-1)n\pi x}{2\ell} dx$$

$$(5.9)$$

$$- A_n \cosh \sqrt{k + \frac{(2n-1)\pi \ell_1}{2\ell}}.$$

Let us find the solution of the problem (5.7). Taking $w(x,y) = X(x) \cdot Y(y)$, we obtain the problem

$$Y''(y) + \lambda Y(y) = 0 \qquad (0 < y < \ell_1), \qquad Y(0) = 0 = Y(\ell_1) = 0,$$

with solutions

$$Y_n(y) = \sin \frac{n\pi y}{\ell_1}.$$

The solution of equation

$$X''(x) - (k - \lambda)X(x) = 0 \qquad (0 < x < \ell),$$

is

$$X_n(x) = A_n \cosh \sqrt{k + \frac{n\pi x}{\ell_1}} + B_n \sinh \sqrt{k + \frac{n\pi x}{\ell_1}} \qquad (n \in \mathbf{N}).$$

The solution of the problem (5.7) is

$$w(x,y) = \sum_{n=1}^{\infty} \left(A_n \cosh \sqrt{k + \frac{n\pi x}{\ell_1}} + B_n \sinh \sqrt{k + \frac{n\pi x}{\ell_1}} \right) \cdot \sin \frac{n\pi y}{\ell_1}.$$

Using the conditions

$$w(0,y) = \varphi(y), \quad \frac{\partial w(\ell, y)}{\partial x} = \psi(y) \quad (0 < y < \ell_1)$$

we obtain

$$\varphi(y) = \sum_{n=1}^{\infty} \left(A_n \cosh \sqrt{k} + B_n \sinh \sqrt{k} \right) \sin \frac{n\pi y}{\ell_1} \quad (0 < y < \ell_1)$$

and

$$\psi(y) = \sum_{n=1}^{\infty} \frac{n\pi}{2\ell_1 \sqrt{k + \frac{n\pi \ell}{\ell_1}}} \left(A_n \sinh \sqrt{k + \frac{n\pi \ell}{\ell_1}} \right.$$

$$\left. + B_n \cosh \sqrt{k + \frac{n\pi \ell}{\ell_1}} \right) \sin \frac{n\pi y}{\ell_1} \quad (0 < y < \ell_1).$$

Therefore the coefficients are given by

$$A_n \cosh \sqrt{k} + B_n \sinh \sqrt{k} = \frac{2}{\ell_1} \int_0^{\ell_1} \varphi(y) \sin \frac{n\pi y}{\ell_1} dy,$$

and

$$B_n \cosh \sqrt{k + \frac{n\pi \ell}{\ell_1}} = \frac{4\sqrt{k + \frac{n\pi \ell}{\ell_1}}}{n\pi} \cdot \frac{2}{\ell} \int_0^{\ell_1} \psi(y) \sin \frac{n\pi y}{\ell_1} dy - A_n \sinh \sqrt{k + \frac{n\pi \ell}{\ell_1}}.$$

Example 5.7 *Find the solution of the Poisson equation*

$$\frac{\partial^2 u}{\partial x^2} + \frac{\partial^2 u}{\partial y^2} = -1 \quad (0 < x < \ell, \ 0 < y < \ell_1),$$

with the conditions

$$u(0,y) = 0, \quad u(\ell, y) = 0 \quad (0 < y < \ell_1),$$
$$u(x,0) = 0, \quad u(x, \ell_1) = 0 \quad (0 < x < \ell).$$

Solution. If we take $u(x,y) = S(x) + v(x,y)$, we obtain

$$S''(x) = -1, \quad S(0) = 0, \quad S(\ell) = 0.$$

The solution of this problem is

$$S(x) = \frac{x}{2}\left(\ell - \frac{x}{2}\right).$$

The function v satisfies the Laplace equation in the problem

$$\frac{\partial^2 v}{\partial x^2} + \frac{\partial^2 v}{\partial y^2} = 0 \quad (0 < x < \ell, \ 0 < y < \ell_1),$$

$$v(0,y) = 0, \quad v(\ell,y) = 0 \quad (0 < y < \ell_1), \tag{5.10}$$

$$v(x,0) = u(x,\ell_1) = -\frac{x}{2}\left(\ell - \frac{x}{2}\right) \quad (0 < x < \ell).$$

The formal solution of this problem is

$$v(x,y) = \sum_{n=1}^{\infty}\left(A_n \cosh\frac{n\pi y}{\ell} + B_n \sinh\frac{n\pi y}{\ell}\right)\sin\frac{n\pi x}{\ell} \quad (0 < x < \ell).$$

Using the boundary conditions we have

$$-\frac{x}{2}\left(\ell - \frac{x}{2}\right) = \sum_{n=1}^{\infty} A_n \sin\frac{n\pi x}{\ell}dx.$$

$$-\frac{x}{2}\left(\ell - \frac{x}{2}\right) = \sum_{n=1}^{\infty}\left(A_n \cosh\frac{n\pi\ell_1}{\ell} + B_n \sinh\frac{n\pi\ell_1}{\ell}\right)\sin\frac{n\pi x}{\ell}.$$

Therefore

$$A_n = -\frac{2}{\ell}\int_0^\ell \frac{x}{2}\left(\ell - \frac{x}{2}\right)\sin\frac{n\pi x}{\ell}dx = \begin{cases} -\dfrac{4\ell^2}{n^3\pi^3}, & n = 1,3,5,\dots \\ \\ 0, & n = 2,4,6\dots. \end{cases}$$

Also, we have

$$A_n \cosh\frac{n\pi\ell_1}{\ell} + B_n \sinh\frac{n\pi\ell_1}{\ell} = -\frac{2}{\ell}\int_0^\ell \frac{x}{2}\left(\ell - \frac{x}{2}\right)\sin\frac{n\pi x}{\ell}dx,$$

wherefrom it follows that

$$B_n = \begin{cases} -\dfrac{4\ell^2}{n^3\pi^3} \cdot \dfrac{1 - \cosh\left(\dfrac{n\pi\ell_1}{\ell}\right)}{\sinh\left(\dfrac{n\pi\ell_1}{\ell}\right)}, & n = 1,3,5,\dots \\ \\ 0, & n = 2,4,6\dots. \end{cases}$$

The solution of the problem (5.10) has the form

$$v(x,t) = -\frac{4\ell^2}{\pi^3} \sum_{n=1}^{\infty} \frac{\sin\left(\frac{(2n-1)\pi x}{\ell}\right)}{\sinh\left(\frac{(2n-1)\pi \ell_1}{\ell}\right)} \left(\sinh\frac{(2n-1)\pi y}{\ell} + \sinh\frac{(2n-1)\pi(\ell_2-y)}{\ell}\right).$$

The solution of the considered problem can be written as

$$u(x,y) = \frac{x}{2}\left(\ell - \frac{x}{2}\right) - \frac{4\ell^2}{\pi^3} \sum_{n=1}^{\infty} \frac{\sin\left(\frac{(2n-1)\pi x}{\ell}\right)}{\sinh\left(\frac{(2n-1)\pi \ell_1}{\ell}\right)}$$

$$\cdot \left(\sinh\frac{(2n-1)\pi y}{\ell} + \sinh\frac{(2n-1)\pi(\ell_2-y)}{\ell}\right).$$

Example 5.8 *Determine the solution of the problem given in previous Example 5.7 by using double Fourier series.*

Solution. Taking $u(x,y) = X(x) \cdot Y(y)$ we obtain the problems

$$X''(x) + \lambda X = 0, \quad X(0) = 0 = X(\ell) = 0,$$

$$Y''(x) - \lambda Y = 0, \quad Y(0) = 0 = Y(\ell_1) = 0.$$

Using the eigenfunctions of the previous problems we can write the formal solution of the consider equation as

$$u(x,y) = \sum_{m=1}^{\infty}\sum_{n=1}^{\infty} A_{mn} \sin\frac{m\pi y}{\ell_1} \cdot \sin\frac{n\pi x}{\ell}.$$

Using

$$\frac{\partial^2 u}{\partial x^2} + \frac{\partial^2 u}{\partial y^2} = -\sum_{m=1}^{\infty}\sum_{n=1}^{\infty} A_{mn} \left(\left(\frac{m\pi}{\ell}\right)^2 + \left(\frac{m\pi}{\ell_1}\right)^2\right) \sin\frac{m\pi y}{\ell} \cdot \sin\frac{n\pi x}{\ell_1}$$

and the Fourier series for the constant 1

$$1 = \frac{4}{\pi}\sum_{m=1}^{\infty}\sum_{n=1}^{\infty} \frac{1}{(2m-1)(2n-1)} \sin\frac{(2m-1)\pi y}{\ell_1} \cdot \sin\frac{(2n-1)\pi x}{\ell}$$

we obtain that $A_{2m\,2n} = 0$, $(m \in \mathbf{N})$, $(n \in \mathbf{N})$, and

$$A_{(2m+1)(2n+1)} = \frac{4}{(2m-1)(2n-1)\pi}\left(\left(\frac{(2m-1)\pi}{\ell}\right)^2 + \left(\frac{(2n-1)\pi}{\ell}\right)^2\right)^{-1}$$

for $m = 1, 2, \ldots, \ (n \in \mathbf{N})$.

So the solution of the consider problem can be written as

$$u(x, y) = \frac{4}{\pi} \sum_{m=1}^{\infty} \sum_{n=1}^{\infty} \frac{4 \sin \dfrac{(2m-1)\pi y}{\ell_1} \cdot \sin \dfrac{(2n-1)\pi x}{\ell}}{(2m-1)(2n-1)\pi \left(\left(\dfrac{(2m-1)\pi}{\ell} \right)^2 + \left(\dfrac{(2n-1)\pi}{\ell} \right)^2 \right)}.$$

Example 5.9 *Determine the solution of the Dirichlet problem for the unit ball* $B(0, R_0) \subset \mathbf{R}^2$

$$\frac{\partial^2 U}{\partial r^2} + \frac{1}{r} \frac{\partial U}{\partial r} + \frac{1}{r^2} \frac{\partial U}{\partial \varphi} = 0 \quad (0 < r < R_0, \ -\pi < \varphi < \pi),$$

$$U(R_0, \varphi) = f(\varphi) \quad (-\pi < \varphi < \pi)$$

for $f \in C^2(\partial B)$.

Solution. Separation of variables $U(r, \varphi) = R(r)H(\varphi)$ leads to

$$0 = \frac{\partial^2 U}{\partial r^2} + \frac{1}{r} \frac{\partial U}{\partial r} + \frac{1}{r^2} \frac{\partial U}{\partial \varphi} = R''(r)H(\varphi) + \frac{1}{r}R'(r)H(\varphi) + \frac{1}{r^2}R(r)H''(\varphi).$$

Dividing by $R(r)H(\varphi)$ and multiplying by r^2 we get

$$r^2 \frac{R'' + \dfrac{1}{r}R'}{R} = -\frac{H''}{H} = \lambda,$$

wherefrom we obtain the problem

$$H'' + \lambda H = 0, \quad H(-\pi) = H(\pi), \quad H'(-\pi) = H'(\pi) \tag{5.11}$$

and the equation

$$r^2 R'' + r R' - \lambda R = 0. \tag{5.12}$$

The solution of equation (5.11) for $\lambda = \lambda_n = n^2, \ n = 0, 1, \ldots,$ is

$$H_n(\varphi) = \begin{cases} \dfrac{1}{2}a_0, & n = 0 \\[2mm] a_n \cos(n\varphi) + b_n \sin(\varphi), & (n \in \mathbf{N}). \end{cases} \tag{5.13}$$

The general solution of Euler equation (5.12), for $\lambda = \lambda_n = n^2, \ n = 0, 1, \ldots,$ is

$$R_n(r) = \begin{cases} C_1 + C_2 \ln r, & n = 0 \\[2mm] C_3 r^n + C_4 r^{-n}, & (n \in \mathbf{N}). \end{cases} \tag{5.14}$$

In order to avoid discontinuities of $\ln r$ and r^{-n}, $n \in \mathbf{N}$ for $r = 0$, we put $C_2 = C_4 = 0$. So the solution of considered equation is

$$U(r, \varphi) = \frac{1}{2}A_0 + \sum_{n=1}^{\infty} \frac{r^n}{R_0^n} \left(A_n \cos n\varphi + B_n \sin n\varphi \right). \qquad (5.15)$$

Using the boundary conditions $U(R_0, \varphi) = f(\varphi)$ i.e.,

$$U(R_0, \varphi) = f(\varphi) = \frac{1}{2}A_0 + \sum_{n=1}^{\infty} \left(A_n \cos n\varphi + B_n \sin n\varphi \right),$$

we shall find A_n and B_n. The last series is uniformly convergent, since $f \in C^2(\partial B)$. The Fourier coefficients can be written as

$$A_0 = \frac{1}{2\pi} \int_{-\pi}^{\pi} f(\theta)d\theta$$

$$R_0^n A_n = \frac{1}{\pi} \int_{-\pi}^{\pi} f(\theta) \cos(n\theta)d\theta \qquad (n \in \mathbf{N}),$$

$$R_0^n B_n = \frac{1}{\pi} \int_{-\pi}^{\pi} f(\theta) \sin(n\theta)d\theta \qquad (n \in \mathbf{N}).$$

The expression $A_n \cos n\varphi + B_n \sin n\varphi$ can be transformed as

$$A_n \cos n\varphi + B_n \sin n\varphi = \frac{1}{R_0^n \pi} \int_{-\pi}^{\pi} f(\theta) \left(\cos(n\theta) \cos(n\varphi) + \sin(n\theta) \sin(n\varphi) \right) d\theta$$

$$= \frac{1}{R_0^n \pi} \int_{-\pi}^{\pi} f(\theta)(\cos(n(\theta - \varphi))d\theta,$$

and substituted in (5.15)

$$U(r, \varphi) = \frac{1}{\pi} \int_{-\pi}^{\pi} f(\theta) \left(\frac{1}{2} + \sum_{n=1}^{\infty} \left(\frac{r}{R_0} \right)^n \cos(n(\theta - \varphi)) \right) d\theta.$$

It can be proved that

$$\frac{1}{2} + \sum_{n=1}^{\infty} \left(\frac{r}{R_0} \right)^n \cos(n(\theta - \varphi)) = \frac{R_0^2 - r^2}{2(R_0^2 + r^2 - 2r R_0 \cos(\theta - \varphi))},$$

wherefrom the Poisson integral formula

$$U(r, \varphi) = \frac{1}{\pi} \int_{-\pi}^{\pi} f(\theta) \frac{R_0^2 - r^2}{2(R_0^2 + r^2 - 2r R_0 \cos(\theta - \varphi))} f(\theta)d\theta, \qquad (5.16)$$

for $0 < r < R$ is obtained.

Exercise 5.10 *Determine the solution of the problem*

$$\frac{\partial^2 U}{\partial r^2} + \frac{1}{r}\frac{\partial U}{\partial r} + \frac{1}{r^2}\frac{\partial U}{\partial \varphi} = 0 \qquad (0 < r < R, \quad -\pi < \varphi < \pi),$$

$$U(R, \varphi) = 1 - \cos 2\varphi \qquad (-\pi < \varphi < \pi).$$

Answer. $U(r, \varphi) = 1 + \left(\dfrac{r}{R}\right)^2 \cos 2\varphi.$

Example 5.11 *Determine the solution of the problem*

$$\frac{\partial^2 U}{\partial r^2} + \frac{1}{r}\frac{\partial U}{\partial r} + \frac{1}{r^2}\frac{\partial U}{\partial \varphi} = 0 \qquad (0 < a < r < b, \quad -\pi < \varphi < \pi),$$

with the following boundary condition

a) $U(a, \varphi) = f(\varphi), \quad U(b, \varphi) = g(\varphi) \qquad (-\pi < \varphi < \pi);$

b) $U(a, \varphi) = T_0, \quad U(b, \varphi) = T_1 \qquad (-\pi < \varphi < \pi).$

Solution. Separating the variables $U(r, \varphi) = R(r)H(\varphi)$, the ordinary differential equations (5.11) and (5.12) with solutions (5.13) and (5.14) are obtained.

So the solution of considered equation is

$$U(r, \varphi) = \frac{1}{2}(A_0 + B_0 \ln r)$$

$$+ \sum_{n=1}^{\infty} \left(\left(A_n r^n + B_n r^{-n} \right) \cos n\varphi + \left(C_n r^n + D_n r^{-n} \right) \right) \sin n\varphi. \tag{5.17}$$

a) After applying the boundary conditions we obtain the following system of equation

$$A_0 + B_0 \ln a = \frac{1}{\pi} \int_{-\pi}^{\pi} f(\varphi) d\varphi,$$

$$A_0 + B_0 \ln b = \frac{1}{\pi} \int_{-\pi}^{\pi} g(\varphi) d\varphi,$$

$$A_n a^n + B_n a^{-n} = \frac{1}{\pi} \int_{-\pi}^{\pi} f(\varphi) \cos(n\varphi) d\varphi$$

$$A_n b^n + B_n b^{-n} = \frac{1}{\pi} \int_{-\pi}^{\pi} g(\varphi) \cos(n\varphi) d\varphi$$

$$(5.18)$$

$$C_n a^n + D_n a^{-n} = \frac{1}{\pi} \int_{-\pi}^{\pi} f(\varphi) \sin(n\varphi) d\varphi$$

$$C_n b^n + D_n b^{-n} = \frac{1}{\pi} \int_{-\pi}^{\pi} g(\varphi) \sin(n\varphi) d\varphi,$$

for $n \in \mathbf{N}$.

b) The boundary conditions are independent on φ and therefore we use the form of the solution as

$$U(r) = A_0 + B_0 \ln r.$$

From boundary conditions we get two equations

$$T_1 = A_0 + B_0 \ln a, \qquad A_0 + B_0 \ln b,$$

with the solutions

$$A_0 = T_1 - (T_2 - T_1) \frac{\ln a}{\ln \dfrac{b}{a}}, \qquad B_0 = \frac{T_2 - T_1}{\ln \dfrac{b}{a}}.$$

The solution has the form

$$U(r) = T_1 + (T_2 - T_1) \frac{\ln \dfrac{r}{a}}{\ln \dfrac{b}{a}}.$$

Example 5.12 *Find the solution of the problem*

$$\frac{\partial^2 u}{\partial x^2} + \frac{\partial^2 u}{\partial y^2} = 0 \qquad (-\infty < x < \infty, \ 0 < y < \infty),$$

with the boundary conditions

a) $\begin{array}{l} u(x,y) \to 0 \quad as \quad x^2 + y^2 \to \infty \\ u(x,0) = f(x) \qquad (-\infty < x < \infty); \end{array}$

b) $\begin{array}{l} u(x,y) \to 0 \quad as \quad x^2 + y^2 \to \infty \\ u(x,0) = T \qquad (-\infty < x < \infty). \end{array}$

Solution. The solution will be constructed by using the method of Fourier integrals.

a) Separation variables leads to the

$$X(x) = A(s)\cos sx + B(s)\sin sx$$

$$Y(y) = D(s)e^{-sy} + E(s)e^{sy},$$

where s are the eigenvalues generated by $\lambda = s^2$. Using the boundary conditions we get

$$f(x) = \int_0^\infty (A(s)\cos sx + B(s)\sin sx)\, ds,$$

wherefrom we get

$$A(s) = \frac{1}{\pi}\int_{-\infty}^\infty f(x)\cos sx\, dx$$

$$B(s) = \frac{1}{\pi}\int_{-\infty}^\infty f(x)\sin sx\, dx.$$

b) $u(x,y) = T$.

Example 5.13 *Solve the following Dirichlet problem on the three-dimensional ball* $B(0,1)$ *(in the spheric coordinate system)*

$$\Delta u = \frac{1}{r^2}\frac{\partial}{\partial r}\left(r^2\frac{\partial u}{\partial r}\right) + \frac{1}{r^2\sin\theta}\frac{\partial}{\partial\theta}(\sin\theta\frac{\partial u}{\partial\theta}) + \frac{1}{r^2\sin^2\theta}\frac{\partial^2 u}{\partial\varphi^2} = 0 \ \ on \ B(0,1),$$

$$u(1,\theta,\varphi) = \cos^2\theta \quad (0 \le \theta \le \pi).$$

Solution. The boundary condition is axially symmetric and therefore we suppose that the solution of the problem is also axially symmetric. We represent the solution by axially symmetric ball functions

$$u(r,\theta) = \sum_{k=0}^{\infty} a_n r^n P_n(\cos\theta),$$

where P_n is the Legendre polynomial. Then we obtain by the boundary condition

$$x^2 = \sum_{k=0}^{\infty} a_n P_n(x).$$

Since

$$P_0(x) = 1,\ P_1(x) = x,\ P_2(x) = \frac{3}{2}x^2 - \frac{1}{2},$$

we obtain

$$a_n = 0\ (n \in \mathbf{N}\setminus\{2\}),\ a_0 = \frac{1}{3},\ a_2 = \frac{2}{3}.$$

The solution is given by

$$u(r,\theta) = \frac{1}{3}(1 - r^2) + r^2 \cos^2\theta,$$

Exercise 5.14 *Solve the exterior Dirichlet problem for $1 < r < 2$ (in the spheric coordinate system)*

$$\Delta u = 0,\ u(1,\theta) = \frac{1}{2}\cos\theta,\ u(2,\theta) = 1 + \cos 2\theta\ \ (0 \le \theta \le 2\pi).$$

Hint. Find the solution in the form of the series

$$u(r,\theta) = \sum_{k=0}^{\infty} (a_n r^n + b_n r^{-(n+1)}) P_n(\cos\theta).$$

Use that

$$P_0(x) = 1,\ P_1(x) = x,\ P_2(x) = \frac{3}{2}x^2 - \frac{1}{2},\qquad P_3(x) = \frac{5}{2}x^3 - \frac{3}{2}x.$$

Answer.

$$u(r,\theta) = \frac{4}{3}(1 - \frac{1}{r}) + \frac{1}{14}(\frac{8}{r^2} - r)P_1(\cos\theta) + \frac{32}{93}(r^2 - \frac{1}{r^3})P_2(\cos\theta).$$

Example 5.15 (Dirichlet problem on cylinder) *Solve the Dirichlet problem on the cylinder $[0, \ell) \times [0, 2\pi) \times (0, H)$ (in the cylindric coordinate system (r, φ, z))*

$$\Delta u = \frac{1}{r}\frac{\partial}{\partial r}(r\frac{\partial u}{\partial r}) + \frac{1}{r^2}\frac{\partial^2 u}{\partial\varphi^2} + \frac{\partial^2 u}{\partial z^2} = 0\ on\ [0, \ell) \times [0, 2\pi) \times (0, H),$$

with the boundary conditions

$$u(\ell, \varphi, z) = 0,\ u(r, \varphi, 0) = 0,\ u(r, \varphi, H) = m(r, \varphi),$$

where m is the given function which is 2π-periodic with respect to φ.

Solution. Applying the Fourier method of separation of variables first in the form

$$u(r, \varphi, z) = V(r, z) \cdot \Phi(\varphi)$$

and then

$$V(r, z) = R(r) \cdot Z(z),$$

where $\Phi : \mathbf{R} \to \mathbf{R}$, $R : (0, \infty) \to \mathbf{R}$, $Z : \mathbf{R} \to \mathbf{R}$ and Φ is a 2π-periodic function, taking in the first step the constant γ and then the constant λ, we obtain the differential equations

$$\varphi'' + \gamma \Phi = 0,$$

$$Z'' - \lambda Z = 0,$$

$$(r R'(r))' + (\lambda r - \frac{n^2}{r}) R(r) = 0.$$

The last equation is the Bessel equation (for $x = r\sqrt{\lambda}$ for $\lambda \geq 0$) with solutions $J_n(r\sqrt{\lambda})$. The boundary condition $R(\ell) = 0$ gives $J_n(\ell\sqrt{\lambda}) = 0$, which implies that the eigenvalues are given by

$$\lambda_{nj} (\frac{x_{nj}}{\ell})^2 \qquad (j \in \mathbf{N}),$$

where $x_{n1} < x_{n2} < \ldots$ are the positive zeros of the function J_n.
Therefore the solution of the considered Dirichlet problem can be represented by the series

$$u(r, \varphi, z) = \sum_{k=0}^{\infty} \sum_{j=1}^{\infty} (a_{kj} \cos k\varphi + b_{kj} \sin k\varphi)(c_{kj} \exp(\frac{x_{kj}z}{\ell}) + d_{kj} \exp(-\frac{x_{kj}z}{\ell})) J_k(\frac{x_{kj}}{\ell} r).$$

By the second boundary conditions we have $c_{kj} = -d_{kj}$. Therefore taking $A_{kj} = 2a_{kj}c_{kj}$ and $B_{kj} = 2b_{kj}c_{kj}$ we obtain the (formal) solution

$$u(r, \varphi, z) = \sum_{k=0}^{\infty} \sum_{j=1}^{\infty} (A_{kj} \cos k\varphi + B_{kj} \sin k\varphi) \sinh(\frac{x_{kj}z}{\ell}) J_k(\frac{x_{kj}}{\ell} r),$$

where A_{kj} and B_{kj} are given by (the last boundary condition and the weighted orthogonality, by the weight r)

$$A_{kj} = \frac{1}{\pi M_{kj} \sinh \frac{x_{kj}H}{\ell}} \int_0^\ell \int_0^{2\pi} r m(r, \varphi) J_k(\frac{x_{kj}}{\ell} r) \cos k\varphi \, d\varphi dr,$$

$$B_{kj} = \frac{1}{\pi M_{kj} \sinh \frac{x_{kj}H}{\ell}} \int_0^\ell \int_0^{2\pi} r m(r, \varphi) J_k(\frac{x_{kj}}{\ell} r) \sin k\varphi \, d\varphi dr,$$

where

$$M_{kj} = \int_0^\ell r J_k(\frac{r}{\ell} x_{kj}) \, dr.$$

5.2 The Maximum Principle

5.2.1 Preliminaries

Theorem 5.1 *Let L be a general elliptic operator from the preliminary on the bounded region Q. If $u \in C^2(Q), L(u) \geq 0$ and $c(x) \leq 0$ $(x \in Q)$, then the function u has no positive maximum in the region Q.*

Theorem 5.2 (Strong maximum principle) *For a function $u \in C^2(Q) \cap C(\overline{Q})$ (where Q is a bounded region) which satisfies $\Delta u \geq 0$ $(x \in Q)$ either $u = const.$ or $u(x) < \max_{y \in \partial Q} u(y)$ $(x \in Q)$.*

Definition 5.3 *The function $E = E(x)$ given by*

$$E(x) = \begin{cases} -\dfrac{1}{(n-2)\sigma_n |x|^{n-2}} & \text{for } n \geq 3, \\ -\dfrac{1}{2\pi} \ln \dfrac{1}{|x|} & \text{for } n = 2 \end{cases}$$

is the fundamental solution of the Laplace equation, where $\sigma_n = \dfrac{2\pi^{n/2}}{\Gamma(n/2)}$ is the surface area of the unit sphere in $\mathbf{R}^n$.

The exterior Dirichlet problem consists in finding a function $u \in C(\overline{Q}_1)$, where $Q_1 = \mathbf{R}^n \setminus \overline{Q}$ is a region for a given bounded region Q, which is harmonic on Q_1 and for $n \geq 3$ we have $\lim_{x \to +\infty} u(x) = 0$ (uniformly), and for $n = 2$ the function u is bounded outside of a ball, and it satisfies the boundary condition $u|_{\partial Q_1} = f$, for a given continuous function f.

5.2.2 Examples and Exercises

Example 5.16 *Prove that the condition $c(x) \leq 0$ $(x \in Q)$ in Theorem 5.1 is necessary.*

Solution. The function $u(x,y) = \sin x \sin y$ is a solution of the equation

$$\frac{\partial^2 u}{\partial x^2} + \frac{\partial^2 u}{\partial y^2} + 2u = 0$$

on the square

$$S = \{(x,y) \mid 0 \leq x, y \leq \pi\},$$

but it attends its maximum inside of S at the point $(\frac{\pi}{2}, \frac{\pi}{2})$.

Example 5.17 *Prove that if for the elliptic operator L from the preliminaries for*
$u \in C^2(Q)$

$$c = 0, \quad L(u) > 0 \quad (\text{ or } L(u) < 0),$$

then the function u has no maximum (minimum) in Q.

Solution. By Theorem 5.1 (see Example 5.23) the function u has no positive
maximum in Q. The negative maximum can be reduced on the preceding case taking
the function $u + k$ for enough big k that for $u + k$ be positive maximum and apply
the preceding result on $u + k$ since $L(u + k) = L(u)$.
For the minimum take instead the function u the function $-u$ and use $\min u = -\max(-u)$.

Example 5.18 *Prove that for a function with the properties*

$$\Delta u = 0 \text{ on } Q, \ u \in C^2(Q) \cap C(\overline{Q}),$$

we have

$$\min_{y \in \partial Q} u(y) \leq u(x) \leq \max_{y \in \partial Q} u(y) \qquad (x \in Q).$$

Solution. Since $u \in C(\overline{Q})$ the function u attends its maximum at some point from
$\overline{Q}$. We introduce for $\varepsilon > 0$ the function u_ε in the following way

$$u_\varepsilon(x) = u(x) + \varepsilon |x|^2,$$

which satisfies

$$\Delta u_\varepsilon(x) = \Delta u(x) + 2n\varepsilon > 0.$$

Therefore by Example 5.17

$$u_\varepsilon(x) \leq \max_{y \in \partial Q} u_\varepsilon(y) \leq \max_{y \in \partial Q} u(y) = \varepsilon \max_{y \in \partial Q} |y|^2.$$

Letting $\varepsilon \to 0$ we obtain

$$\sup_{x \in Q} u(x) \leq \max_{y \in \partial Q} u(y).$$

We can prove in an analogous way the left part of the desired inequality.

   Remark 5.18.1. It is true the strong maximum principle - see Example 5.32.

Example 5.19 *There can exist only one solution for the Dirichlet problem for the
Laplace equation.*

Solution. Suppose contrary, i.e., that there are two solutions u_1 and u_2. Then the
function $u = u_1 - u_2$ is the solution of Dirichlet problem for the Laplace equation
with the boundary condition $u|_{\partial Q} = 0$. Therefore by Example 5.18 we obtain $u = 0$,
i.e., $u_1 = u_2$.

Example 5.20 *Let $\{u_n\}_{n\in\mathbf{N}}$ be a sequence of continuous functions on $\overline{Q}$ and harmonic on bounded region Q. Prove that if the sequence $\{u_n\}_{n\in\mathbf{N}}$ converges uniformly on ∂Q then it converges uniformly also on the whole $\overline{Q}$.*

Solution. Since the sequence $\{u_n\}_{n\in\mathbf{N}}$ uniformly converges on ∂Q it is also a Cauchy sequence, i.e., for every $\varepsilon > 0$ there exists $n_0 \in \mathbf{N}$ such that for every $m,n > n_0$

$$|u_n(x) - u_m(x)| < \varepsilon \qquad (x \in \partial Q),$$

i.e.,

$$-\varepsilon < u_n(x) - u_m(x) < \varepsilon \qquad (x \in \partial Q).$$

Therefore applying Example 5.18 on the function $u_n - u_m$ we obtain

$$|u_n(x) - u_m(x)| < \varepsilon \qquad (x \in \overline{Q}).$$

Since the space $C(\overline{Q})$ is complete with respect to sup-norm, we obtain that the sequence $\{u_n\}_{n\in\mathbf{N}}$ uniformly converges on $\overline{Q}$.

Exercise 5.21 *Prove for the Dirichlet problem for the two dimensional Laplace equation on the unit circle with the boundary condition $u(1,\theta) = A(\theta)$, $-\pi \le \theta \le \pi$*

a) the Poisson formula for $A \in C^2[-\pi,\pi]$;

b) $\lim_{r\to 1-0} u(r,\theta) = A(\theta)$ $(\theta \in [-\pi,\pi])$ for $A \in C^2[-\pi,\pi]$;

c) the Poisson formula for A is continuous on $[-\pi,\pi]$.

Hints.

a) The Fourier series of the function A uniformly converges to A on $[-\pi,\pi]$. Use the Fourier method of separation of variables for the Laplace equation in the polar coordinate system (See Example 5.9.)

b) Use the fact that the function A is uniformly continuous on the interval $[-\pi,\pi]$, i.e., for every $\varepsilon > 0$ there exists $\delta > 0$ such that $|\theta - t| < \delta$ implies $|A(\theta) - A(t)| < \varepsilon$. For the case $|\theta - t| \ge \delta$ use that

$$\lim_{r\to 1-0} \frac{1 - r^2}{1 - 2r\cos(\theta - t) + r^2} = 0.$$

c) Use the Weierstrass theorem on the uniform approximation of continuous function with trigonometric polynomials

$$T_n(\theta) = \frac{a_0}{2} \sum_{k=1}^{n}(a_k \cos k\theta + b_k \sin k\theta),$$

and consider Dirichlet problems with boundaries T_n $(n \in \mathbf{N})$. Then apply a) and b) on the sequence $\{T_n(\theta)\}_{n\in\mathbf{N}}$ and the corresponding sequence of solutions $\{u_n\}_{n\in\mathbf{N}}$ and use Example 5.20.

Example 5.22 *Prove that the solution u of the Laplace equation in* $\mathbf{R}^n \setminus \{0\}$ *which is spheric symmetric, i.e., with the property that it has the same value in the points* x *which are on the same distance of the origin is the fundamental solution.*

Solution. Put
$$u = V(r) \text{ for } r = |x| > 0.$$

Since we have
$$\frac{\partial u}{\partial x_i} = V'(r)\frac{\partial r}{\partial x_i} = V'(r)\frac{x_i}{r}$$

and so
$$\frac{\partial^2 u}{\partial x_i^2} = V''(r)\frac{x_i^2}{r^2} + V'(r)\frac{1}{r} - V'(r)\frac{x_i^2}{r^3}$$

we obtain that V satisfies the ordinary differential equation
$$\Delta u = V''(r) + \frac{n-1}{r}V'(r) = 0,$$

whose solution is
$$V(r) = C_1 \int \frac{dr}{r^{n-1}} + C_2,$$

where C_1 and C_2 are arbitrary real constants. Taking $C_1 = \frac{1}{\sigma_n}$ and $C_2 = 0$ we obtain

$$V(r) = \begin{cases} \dfrac{C_1 r^{2-n}}{\sigma_n(2-n)} & \text{for } n \geq 3, \\ C_1 \ln r & \text{for } n = 2. \end{cases}$$

The function $u(x) = V(r)$ for $|x| = r, 0 < r < \infty$ is the fundamental solution of the Laplace equation.

Exercise 5.23 *Prove Theorem 5.1 with an additional supposition (which is not necessary!):*
Let L be a general elliptic operator from the preliminaries on the bounded region Q. If $u \in C(Q), L(u) \geq 0$ and

$$c(x) \leq 0 \quad \text{and} \quad (L(u))^2 + c^2 > 0 \qquad (x \in Q),$$

then the function u has no positive maximum in the region Q.

Hints. Suppose the contrary, i.e., that the function has its positive maximum at some point $x^0 \in Q$. Therefore

$$u(x^0) > 0, \qquad \frac{\partial u(x^0)}{\partial x_i} = 0 \quad (i = 1, \ldots, n),$$

$$u(x^0 + tp) \leq u(x^0)$$

for enough small t and fixed $p = (p_1, \ldots, p_n)$. Apply Taylor formula on u at the point $x^0 + tp$ and take $t \to 0$, which will imply

$$\sum_{i=1}^{n} \sum_{j=1}^{n} \frac{\partial^2 u(x^0)}{\partial x_i \partial_j} p_i p_j \leq 0.$$

Using that $L(u(x^0)) \geq 0$ and applying the given conditions we can obtain a contradiction with the preceding inequality taking special coordinates $p_i = \alpha_{k_0 i}$, where α_{ij} are the coefficients in the transformation to a canonical form:

$$q_i = \sum_{j=1}^{n} \alpha_{ij} p_j,$$

and $k = k_0$ is chosen such that

$$\sum_{i=1}^{n} \sum_{j=1}^{n} \alpha_{k_0 i} \alpha_{k_0 j} \frac{\partial^2 u(x^0)}{\partial x_i \partial x_j} > 0.$$

Example 5.24 *Prove that the exterior Dirichlet problem*

$$\Delta u = 0 \quad (x \in Q_1) \quad and \quad u|_{\partial Q_1} = 1,$$

where $Q = B(0,1) \subset \mathbf{R}^3$ and $Q_1 = \mathbf{R}^3 \setminus B(0,1)$, has not a unique solution.

Solution. The functions $u = 1$ and $u = \dfrac{1}{|x|}$ are solutions of the given problem.

5.3 The Green Function

5.3.1 Preliminaries

Definition 5.4 *A function $G = G(x,y)$ defined on $\overline{Q} \times Q$ for a bounded region $Q \subset \mathbf{R}^3$ is Green function for the Dirichlet problem if for an arbitrary but fixed $y \in Q$ has the form*

$$G(x,y) = \frac{1}{4\pi |x - y|} + g(x,y),$$

where the function g is harmonic with respect to x in Q and continuous on $\overline{Q}$ and

$$G|_{\partial Q} = 0.$$

5.3.2   Examples and Exercises

Example 5.25 *Prove that the functions g and G in Definition 5.4 satisfy*

$$\text{a)} \qquad g(x,y) = -\frac{1}{4\pi|x-y|} \qquad (x \in \partial Q, y \in Q);$$

$$\text{b)} \qquad \text{the function } g \text{ is continuous on } \overline{Q} \times Q;$$

$$\text{c)} \qquad 0 < G(x,y) < \frac{1}{4\pi|x-y|} \qquad (x \in Q, x \neq y).$$

Solution.

a) Follows by Definition 5.4.

b) Take $x^0 \in \overline{Q}$ and $y^0 \in Q$. Applying the Maximum Principle we obtain for $x \in \overline{Q}$ and $y \in Q$

$$|g(x,y) - g(x^0,y^0)| \;\leq\; |g(x,y^0) - g(x^0,y^0)| + |g(x,y) - g(x,y^0)|$$

$$\leq\; |g(x,y^0) - g(x^0,y^0)| + \max_{x' \in \partial Q} \frac{1}{4\pi}\Big|\frac{1}{|x'-y^0|} - \frac{1}{|x'-y|}\Big|.$$

Since the function $g = g(x,y)$ is continuous with respect to x the last inequality easily implies the desired continuity of g at (x^0, y^0).

c) The function $G = G(x,y)$ with respect to variable x is harmonic on $Q \setminus \{y\}$ and continuous on $\overline{Q} \setminus \{y\}$. Therefore by the Maximum Principle we have

$$G(x,y) > 0 \qquad (x, y \in Q).$$

Since by a) $g(x,y) < 0$ $(x \in \partial Q, y \in Q)$ we obtain again by the Maximum principle $g(x,y) < 0$ $(x \in \overline{Q}, y \in Q)$. Then by Definition 5.4 follows the desired conclusion.

Remark 5.25.1. It can be proved that for enough smooth boundary ∂Q always exists the Green function, exists $\dfrac{\partial G(x,y)}{\partial n}$ for every fixed $y \in Q$ and G is symmetric, i.e., $G(x,y) = G(y,x)$ $(x,y \in Q)$.

Example 5.26 *Find the Green function on the ball $B = B(0,R) = \{x \mid |x| < R\}$.*

Solution. We correspond to a point $y \neq 0$ from the ball $B = B(0,R) = \{x \mid |x| < R\}$ a point y^* with the spheric inversion, i.e.,

$$y^* = y\frac{R^2}{|y|^2}, \qquad |y| \cdot |y^*| = R^2.$$

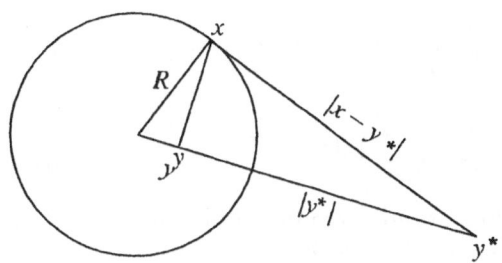

Figure 5.1

We shall find the Green function in the following form

$$G(x,y) = \frac{1}{4\pi|x-y|} - \frac{A}{4\pi|x-y^*|}.$$

Using the condition $G(x,y)|_{x\in\partial B} = 0$ we obtain $A = \dfrac{R}{|y|}$. We obtain from Figure 5.1 that the triangles Oxy^* and Oxy are similar. Hence the sides are proportional, i.e.,

$$\frac{R}{|y|} = \frac{|x-y^*|}{|x-y|},$$

where $|x| = R$. Therefore we obtain

$$G(x,y) = \frac{1}{4\pi|x-y|} - \frac{R|y|}{4\pi|x|y| - yR^2|}.$$

Exercise 5.27 *Find the Green functions on the following regions in* $\mathbf{R}^3$

a) *the semi-ball* $B(0,R) \cap \{(x,y,z) \mid z > 0\}$;

b) *the part of the ball in the first octant*

$$B(0,R) \cap \{(x,y,z)\mid x,y,z > 0\}.$$

Hint. Use the spheric inversion and remember that for the ball $B(0,R)$ by Exercise 5.26

$$\frac{1}{4\pi}\left(\frac{1}{|x-y|} - \frac{R}{|y||x-y'|}\right)$$

for $y' = \dfrac{R^2}{|y|^2} y$.

Answers. Let

$$y_{ijs} = ((-1)^i y_1, (-1)^j y_2, (-1)^s y_3) \text{ and } y'_{ijs} = \dfrac{R^2}{|y|^2} y_{ijs}.$$

a) $\dfrac{1}{4\pi} \displaystyle\sum_{s=0}^{1} (-1)^s \left(\dfrac{1}{|x - y_{00s}|} - \dfrac{R}{|y||x - y_{00s}^*|} \right);$

b) $\dfrac{1}{4\pi} \displaystyle\sum_{i=0}^{1}\sum_{j=0}^{1}\sum_{s=0}^{1} (-1)^{i+j+s} \left(\dfrac{1}{|x - y_{ijs}|} - \dfrac{R}{|y||x - y_{ijs}^*|} \right).$

Example 5.28 (The Green formula) *Prove that if $u \in C^1(\overline{Q})$ is such that $u(x) = 0$ for $x \notin \overline{Q}$, then we have for $x \notin \partial Q$*

$$u(x) = -\frac{1}{(n-2)\sigma_n} \int_Q \frac{\Delta u(y)}{|x-y|^{n-2}}\, dy + \frac{1}{(n-2)\sigma_n} \int_{\partial Q} \left(\frac{1}{|x-y|^{n-2}} \frac{\partial u(y)}{\partial n} \right.$$

$$\left. - u(y)\frac{\partial}{\partial n_y}\left(\frac{1}{|x-y|^{n-2}} \right) \right) dS_y \quad (n \geq 3), \tag{5.19}$$

where $\sigma_n = \dfrac{2\pi^{n/2}}{\Gamma\left(\dfrac{n}{2}\right)}$ is the surface area of the unit sphere in $\mathbf{R}^n$, and

$$u(x) = -\frac{1}{2\pi} \int_Q \Delta u(y) \ln \frac{1}{|x-y|}\, dy + \frac{1}{2\pi} \int_{\partial Q} \left(\ln \frac{1}{|x-y|} \cdot \frac{\partial u(y)}{\partial n} \right.$$

$$\left. - u(y)\frac{\partial}{\partial n_y}\left(\ln \frac{1}{|x-y|} \right) \right) dS_y \quad (n = 2). \tag{5.20}$$

Solution. For an arbitrary but fixed $x \in Q$ we take a ball $B(x,\varepsilon)$ such that $\overline{B(x,\varepsilon)} \subset Q$. Applying the classical Green formula on the fundamental solution $E(y-x)$ and the function $u \in C^2(\overline{Q})$ on the region $Q_\varepsilon = Q \setminus \overline{B(x,\varepsilon)}$ we obtain

$$\int_{Q_\varepsilon} \Delta u(y) E(x-y)\, dy = \int_{\partial Q} \left(E(y-x)\frac{\partial u(y)}{\partial n} - u(y)\frac{\partial E(y-x)}{\partial n_y} \right) dS_y$$

$$+ \int_{\partial B(x,\varepsilon)} \left(E(y-x)\frac{\partial u(y)}{\partial n} - u(y)\frac{\partial u(y-x)}{\partial n_y} \right) dS_y. \tag{5.21}$$

The second summand on the right side in (5.21) using the equality

$$\frac{\partial}{\partial n_y} = -\frac{\partial}{\partial |y-x|} \qquad (y \in \partial B(x,\varepsilon))$$

can be written for $n \geq 3$ in the form

$$-\frac{1}{(n-2)\sigma_n\varepsilon^{n-2}} \int\limits_{\partial B(x,\varepsilon)} \frac{\partial u(y)}{\partial \mathbf{n}}\, dS_y + \frac{1}{\sigma_n\varepsilon^{n-1}} \int\limits_{\partial B(x,\varepsilon)} u(y)\, dS_y$$

$$= u(x) + \frac{1}{\sigma_n\varepsilon^{n-1}} \int\limits_{\partial B(x,\varepsilon)} (u(y) - u(x))\, dS_y - \frac{1}{(n-2)\sigma_n\varepsilon^{n-2}} \int\limits_{\partial B(x,\varepsilon)} \frac{\partial u(y)}{\partial \mathbf{n}}\, dS_y.$$

Putting this in the equality (5.21) and using the facts that the surface area of $\partial B(x,\varepsilon)$ is equal to $\sigma_n\varepsilon^{n-1}$ and $u(y) - u(x) \leq k\varepsilon$ and $|\frac{\partial u(y)}{\partial \mathbf{n}}| \leq S$ ($k > 0$ and $S > 0$ are constants) and finally letting $\varepsilon \to 0$ we obtain the desired equality (5.19) for $n \geq 3$.
Using again (5.21) we prove analogously the equality (5.20) for $n = 2$.

Exercise 5.29 *Prove that for $n = 3$ the boundary problem for the Poisson equation*

$$\Delta u = -F$$

for $F \in C^1(Q) \cap C(\overline{Q})$ can be reduced on the boundary problem for a Laplace equation $\Delta v = 0$.

Hint. Put

$$u = v + V, \text{where } V(x) = \frac{1}{4\pi} \int\limits_Q \frac{F(y)}{|x - y|}\, dy.$$

Then the function $v \in C^2(Q) \cap C^1(\overline{Q})$ satisfies the Poisson equation $\Delta V = -F$ and v satisfies the Laplace equation.

Example 5.30 *Supposing that the boundary ∂Q of the bounded region Q is enough regular and that the solution $u \in C^2(Q) \cap C(\overline{Q})$ of the Dirichlet problem*

$$\Delta u = -F \; , \; u|_{\partial Q} = f,$$

where $\int\limits_Q F^2(x)\, dx < \infty$ and $f \in C(\partial Q)$, and there exists the derivative with respect to the normal on the surface ∂Q, prove that u can be represented by the Green function G in the following form

$$u(x) = -\int\limits_{\partial Q} \frac{\partial G(x,y)}{\partial \mathbf{n}_y} f(y)\, dS_y + \int\limits_Q G(x,y)F(y)\, dy \quad (x \in Q).$$

Solution. Applying the Green formula from Example 5.28 on the function u for $n \geq 3$ we obtain

$$u(x) = \frac{1}{4\pi} \int_{\partial Q} \left(\frac{\partial u(y)}{\partial n_y} \frac{1}{|x-y|} - f(y) \frac{\partial}{\partial n_y} \left(\frac{1}{|x-y|} \right) \right) dS_y + \frac{1}{4\pi} \int_Q \frac{F(y)}{|x-y|} \, dy \quad (x \in Q).$$

(5.22)

By the definition of the Green function the function $g = g(x,y)$ for a fixed $x \in Q$ is harmonic with respect to y, continuous on $\overline{Q}$ and on ∂Q there exists $\dfrac{\partial g(x,y)}{\partial n_y}$.

Applying the classical Green formula we obtain

$$\int_{\partial Q} \left(\frac{\partial u(y)}{\partial n_y} g(x,y) - f(y) \frac{\partial g(x,y)}{\partial n_y} \right) dS_y + \int_Q F(y) g(x,y) \, dy = 0 \quad (x \in Q). \quad (5.23)$$

Adding the equalities (5.22) and (5.23) and using the definition of the Green function G and Example 5.25 a) we obtain the desired representation of u by the Green function.

Example 5.31 *Prove that the function u for $n = 3$, given by*

$$u(x) = \frac{1}{4\pi R} \int_{B(0,R)} \frac{R^2 - |x|^2}{|x-y|^3} f(y) \, dS_y \quad (x \in B(0,R)),$$

is the unique solution of the Dirichlet problem for $n = 3$

$$\Delta u = 0, \qquad u|_{\partial B} = f,$$

for $f \in C(\partial B)$, where B is the ball $B(0,R) = \{x \mid |x| < R\}$.

Solution. The uniqueness follows by Example 5.19.
If u is the solution of the Dirichlet problem on B, then its restriction on $\overline{B(0,r)}$ for $r < R$ is the solution of the Dirichlet problem on the ball $B(0,r)$ with the boundary condition $u|_{\partial B(0,r)}$. Then this solution can be represented by Example 5.30 (since $F = 0$) in the form

$$u(x) = \int_{\partial B(0,r)} \frac{\partial G(x,y)}{\partial n_y} u(y) \, dS_y. \quad (5.24)$$

On the other side by Example 5.26 we have

$$\left. \frac{\partial G(x,y)}{\partial n_y} \right|_{\partial B(0,r)} = \left. \frac{\partial}{\partial |y|} \left(\frac{1}{4\pi|x-y|} - \frac{r|y|}{4\pi|x|y|^2 - yr^2|} \right) \right|_{|y|=r}$$

$$= \frac{1}{4\pi} \frac{\partial}{\partial \rho} \left(\frac{1}{(|x|^2 + \rho^2 - 2|x|\rho \cos \gamma)^{1/2}} \right.$$

$$\left. - \frac{1}{(r^4 + |x|^2\rho^2 - 2r^2|x|\rho \cos \gamma)^{1/2}} \right) \Bigg|_{\rho=r}$$

$$= \frac{|x|^2 - r^2}{4\pi r |x - y|^3} \Bigg|_{y \in \partial B(0,r)}$$

Putting the obtained equality in (5.24) we obtain

$$u(x) = \frac{1}{4\pi r} \int_{\partial B(0,r)} \frac{r^2 - |x|^2}{|x - y|^3} u(y) \, dS_y \qquad (x \in B(0,r)).$$

Letting $r \to R$ we obtain by the continuity of the function u in $B(0, R)$ the desired conclusion.

Example 5.32 (Strong maximum principle) *Prove that for a function $u \in C^2(Q) \cap C(\overline{Q})$ (where Q is a bounded region) which satisfies $\Delta u \geq 0$ $(x \in Q)$ either $u = const.$ or $u(x) < \max_{y \in \partial Q} u(y)$ $(x \in Q)$.*

Solution. Since $u \in C^2(Q)$ and $\Delta u(x) \geq 0$ $(x \in Q)$ we can apply the Green formula from Example 5.28. Therefore we have on the ball $B(x^0, R_0)$

$$u(x^0) \leq \frac{1}{\sigma_n R_0^{n-1}} \int_{\partial B(x^0, R_0)} u(x) \, dS_x.$$

We consider now the partition of Q on two sets Q_1 and Q_2 such that

$$Q_1 = \{x \mid x \in Q, u(x) = \max_{y \in \partial Q} u(y)\},$$

$$Q_2 = \{x \mid x \in Q, u(x) < \max_{y \in \partial Q} u(y)\}.$$

Both sets Q_1 and Q_2 are open sets. The first by the continuity of the function u and the second by the inequality from the beginning. Since Q is connected either Q_1 or Q_2 have to be empty set.

5.4 The Harmonic Functions

5.4.1 Examples and Exercises

Exercise 5.33 *Find the harmonic function u on the ring*

$$\{(x, y) \mid x^2 + y^2 = r^2, 1 < r < 2\}$$

which satisfies the conditions

$$v(1,t) = 0, \ v(2,t) = 2\sin t \quad (0 < t < 2\pi),$$

where $u(x,y) = v(r,t)$.

Answer. $v(r,t) = \dfrac{8}{3} \cdot \sinh(\ln r) \cdot \sin t.$

Example 5.34 (Mean value theorem) *Prove that if a function u is harmonic on the ball $B = B(0, R_0)$ and continuous on $\overline{B}$, then we have*

a) for $x \notin \partial B(0,r), \quad r < R_0$,

$$u(x) = \frac{1}{(n-2)\sigma_n} \int_{\partial B(0,r)} \left(\frac{1}{|x-y|^{n-2}} \frac{\partial u(y)}{\partial n} \right.$$

$$\left. - u(y) \frac{\partial}{\partial n_y} \left(\frac{1}{|x-y|^{n-2}} \right) \right) dS_y \quad (n \geq 3), \tag{5.25}$$

where $\sigma_n = \dfrac{2\pi^{n/2}}{\Gamma(\frac{n}{2})}$ is the surface area of the unit sphere of $\mathbf{R}^3$, and

$$u(x) = \frac{1}{2\pi} \int_{\partial B(0,r)} \left(\ln \frac{1}{|x-y|} \cdot \frac{\partial u(y)}{\partial n} \right.$$

$$\left. - u(y) \frac{\partial}{\partial n_y} \left(\ln \frac{1}{|x-y|} \right) \right) dS_y \quad (n = 2). \tag{5.26}$$

b) the equality

$$u(0) = \frac{1}{\sigma_n R_0^{n-1}} \int_{\partial B} u(y) \, dS_y; \tag{5.27}$$

c) the equality

$$u(0) = \frac{1}{\sigma_n R_0^{n-1}} \int_B u(y) \, dy.$$

Solution.

a) Since $u \in C^1(\overline{B(0,r)})$ and u is harmonic on $B(0,r)$ we obtain by Example 5.28 the equalities (5.25) and (5.26).

b) For $n \geq 3$ taking $x = 0$ we obtain by (5.25)

$$u(x) = \frac{1}{(n-2)\sigma_n}(\frac{1}{r^{n-2}} \int_{\partial B(0,r)} \frac{\partial u(y)}{\partial n} dS_y$$

$$- \int_{\partial B(0,r)} u(y)\frac{\partial}{\partial n_y}(\frac{1}{|y|^{n-2}}) dS_y). \tag{5.28}$$

On the other hand by the classical Green formula we obtain

$$\int_{\partial B(0,r)} \frac{\partial u(y)}{\partial n} dS_y = 0.$$

Putting this in (5.28) we have

$$u(0) = \frac{1}{r^{n-1}\sigma_n} \int_{\partial B(0,r)} u(y) dS_y. \tag{5.29}$$

Letting in (5.29) $r \to R_0$ we obtain by the continuity of the function u (5.27). Using (5.26) we can prove in an analogous way (5.27) for $n = 2$.

c) Apply the integral with respect to r on (5.29) from 0 to R_0.

Example 5.35 *Prove that if a function u has the mean value property on simple connected region $Q \subset \mathbf{R}^n$, i.e., for every ball $B(x_0, R) = \{y \mid |x_0 - y| < R\}$ such that $\overline{B(x_0, R)} \subset Q$ we have*

$$u(x_0) = \frac{1}{2\pi R} \int_{\partial B(x_0,R)} u(x) dS_x,$$

and u is continuous on Q then:

a) the function takes its maximum and minimum on ∂Q;

b) the function u is harmonic.

Solution. a) We shall prove that either $u = const.$ or

$$\min_{z \in \partial Q} u(z) < u(x) < \max_{z \in \partial Q} u(z) \qquad (x \in Q). \tag{5.30}$$

Let $M = \max_{x \in \overline{Q}} u(x)$. We shall prove that if there exists a point from Q where the right part in the inequality (5.30) is not satisfied, then it have to be $u = M = const.$ on Q. Suppose that there exists such point. Hence there exists $x^0 \in Q$ such that $u(x^0) = M$. We connect an arbitrary but fixed point $y \in Q$ with x^0 by a broken line L which does not intersect itself and which whole lies in Q. Let $d > 0$ be the distance between L and ∂Q. We cover the line L with a finite number of balls $B\left(x^i, \frac{d}{2}\right)$ $(i = 0, 1, \ldots, N)$ with centers $x^i \in L \cap B(x^i, \frac{d}{2})$ $(i = 1, 2, \ldots, N)$ and $y \in B(x^N, \frac{d}{2})$. We have

$$u(x^0) = \frac{1}{\pi d} \int\limits_{|x-x_0|<d/2} u(x)\, dx,$$

i.e.,

$$\int\limits_{|x-x_0|<d/2} (u(x) - u(x^0))\, dx.$$

Since the function under the integral is not positive, we have $u(x) = u(x^0)$ on $B(x^0, \frac{d}{2})$ and specially $u(x^1) = M$. We conclude that $M = u(x^1) = u(x^2) = \cdots = u(y)$.

We prove in analogous way that either the left part in the inequality (5.30) is true or $u = const.$ on Q.

Example 5.36 *Let a function u be harmonic on a bounded region Q in $\mathbf{R}^n$ and bounded on it, i.e., there exists $M > 0$ such that $|u(x)| \le M$ $(x \in Q)$. Prove that for arbitrary $x \in Q$*

$$|D^\alpha u(x)| \le M(\frac{n}{r})^{|\alpha|}|\alpha|^{|\alpha|}, \tag{5.31}$$

where $r = d(x, \partial Q)$ is the distance of the point x from ∂Q.

Solution. The proof goes by induction.

Suppose that $|\alpha| = 1$. Since $\frac{\partial u}{\partial x_i}$ $(i = 1, \ldots, n)$ are harmonic functions on Q we have by Example 5.34 c) (it holds also for the ball $B(x^0, R_0)$ for $x^0 \ne 0$ since the property of the harmonicity is translation invariant)for every $r' < r$

$$\frac{\partial u(x)}{\partial x_i} = \frac{n}{\sigma_n r'^n} \int\limits_{B(x,r')} \frac{\partial u(y)}{\partial y_i}\, dy = \frac{n}{\sigma_n r'^n} \int\limits_{\partial B(x,r')} u(y) \cos \varphi_i\, dS_y,$$

where φ_i is the angle between $y - x$ and the coordinates line y_i. Therefore we have

$$\left|\frac{\partial u(x)}{\partial x_i}\right| \le \frac{n}{\sigma_n r'^n} \int\limits_{\partial B(x,r')} |u(y)|\, dS_y$$

$$\le M\frac{n}{\sigma_n r'^n}\sigma_n r'^{n-1} = \frac{Mn}{r'}.$$

Letting $r' \to r$ we obtain (5.31) for $|\alpha| = 1$.

Suppose now that the conclusion holds for every α such that $|\alpha| \leq k - 1, k \geq 2$. We shall prove (5.31) for $|\alpha| = k$. By the inductive supposition we have for every $y \in B(x, \frac{r'}{k})$, $0 < r' < r$, and every β, $|\beta| = k - 1$,

$$|D_y^\beta u(y)| \leq M(\frac{n}{r' - \frac{r'}{k}})^{k-1}(k-1)^{k-1} = M',$$

where $M' = M(\frac{n}{r'})^{k-1}k^{k-1}$.

By the first inductive step for the first derivative of $D_y^\beta u$ we obtain

$$|D_{y_i}D_y^\beta u(y)| \leq M'\frac{n}{\frac{r'}{k}} = M(\frac{n}{r'})^k k^k \quad (i = 1, \ldots, n).$$

Letting $r' \to r$ we obtain (5.31).

Example 5.37 *Prove that a harmonic function on a bounded region $Q \subset \mathbf{R}^n$ is a (real) analytic function.*

Solution. Let u be a harmonic function on Q. Let x^0 be an arbitrary but fixed point from Q. We denote by

$$\delta = d(x^0, \partial Q) > 0 \text{ and } B(x^0, \frac{\delta}{4}) = \{x \mid |x - x_0| < \frac{\delta}{4} |\}.$$

We have $u \in C(\overline{Q}_{\delta/2})$, where $Q_\varepsilon = \{x \mid d(x, Q) < \varepsilon\}$. Hence

$$M = \max_{x \in \overline{Q}_{\delta/2}} |u(x)| \in \mathbf{R}.$$

Since

$$d(\overline{B(x^0, \delta/4)}, \delta Q_{\delta/2}) \geq \frac{\delta}{4} \quad (x \in B(x^0, \frac{\delta}{4}))$$

and for every multi-index α we have

$$|D^\alpha u| \leq M(\frac{4n}{\delta})^{|\alpha|}|\alpha|^{|\alpha|},$$

see Example 5.36. We obtain by the Stirling formula

$$\lim_{k \to \infty} \frac{k^{(k+1)/2}}{k!e^k} = \frac{1}{\sqrt{2\pi}},$$

that there exists $c > 0$ such that $|\alpha|^{|\alpha|} \leq ce^{|\alpha|}|\alpha|!$.

Taking in the equality

$$(x_1 + \cdots + x_n)^k = \sum_{|\alpha|=k} \frac{|\alpha|!}{\alpha!}x^\alpha \quad (k \in \mathbf{N}).$$

Putting $x_i = 1$ for all i we can obtain the estimation $\dfrac{|\alpha|!}{\alpha!} \leq n^{|\alpha|}$. Therefore for every $x \in B(x^0, \dfrac{\delta}{4})$ and every $\alpha \in \mathbf{Z}_+^n$ we have

$$|D^\alpha u| \leq cM \left(\frac{4n^2 e}{\delta}\right)^{|\alpha|} \alpha!. \qquad (5.32)$$

Hence the series

$$\sum_\alpha \frac{D^\alpha u(x^0)}{\alpha!}(x - x^0)^\alpha \qquad (5.33)$$

converges absolutely on $B(x^0, \dfrac{\delta}{4n^2 e})$. Therefore the sum (5.33) is an analytical function on $B(x^0, \dfrac{\delta}{4n^2 e})$. We shall prove that the series (5.32) converges to the function u on the ball $B' = B(x^0, \dfrac{\delta}{8n^2 e})$, i.e.,

$$R_N(x) = u(x) - \sum_{k=0}^{N-1} \sum_{|\alpha|=k} \frac{D^\alpha u(x^0)}{\alpha!}(x - x^0)^\alpha = \sum_{|\alpha|=N} \frac{D^\alpha u(x^0 + \theta(x - x^0))}{\alpha!}(x - x^0)^\alpha,$$

where $|\theta| < 1$, converges to zero as $N \to \infty$ for every $x \in B'$. Since $x^0 + \theta(x - x^0) \in B' \subset B(x^0, \dfrac{\delta}{4})$ we obtain by (5.32) for every $x \in B'$

$$
\begin{aligned}
|R_N(x)| &\leq \sum_{|\alpha|=N} cM \left(\frac{4n^2 e}{\delta}\right)^N \left(\frac{\delta}{8n^3 e}\right)^N \\
&\leq \frac{cM}{(2n)^N} n^N = \frac{cM}{2^N}.
\end{aligned}
$$

Therefore

$$\lim_{N \to \infty} R_N(x) = 0 \qquad (x \in B').$$

Since x^0 was an arbitrary element of Q we conclude that the function u is analytical on the whole Q.

Example 5.38 *Prove that a function u is harmonic on simple connected region $Q \subset \mathbf{R}^2$ if and only if there exists an analytical function f on Q (taking $z = x + iy$) such that $u = \Re f$.*

Solution. If f is an analytical function and $u = \Re f$, then $\Delta u = 0$ by Example 1.6. Suppose now that the function u is harmonic on Q. We define a function v by

$$v(x, y) = \int_{L((x^0, y^0), (x,y))} \left(-\frac{\partial u(x, y)}{\partial y} \, dx + \frac{\partial u(x, y)}{\partial x}\right) dy,$$

where (x^0, y^0) is an arbitrary but fixed point from Q, $(x, y) \in Q$ and $L((x^0, y^0), (x, y))$ is an arbitrary path which connect the points (x^0, y^0) and (x, y) and completely lies in Q. The function is independent of the choice of curve L since the region Q is simple connected and by the classical Green formula. Since $u \in C^2(Q)$ we obtain that $v \in C^1(Q)$. Since v satisfies the Cauchy-Riemann equations (check it!) we have that the function $f = u + iv$ is an analytical function with respect to $x + iy$, where $(x, y) \in Q$.

Example 5.39 *Prove that from every infinite family of harmonic functions on a bounded region Q, which are uniformly bounded we can select a sequence of harmonic functions which uniformly converges on a subregion Q' of Q such that $\overline{Q'} \subset Q$.*

Solution. Let $\mathcal{H}$ be an infinite family of harmonic functions on region Q uniformly bounded by a constant $M > 0$. We choose a sequence of regions $Q_1, Q_2, \ldots$ in the following way: $Q_1 \subset Q_2 \subset \ldots$ and Q_i are relative compact in Q for every $i \in \mathbf{N}$ and $\cup_{i=1}^{\infty} Q_i = Q$. Since $\mathcal{H} \subset C(\overline{Q})$ and the functions from $\mathcal{H}$ are uniformly bounded on Q_1 there exists a constant $C > 0$ which depends on Q_1 such that for every $u \in \mathcal{H}$ we have

$$| \nabla u| \leq C \qquad (x \in Q_1),$$

see Example 5.36. Hence $\mathcal{H}$ is equicontinuous on $\overline{Q}_1$. Therefore by Arzela- Ascoli theorem we can select from the family $\mathcal{H}$ a sequence $u_{11}, u_{12}.u_{13}, \ldots$ which is uniformly convergent on $\overline{Q}_1$. Since this sequence satisfies also the conditions of Arzel-Acsoli theorem on $\overline{Q}_2$ there exists its subsequence $u_{21}, u_{22}, u_{23}, \ldots$ which is uniformly convergent on $\overline{Q}_2$. Continuing this procedure we obtain at the i-th step a subsequence $u_{i1}, u_{i2}, u_{i3}, \ldots$ which is uniformly convergent on $\overline{Q}_i$. Then the diagonal sequence $u_{11}, u_{22}, u_{33}, \ldots$ is the desired sequence.

Example 5.40 *Let $\{u_k\}_{k \in \mathbf{N}}$ be a sequence of harmonic functions on the region Q which is uniformly convergent on every region $Q' \subset Q$ such that $\overline{Q'} \subset Q$. Prove that the limit function u is also harmonic on Q and that for every multi-index $\alpha = (\alpha_1, \ldots, \alpha_n)$ the sequence $\{D^\alpha u_k\}_{k \in \mathbf{N}}$ uniformly converges to $D^\alpha u$ on every region $Q' \subset Q$ such that $\overline{Q'} \subset Q$.*

Solution. First we have $u \in C(\overline{Q'})$. Let Q'' be a region such that $Q' \subset\subset Q'' \subset\subset Q$. Since every function u_m is bounded on Q'' for every multi-index α there exists a constant $C > 0$ which depends only from Q', Q'' and $|\alpha|$ such that

$$\|D^\alpha(u_m - u_s)\|_{C(\overline{Q'})} \leq C \|u_m - u_s\|_{C(\overline{Q''})} \qquad (s, m = 1, 2, \ldots).$$

Since

$$\lim_{m,s \to \infty} \|u_m - u_s\|_{C(\overline{Q''})} = 0$$

we obtain that all sequences $\{D^\alpha u_m\}_{m\in N}$ are Cauchy sequences in the space $C(\overline{Q'})$. Therefore $u \in C^\infty(\overline{Q'})$ and for every α we have

$$\lim_{m\to\infty} \|D^\alpha u_m - D^\alpha U\|_{C(\overline{Q'})} = 0.$$

Taking $m \to \infty$ in $\Delta u_m = 0$ $(x \in Q')$ we obtain $\Delta u = 0$ $(x \in Q')$.

Example 5.41 (Harnack first theorem) *Let $\{u_k\}_{k\in N}$ be a sequence of harmonic functions on the region Q which belongs to $C(\overline{Q})$. Prove that if the sequence $\{u_k|_{\partial Q}\}_k \in$ N is uniformly convergent on the border ∂Q, then the sequence uniformly converges on $\overline{Q}$ to a harmonic function.*

Solution. The first part follows from Example 5.20.
    The function u is harmonic by Example 5.40.

Exercise 5.42 (Harnack second theorem) *Let $\{u_k\}_{k\in N}$ be a sequence of harmonic functions on the region $Q \subset \mathbf{R}^2$ which belongs to $C(\overline{Q})$. Prove that if the sequence $\{u_k|_{\partial Q}\}_{k\in N}$ satisfies the inequality*

$$u_k(x_i) \leq u_{k+1}(x_i) \qquad (x_i \in Q, i \in \mathbf{N}),$$

then the convergence of the sequence $\{u_k\}_{k\in N}$ in a point of the region Q implies its uniform convergence on every compact set in Q.

Hints. Prove that u is a harmonic and non-negative function and using the Poisson formula prove the inequalities on the disc of radius R

$$\frac{R-r}{R+r}u(0,0) \leq u(r,\theta) \leq \frac{R+r}{R-r}u(0,0) \quad (r \in \mathbf{R}).$$

Exercise 5.43 *Prove that a function $u \in C^2(Q)$ for a bounded region $Q \subset \mathbf{R}^n$ which is a solution of the Helmholtz equation*

$$\Delta u + \lambda u = 0,$$

where λ is a constant, is an analytical function on Q, i.e., all eigenfunctions for a Laplace operator are analytical functions.

Example 5.44 *We correspond to a point $x \neq 0$ from the ball $B = B(0,R) = \{y \mid |y| < R\}$ a point x^* with the spheric inversion, i.e., $x^* = x\dfrac{R^2}{|x|^2}$. Prove that if a function u is harmonic on $\mathbf{R}^n \setminus \overline{B}$, then the function u^* defined by (Kelvin transformation)*

$$u^*(x^*) = \frac{R}{|x^*|^{n-2}}u(\frac{R^2x^*}{|x^*|^2}) \tag{5.34}$$

is harmonic on $B \setminus \{0\}$.

Solution. Suppose $n > 2$ (for $n = 2$ we proceed analogously). Let B_1' be a region strictly inside of $B \setminus \{0\}$ and B_1 is its original with respect to the inversion. Then B_1 is a bounded region strictly inside of $\mathbf{R}^n \setminus \overline{B}$. The function u is harmonic and belongs to $C^2(\overline{Q}_1)$ and therefore

$$u(x) = \int_{\partial B_1} \left(\frac{\mu(z)}{|x - z|^{n-2}} + \nu(z) \frac{\partial}{\partial n_z} \left(\frac{1}{|x - z|^{n-2}} \right) \right) dS_z,$$

where

$$\mu(z) = \frac{1}{(n-2)\sigma_n} \cdot \frac{\partial u(z)}{\partial n} \Big|_{\partial B_1}, \quad \nu(z) = -\frac{u(z)}{(n-2)\sigma_n} \Big|_{\partial B_1}$$

and these are continuous functions on ∂B_1. By (5.34) we have for all $x \in B_1'$

$$u^*(x^*) = \int_{\partial B_1} \left(\frac{R\mu(z)}{|x^*|^{n-2} \left| \frac{R^2 x^*}{|x^*|^2} - z \right|^{n-2}} \right. \tag{5.35}$$

$$\left. + \nu(z) \frac{\partial}{\partial n_z} \left(\frac{R}{|x^*|^{n-2} \left| \frac{R^2 x^*}{|x^*|^2} - z \right|^{n-2}} \right) \right) dS_z.$$

Since the function

$$\frac{|x^*|^{2-n}}{R} \left| \frac{R^2 x^*}{|x^*|^2} - z \right|^{2-n}$$

is harmonic we have that

$$\frac{\partial}{\partial n_z} \left(\frac{|x^*|^{2-n}}{R} \left| \frac{R^2 x^*}{|x^*|^2} - z \right|^{2-n} \right)$$

is harmonic function with respect to x^* when $\frac{R^2 x^*}{|x^*|^2} \neq z$. Therefore the function under integral in (5.35) is continuous with respect to x^* and z as well as its derivatives with respect to x^* and it is harmonic with respect to x^* for $z \in \partial B$ and $x^* \in B \setminus \{0\}$.

Exercise 5.45 *Prove that the exterior Dirichlet problem has maximally one solution.*

Hint. For two solutions u_1 and u_2 consider the function $u = u_1 - u_2$ which satisfies the homogeneous Dirichlet problem. Apply the inversion transformation and use Example 5.44.

Exercise 5.46 *The Poisson equation $\Delta u = -F$ has has not more than one regular solution in the infinite for $n \geq 3$. For $n = 2$ regular solutions in the infinite differ up to a constant.*

Hint. For $n \geq 3$ use the maximum principle on the ball $B(0, R)$, and for $n = 2$ the inversion and the maximum principle.

Exercise 5.47 *Prove that the solution of the Dirichlet problem in* $\mathbf{R}^2, \mathbf{R}^3$

$$\Delta u = -F \quad (|x| < R) \quad and \quad u|_{\partial B(0,R)} = f$$

is given by

$$u(x) = \frac{1}{4\pi R} \int_{|y|=R} \frac{R^2 - |x|^2}{|x-y|^2} f(y)\, dS_y + \frac{1}{4\pi} \int_{|y|\leq R} \left(\frac{1}{|x-y|} - \frac{R}{|y||x-y'|} \right) F(y)\, dy,$$

where $y' = y\dfrac{R^2}{|y|^2}$ *is obtained by inversion.*

5.5 Gravitational Potential

Let the function

$$u(x,y,z) = \frac{c}{r}, \tag{5.36}$$

defines the gravitational potential. In equation (5.36) $c = \gamma Mm$, and γ is gravitational constant; M is the mass of a particle at a fixed point (ξ, η, ζ), m is the mass of a particle at a fixed point, and r is the distance between two points

$$r = \sqrt{(x-\xi)^2 + (y-\eta)^2 + (z-\zeta)^2}.$$

If P_1 is particle of mass M and P_2 is particle of mass m, then P_1 attract P_2 with a gravitational force that is the gradient of the function $\dfrac{c}{r}$.

Let $u(x,y,z)$ be a potential u of continuous body at a point (x,y,z) outside the body. Then the potential u is defined by

$$u(x,y,z) = k \int\!\!\int_V\!\!\int \frac{\rho\, d\xi\, d\eta\, d\zeta}{r}$$

where ρ is the density of a mass at $(\xi, \eta\zeta)$ and k is a positive constant. The function u satisfies the Laplace equation

$$\frac{\partial^2 u}{\partial x^2} + \frac{\partial^2 u}{\partial y^2} + \frac{\partial^2 u}{\partial z^2} = 0.$$

In Laplace equation, u may represent electric or magnetic potential functions at points not filled with electric charges or magnetic poles. Laplace equation is associated with incompressible fluid flow problems, too.

Chapter 6

Parabolic Equations

6.1 Cauchy Problem

6.1.1 Preliminaries

The problem for heat equation given as

$$\frac{\partial u}{\partial t} = a^2 \Delta u + F(x,t) \qquad (x \in \mathbf{R}^n \quad t > 0), \tag{6.1}$$

with condition

$$u(x,0) = f(x) \qquad (x \in \mathbf{R}^n), \tag{6.2}$$

where $a > 0$, f is bounded and $f \in C(\mathbf{R}^n)$, $F \in C^2(\mathbf{R}^n \times [0,\infty))$, and all its second order partial derivatives are bounded on every set of the form $\mathbf{R}^n \times [0,T]$ is called the *Cauchy problem*.

If $F = 0$, then we are working with *homogeneous heat equation* otherwise this is *nonhomogeneous heat equation*. The classical solution of problem (6.1), (6.2) is the function $u = u(x,t) \in C^2(\mathbf{R}^n \times [0,T]) \cup C(\mathbf{R}^n \times [0,T])$ given by

$$
\begin{aligned}
u(x,t) \quad = \quad & (4\pi a^2 t)^{-n/2} \int\limits_{\mathbf{R}^n} \exp\left(-\frac{|s-x|^2}{4a^2 t}\right) f(s)\, ds \\[2mm]
& + \int\limits_0^t \int\limits_{\mathbf{R}^n} \frac{F(\xi,\tau)}{(4\pi a^2(t-\tau))^{n/2}} \exp\left(-\frac{|x-\xi|^2}{4a^2(t-\tau)}\right) d\xi\, d\tau.
\end{aligned}
\tag{6.3}
$$

The Maximum Principle

For an arbitrary but fixed $T > 0$ we introduce the set Q by

$$Q = \{(x,t) \mid x \in O, 0 < t < T\},$$

183

where O is an open bounded set of $\mathbf{R}^n$. The boundary ∂Q we can divide on two disjoint parts "lower" boundary $\partial' Q$ and "upper" boundary $\partial'' Q$ given by

$$\partial' Q = \{(x,t)| \text{ either } x \in \partial O, 0 \le t \le T \text{ or } x \in O, t = 0\},$$

$$\partial'' Q = \{(x,t)| \ x \in O, \ t = T\}.$$

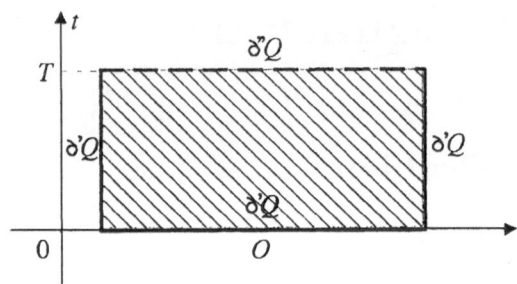

Figure 6.1 For $n = 1$ and O an open interval

Theorem 6.1 (Maximum Principle) *Let*

$$u \in C(\overline{Q}), \quad \frac{\partial u}{\partial t}, \frac{\partial^2 u}{\partial x_i \partial x_j} \in C(Q)$$

and

$$\frac{\partial u}{\partial t} - \Delta u \le 0.$$

Then

$$\max_{(x,t) \in \overline{Q}} u(x,t) = \max_{(x,t) \in \partial' Q} u(x,t).$$

6.1.2   Examples and Exercise

Example 6.1 *Suppose that the function $f : \mathbf{R} \to \mathbf{R}$ is continuous and bounded, i.e.,*

$$|f(x)| \le M \quad (-\infty < x < \infty).$$

Prove that the function

$$u(x,t) = \frac{1}{2a\sqrt{\pi t}} \int_{-\infty}^{\infty} f(y) \exp\left(-\frac{(x-y)^2}{4a^2 t}\right) dy, \tag{6.4}$$

is bounded and satisfy the equation in the problem

$$\frac{\partial u}{\partial t} = a^2 \frac{\partial^2 u}{\partial x^2} \qquad (x \in \mathbf{R}, \quad t > 0),$$

$$u(x, 0) = f(x) \qquad (x \in \mathbf{R}).$$

Solution. The integral (6.4) converges absolutely for arbitrary $(x, t) \in \mathbf{R} \times \mathbf{R}^+$, because

$$|u(x, t)| \leq \frac{M}{2a\sqrt{\pi t}} \int\limits_{-\infty}^{\infty} \exp\left(-\frac{(x - y)^2}{4ta^2}\right) dy,$$

where $M = \sup\limits_{x \in \mathbf{R}} |f(x)|$. The change of variables, $z = \dfrac{x - y}{2a\sqrt{t}}$, gives

$$|u(x, t)| \leq \frac{M}{\sqrt{\pi}} \int\limits_{-\infty}^{\infty} e^{-z^2} dz = M.$$

Differentiating the integral (6.4) by x and t under the integral the obtained integrals converge absolutely. In an arbitrary compact neighbourhood of a point $(x, t) \in \mathbf{R} \times \mathbf{R}^+$ the integral (6.4) with its partial derivatives converge uniformly. The partial derivatives are

$$\frac{\partial u}{\partial t} = \frac{1}{2a\sqrt{\pi}} \left(-\frac{1}{2\sqrt{t^3}} \int\limits_{-\infty}^{\infty} f(y) \exp\left(-\frac{(x - y)^2}{4ta^2}\right) dy \right.$$

$$+ \frac{1}{\sqrt{t}} \int\limits_{-\infty}^{\infty} f(y) \frac{(x - y)^2}{4a^2t^2} \exp\left(-\frac{(x - y)^2}{4ta^2}\right) dy \Bigg)$$

$$= \frac{1}{2a\sqrt{\pi}} \cdot \frac{1}{4a^2\sqrt{t^5}} \int\limits_{-\infty}^{\infty} f(y) \exp\left(-\frac{(x - y)^2}{4ta^2}\right) ((x - y)^2 - 2ta^2) dy;$$

$$\frac{\partial u}{\partial x} = \frac{1}{2a\sqrt{\pi t}} \left(\int\limits_{-\infty}^{\infty} f(y) \exp\left(-\frac{(x - y)^2}{4ta^2}\right) \cdot \frac{-(x - y)}{2a^2t} dy\right);$$

$$\frac{\partial^2 u}{\partial x^2} = \frac{1}{2a\sqrt{\pi t}} \left(\int\limits_{-\infty}^{\infty} \frac{f(y)}{2a^2t} \exp\left(-\frac{(x - y)^2}{4ta^2}\right) \cdot \left(-1 + \frac{(x - y)^2}{2a^2t}\right) dy\right),$$

wherefrom it follows that

$$\frac{\partial u}{\partial t} = a^2 \frac{\partial^2 u}{\partial x^2} \qquad (-\infty < x < \infty, \, t > 0).$$

Example 6.2 *Determine the solution of the problem*

$$\frac{\partial u}{\partial t} = a^2 \frac{\partial^2 u}{\partial x^2} \qquad (x \in \mathbf{R}, \quad t > 0),$$

$$u(x,0) = \begin{cases} 0, & for & x < c, \\ T, & for & c < x < d, \\ 0, & for & x > d. \end{cases}$$

Solution. In this case formula (6.4) reduces to

$$u(x,t) = \frac{T}{2a\sqrt{\pi t}} \int_c^d \exp\left(-\frac{(x-y)^2}{4a^2 t}\right) dy.$$

After changing of variables, $\xi = \dfrac{x-y}{a\sqrt{2t}}$, we get

$$u(x,t) = T\frac{1}{\sqrt{2\pi}} \int_{(x-c)/a\sqrt{2t}}^{(x-d)/a\sqrt{2t}} e^{-\xi^2/2} d\xi.$$

Remark 6.2.1 Since the *normal distribution function* Φ is defined by

$$\Phi(x) = \int_{-\infty}^x e^{-t^2/2} dt,$$

we have shown that the solution of this problem can be written as

$$u(x,t) = T\left(\Phi\left(\frac{x-d}{\sqrt{2at}}\right) - \Phi\left(\frac{x-c}{\sqrt{2at}}\right)\right).$$

Exercise 6.3 *Determine the solution of the problem*

$$\frac{\partial u}{\partial t} = a^2 \frac{\partial^2 u}{\partial x^2} \qquad (-\infty < x < \infty, \ t > 0),$$

$$u(x,0) = T_1 \qquad (x < 0),$$

$$u(x,0) = T_2 \qquad (x > 0).$$

Answer.

$$u(x,t) = T_1\left(1 - \Phi\left(\frac{x}{a\sqrt{2t}}\right)\right) + T_2\Phi\left(\frac{x}{a\sqrt{2t}}\right).$$

It holds $\lim_{x \to 0} \Phi(x) = \Phi(0) = \dfrac{1}{2}$.

Exercise 6.4 *Show that the solution of Cauchy problem*

$$\frac{\partial u}{\partial t} = a^2 \frac{\partial^2 u}{\partial x^2} \qquad (a > 0, \quad x \in \mathbf{R}, \quad t > 0),$$

$$u(x,0) = f(x) \qquad (x \in \mathbf{R}),$$

where the function $f \in C^2(\mathbf{R})$, satisfies the following

a) $u(0,t) = 0$, *if the function f is odd;*

b) $\dfrac{\partial u(0,t)}{\partial x} = 0$, *if the function f is even;*

Hint. It can be done similarly as in Example 4.3.

Example 6.5 *Solve the following problem*

$$\frac{\partial u}{\partial t} = a^2 \frac{\partial^2 u}{\partial x^2} \qquad (x > 0, \quad t > 0),$$

$$u(0,t) = 0 \qquad (t > 0), \quad u(x,0) = f(x) \qquad (x > 0). \tag{6.5}$$

Solution. Let us define the function Ψ as

$$\Psi(x) = \begin{cases} f(x), & \text{for} \quad x > 0, \\ -f(-x), & \text{for} \quad x < 0, \end{cases}$$

and consider the problem

$$\frac{\partial U}{\partial t} = a^2 \frac{\partial^2 U}{\partial x^2} \qquad (x \in \mathbf{R}, \ t > 0),$$

$$U(x,0) = \Psi(x) \qquad (x \in \mathbf{R}).$$

The solution of this problem is

$$U(x,t) = \frac{1}{2a\sqrt{\pi t}} \int\limits_{-\infty}^{\infty} \Psi(y) \exp\left(-\frac{(x-y)^2}{4ta^2}\right) dy,$$

wherefrom it follows that the solution of the considered problem (6.5) is

$$u(x,t) = \frac{1}{2a\sqrt{\pi t}} \int\limits_{0}^{\infty} f(y) \left(\exp\left(-\frac{(x-y)^2}{4a^2 t}\right) - \exp\left(-\frac{(x+y)^2}{4a^2 t}\right)\right) dy.$$

Exercise 6.6 *Solve the following problem*

$$\frac{\partial u}{\partial t} = a^2 \frac{\partial^2 u}{\partial x^2} \qquad (x > 0, \quad t > 0),$$

$$\frac{\partial u(0,t)}{\partial x} = 0 \qquad (t > 0), \qquad u(x,0) = f(x) \qquad (x > 0).$$

(6.6)

Answer.

$$u(x,t) = \frac{1}{2a\sqrt{\pi t}} \int_0^\infty f(y) \left(\exp\left(-\frac{(x-y)^2}{4a^2 t} \right) + \exp\left(-\frac{(x+y)^2}{4a^2 t} \right) \right) dy.$$

Exercise 6.7 *Introducing the function v such that $u(x,t) = e^{ht} v(x,t)$, determine the solution of the equation*

$$\frac{\partial u}{\partial t} = a^2 \frac{\partial^2 u}{\partial x^2} - hu \qquad (x > 0, \, t > 0),$$

with the conditions

a) $u(0,t) = 0 \quad (t > 0), \quad u(x,0) = f(x) \quad (x > 0)$;

b) $\dfrac{\partial u(0,t)}{\partial x} = 0 \quad (t > 0), \quad u(x,0) = f(x) \quad x > 0$;

where h is a constant and the function f is continuous and bounded on the interval $(0,\infty)$.

Answer.

a) $\quad u(x,t) = \dfrac{e^{-ht}}{2a\sqrt{\pi t}} \displaystyle\int_0^\infty f(y) \left(\exp\left(-\dfrac{(x-y)^2}{4a^2 t} \right) - \exp\left(-\dfrac{(x+y)^2}{4a^2 t} \right) \right) dy.$

b) $\quad u(x,t) = \dfrac{e^{-ht}}{2a\sqrt{\pi t}} \displaystyle\int_0^\infty f(y) \left(\exp\left(-\dfrac{(x-y)^2}{4a^2 t} \right) + \exp\left(-\dfrac{(x+y)^2}{4a^2 t} \right) \right) dy.$

Exercise 6.8 *Let u_i be the solution of the Cauchy problem on the set $\{(x,t)| x \in \mathbf{R}^n, \, t > 0\}$,*

$$\frac{\partial u}{\partial t} = a^2 \Delta u,$$

$$u(x,0) = f_i(x_i), \qquad for \qquad f_i : \mathbf{R} \to \mathbf{R}, i = 1,2,\ldots,n.$$

Prove that then the function u given by

$$u(x,t) = u_1(x,t) \ldots u_n(x,t)$$

represent the solution of the Cauchy problem on the set $\{(x,t)|x \in \mathbf{R}^n,\ t > 0\}$,

$$\frac{\partial u}{\partial t} = a^2 \Delta u,$$

$$u(x,0) = f_1(x_1)\ldots f_n(x_n).$$

Exercise 6.9 *Determine the solution of the Cauchy problem on the set $\{(x,t)|x \in \mathbf{R}^n,\ t > 0\}$,*

$$\frac{\partial u}{\partial t} = a^2 \Delta u, \qquad u(x,0) = f(x),$$

for

a) $f(x) = \cos(x_1 + x_2 + \ldots + x_n)$

b) $f(x) = e^{-|x|^2}$.

Answers.

a) $u(x,t) = e^{nt}\cos(x_1 + x_2 + \ldots x_n)$

b) $u(x,t) = (1 + 4t)^{-n/2} \exp\left(\dfrac{-|x|^2}{1+4t}\right).$

Exercise 6.10 *Show that the function $u(x,t) = \displaystyle\int_0^t v(x,t,\tau)d\tau$, where*

$$v(x,t,\tau) = \frac{1}{2\sqrt{\pi(t-\tau)}}\int_{-\infty}^{\infty} g(x,y)\exp\left(-\frac{(x-y)^2}{4(t-\tau)}\right)dy,$$

and the function $g(x,y)$, $(x,y) \in \mathbf{R}^2$, is continuous and bounded on $\mathbf{R}^2$ and satisfies the nonhomogeneous equation

$$\frac{\partial u}{\partial t} = \frac{\partial^2 u}{\partial x^2} + g(x,t).$$

Exercise 6.11 *Suppose that the functions $f \in C(\mathbf{R}^n)$ and $F \in C^2(\mathbf{R}^n \times \{t|t \geq 0\})$ are bounded and the derivatives of F are bounded on $\mathbf{R}^n \times \{t|0 \leq t \leq T\})$. Prove that the function*

$$u(x,t) = \frac{1}{(2a\sqrt{\pi t})^n}\int_{\mathbf{R}^n} f(y)\exp\left(-\frac{|x-y|^2}{4a^2 t}\right)dy$$

$$+ \int_{-\infty}^{\infty}\int_{\mathbf{R}^n} \frac{F(y,s)}{\left(2a\sqrt{\pi(t-s)}\right)^n}\exp\left(-\frac{|x-y|^2}{4a^2(t-s)}\right)dyds,$$

(6.7)

represent the solution of the problem

$$\frac{\partial u}{\partial t} = a^2 \Delta u + F \qquad (x \in \mathbf{R}^n \quad t > 0),$$

$$u(x,0) = f(x) \quad (x \in \mathbf{R}^n).$$

Exercise 6.12 *Determine the solution of the following Cauchy problems*

a) $\dfrac{\partial u}{\partial t} = \dfrac{\partial^2 u}{\partial x^2} + 3t^2 \quad (x \in \mathbf{R}, \ t > 0), \qquad u(x,0) = \sin x \quad (x \in \mathbf{R}),$

b) $\dfrac{\partial u}{\partial t} = \Delta u + \cos t \quad (x \in \mathbf{R}^2, t > 0), \qquad v(x,0) = x_1 x_2 \exp\left(-x_1^2 - x_2^2\right) \ (x \in \mathbf{R}^2)$

c) $\dfrac{\partial u}{\partial t} = 3\Delta u + e^t \qquad (x \in \mathbf{R}^3, \ t > 0), \quad u(x,0) = \sin(x_1 - x_2 - x_3) \ (x \in \mathbf{R}^3).$

Answers.

a) $\qquad u(x,t) = t^3 + e^{-t} \sin x$

b) $\qquad u(x,t) = \sin t + \dfrac{x_1 x_2}{(1+4t)^3} \exp\left(\dfrac{x_1^2 + x_2^2}{1+4t}\right)$

c) $\qquad u(x,y,t) = e^t - 1 + e^{-9t} \sin(x_1 - x_2 - x_3).$

Exercise 6.13 *Determine the solution of the following problems*

a) $\dfrac{\partial u}{\partial t} = \dfrac{\partial^2 u}{\partial x^2} \qquad (x \in \mathbf{R}, \ t > 0), \qquad\qquad u(x,0) = xe^{-x^2} \qquad (x \in \mathbf{R})$

b) $\dfrac{\partial u}{\partial t} = \Delta u \qquad (x \in \mathbf{R}^2, \ t > 0), \qquad\qquad u(x,0) = \cos x_1 x_2 \quad (x \in \mathbf{R}^2),$

c) $\dfrac{\partial u}{\partial t} = a\Delta u \qquad (x \in \mathbf{R}^3, \ t > 0), \qquad\qquad u(x,0) = \cos(x_1 x_2) x_3 \quad (x \in \mathbf{R}^3).$

Answers.

a) $\qquad u(x,t) = x(1+4t)^{-3/2} \exp\left(\dfrac{-x^2}{1+4t}\right)$

b) $\qquad u(x_1, x_2, t) = (1+t^2)^{-1/2} \cos\dfrac{x_1 x_2}{1+t^2} \exp\left(\dfrac{-t(x_1^2 + x_2^2)}{2(1+t^2)}\right),$

c) $\qquad u(x_1, x_2, x_3, t) = \sin x_3 \cdot (1+4t^2)^{-1/2} \cos\dfrac{x_2 x_2}{1+4t^2} \exp\left(-t - t\dfrac{x_1^2 + x_2^2}{2(1+t^2)}\right).$

Exercise 6.14 *Determine the solution of the Dirichlet problem for parabolic equation*

$$t^2 \frac{\partial^2 u}{\partial x^2} - 2xt \frac{\partial^2 u}{\partial x \partial t} + x^2 \frac{\partial^2 u}{\partial t^2} - x \frac{\partial u}{\partial x} - tx \frac{\partial u}{\partial t} - u = 0,$$

$$u(x, -1) = u(x, 1) = x^2, \quad u(-1, t) = u(1, t) = t^2.$$

Answer. The given equation in polar coordinates (r, φ) has the form

$$\frac{\partial^2 U}{\partial \varphi^2} - U = 0,$$

and the solution can be written as

$$u(x, t) = 0, \quad \text{for} \quad x^2 + t^2 \leq 1,$$

$$u(x, t) = (x^2 + t^2 - 1) \left(\frac{\exp\left(\arcsin \dfrac{|x|}{\sqrt{x^2 + t^2}}\right) + \exp\left(\arccos \dfrac{|x|}{\sqrt{x^2 + t^2}}\right)}{\exp\left(\arcsin \dfrac{1}{\sqrt{x^2 + t^2}}\right) + \exp\left(\arccos \dfrac{1}{\sqrt{x^2 + t^2}}\right)} \right),$$

for $x^2 + t^2 > 1$.

Example 6.15 *Let*

$$u \in C(\overline{Q}), \quad \frac{\partial u}{\partial t}, \frac{\partial^2 u}{\partial x_i \partial x_j} \in C(Q).$$

Prove that u is determined uniquely in Q by the value of

$$\frac{\partial u}{\partial t} - \Delta u \text{ on } Q$$

and of u on the boundary $\partial' Q$ (see Preliminaries).

Solution. It is sufficient to consider the case

$$\frac{\partial u}{\partial t} - \Delta u = 0 \text{ on } Q$$

and $u|_{\partial' Q} = 0$. Then applying the Maximum Principle Theorem 6.1 on u and $-u$ we obtain

$$\max_{(x,t) \in \overline{Q}} u(x, t) = \max_{(x,t) \in \overline{Q}} (-u(x, t)) = 0.$$

Therefore $u = 0$ on $\overline{Q}$.

Exercise 6.16 (Widder) *Prove that if*

$$u \in C(\{(x,t)| \ x \in \mathbf{R}, 0 \leq t \leq T\}), \qquad \frac{\partial u}{\partial t}, \frac{\partial u}{\partial x}, \frac{\partial^2 u}{\partial x^2} \in C(V),$$

where $V = \{(x,t)| \ x \in \mathbf{R}, 0 < t < T\}$ and

$$\frac{\partial u}{\partial t} - \frac{\partial^2 u}{\partial x^2} = 0 \ on \ V,$$

$$u(x,0) = f(x) \quad (x \in \mathbf{R}), \qquad u(x,t) \geq 0 \quad ((x,t) \in V),$$

then $u = u(x,t)$ is determined uniquely on V and it is a real analytic function given by

$$u(x,t) = \frac{1}{(4\pi t)^{-n/2}} \int_{\mathbf{R}^n} \exp\left(-\frac{|s-x|^2}{4t}\right) f(s) \, ds.$$

Hints. We define for $a > 1$ the functions z^a and v^a in the following way

$$z^a(x) = \begin{cases} 1 & \text{for } |x| \leq a-1, \\ a - |x| & \text{for } a-1 < |x| < a, \\ 0 & \text{for } |x| \geq a, \end{cases}$$

and

$$v^a(x,t) = \frac{1}{(4\pi t)^{n/2}} \int_{\mathbf{R}^n} \exp\left(-\frac{|s-x|^2}{4t}\right) z^a(s) f(s) \, ds.$$

Using the Maximum Principle prove that for $\varepsilon > 0$ we have

$$v^a(x,t) \leq \varepsilon + u(x,t)$$

for $|x| < r$, $0 \leq t < T$, where

$$r > a + \frac{2Ma}{\varepsilon\sqrt{2\pi e}}$$

for $M = \max\limits_{|x| \leq a} f(x)$. Letting $r \to \infty$ and $\varepsilon \to 0$ prove that

$$0 \leq v(x,t) \leq u(x,t) \quad ((x,t) \in V),$$

where

$$v(x,t) = \lim_{a \to \infty} v^a(x,t) = \frac{1}{(4\pi t)^{n/2}} \int_{\mathbf{R}^n} \exp\left(-\frac{|s-x|^2}{4t}\right) f(s) \, ds.$$

Prove the regularity of $v = v(x,t)$ using v^a and that

$$\frac{\partial v}{\partial t} - \frac{\partial^2 v}{\partial x^2} = 0, \qquad v(x,0) = f(x).$$

Introduce a function h by $h = u - v$, and prove that

$$\frac{\partial h}{\partial t} - \frac{\partial^2 h}{\partial x^2} = 0 \qquad \text{on } V,$$

$$h(x, 0) = f(x), \quad (x \in \mathbf{R}), \qquad h(x, t) \geq 0 \quad ((x, t) \in V).$$

Then prove that $h = 0$, using the function H given by

$$H(x, t) = \int_0^t h(x, s)\, ds,$$

which has the same properties as h and which satisfies additionally the inequality

$$H(x, s) \leq \frac{\pi (t - s)^{1/2}}{x^2} \exp\left(\frac{x^2}{t - s}\right) H(0, t).$$

Prove by this inequality that $H = 0$.

6.2 Mixed Type Problem

6.2.1 Preliminaries

The *mixed* (initial and boundary) value problem (for $a > 0$ and $\ell > 0$) is given by the equation

$$\frac{\partial u}{\partial t} = a^2 \frac{\partial^2 u}{\partial x^2}, \quad (0 < x < \ell,\ t > 0),$$

with boundary conditions

$$u(0, t) = g_1(t), \qquad u(\ell, t) = g_2(t), \quad (t > 0),$$

and initial condition

$$u(x, 0) = f(x), \quad (0 < x < \ell),$$

where the function f is defined on $(0, \ell)$, g_1, g_2 are functions defined for $t > 0$. This is one dimensional heat equation (with two variables).

Remark. The solution of this equation $u(x, t)$ means the temperature distribution on a finite thin rod or wire of length ℓ. Usually it is supposed that this rod made of uniform material, has a uniform cross section and it coincides with a part of $x-$axis.

The homogeneous boundary ($g_1(x) = g_2(x) = 0$) conditions express the situation that the both ends of the rod are maintained all the time at the temperature 0^0C.

The initial conditions express that the initial temperature of the rod given by f depends on x namely on the distance from one end of the rod ($x = 0$).

The solution of this problem is given by applying Fourier method of separation of variables. It has the following form

$$u(x,t) = \sum_{n=1}^{\infty} C_n \exp\left(-t\left(\frac{\pi n a}{\ell}\right)^2\right) \sin(\frac{\pi n x}{\ell}),$$

where

$$C_n = \frac{2}{\ell} \int_0^\ell f(x) \sin\left(\frac{\pi n x}{\ell}\right) dx \quad (n \in \mathbf{N}).$$

6.2.2   Examples and Exercises

Example 6.17 *Find the solution of the problem*

$$\frac{\partial^2 u}{\partial x^2} - \frac{1}{a^2}\frac{\partial u}{\partial t} = 0 \qquad (0 < x < \ell, \ t > 0),$$

where a > 0 is a constant, with

(i) boundary conditions u(0,t) = 0,   u(\ell,t) = 0   t > 0,

(ii) initial conditions u(x,0) = f(x)   (0 < x < \ell).

Solution. Let us suppose that the solution of given equation has the form

$$u(x,y) = X(x) \cdot T(t),$$

then we obtain the problems

$$X''(x) + \lambda X(x) = 0, \quad X(0) = 0, \quad X(\ell) = 0, \tag{6.8}$$

and the equation

$$T'(t) + \lambda \cdot a^2 T(t) = 0. \tag{6.9}$$

The eigenvalues of the problem (6.8) is the considered problem are

$$\lambda = \lambda_n = \frac{n^2\pi^2}{\ell^2} \quad (n \in \mathbf{N}) \tag{6.10}$$

and the corresponding eigenfunctions have the forms

$$X_n(x) = \sin\frac{n\pi x}{\ell} \quad (n \in \mathbf{N}),$$

where we have taken $C_2 = 1$.

For λ given by (6.10) the solution of equation (6.9) has the form

$$T(t) = C_n \exp\left(-t\frac{a^2 n^2 \pi^2}{\ell^2}\right) \qquad (n \in \mathbf{N}),$$

where C_n are arbitrary constants. The solution of the considered problem is

$$u(x, y) = \sum_{n=1}^{\infty} C_n \sin \frac{n\pi x}{\ell} \cdot \exp\left(-t\frac{a^2 n^2 \pi^2}{\ell^2}\right).$$

This solution has to satisfy the initial condition and therefore for $t = 0$, we obtain the Fourier series

$$u(x, 0) = f(x) = \sum_{n=1}^{\infty} C_n \sin \frac{n\pi x}{\ell}.$$

The constants C_n can be found as the coefficients of Fourier series

$$C_n = \frac{2}{\ell} \int_0^{\ell} f(x) \sin \frac{n\pi x}{\ell} dx, \qquad (n \in \mathbf{N}).$$

Example 6.18 *Solve the following problem*

$$\frac{\partial u}{\partial t} = a^2 \frac{\partial^2 u}{\partial x^2} \qquad (0 < x < \pi,\ t > 0),$$

$$u(0, t) = 0, \qquad u(\pi, t) = 0 \qquad (t > 0),$$

$$u(x, 0) = T_0 \qquad (0 < x < \pi),$$

where T_0 is a constant.

Solution. By the method of separation of variables we obtain two differential equations

$$X'' + \lambda X = 0, \quad X(0) = 0, \quad X(\pi) = 0, \tag{6.11}$$

$$T' + \lambda a^2 T = 0. \tag{6.12}$$

Eigenvalues for (6.11) are $\lambda_n = n^2$ and eigenfunctions are $X_n(x) = \sin x$. Therefore solutions of the equation (6.12) are of the form

$$T_n(t) = A_n \exp(-a^2 n^2 t) \qquad (n \in \mathbf{N}).$$

Then the solution of the problem is of the form

$$u(x, t) = \sum_{n=1}^{\infty} A_n \sin(nx) \exp(-a^2 n^2 t).$$

By the initial condition

$$u(x, 0) = T_0 = \sum_{n=1}^{\infty} A_n \sin(nx),$$

therefore one can obtain coefficients A_n as coefficients of a Fourier's series

$$A_n = \frac{2}{\pi} \int_0^{\pi} T_0 \sin(nx) = \frac{2T_0}{n\pi}(1 - (-1)^n)$$

$$= \begin{cases} \dfrac{4T_0}{n\pi}, & n = 1, 3, 5, \ldots \\[2mm] 0 & n = 2, 4, 6, \ldots, \end{cases}$$

and the solution is

$$u(x, t) = \frac{4T_0}{\pi} \sum_{n=1}^{\infty} \frac{\sin((2n-1)x)}{2n-1} \exp(-a^2(2n-1)^2 t).$$

Exercise 6.19 *Determine the solution of the following problem*

$$\frac{\partial u}{\partial t} = a^2 \frac{\partial^2 u}{\partial x^2} \qquad (0 < x < \ell, \ t > 0),$$

$$u(0, t) = u(\ell, t) = 0 \quad (t > 0), \qquad u(x, 0) = f(x) \qquad (0 < x < \ell),$$

where

a) $f(x) = x$;

b) $f(x) = \begin{cases} x, & 0 < x \le \dfrac{\ell}{2} \\[2mm] \ell - x, & \dfrac{\ell}{2} \le x < \ell. \end{cases}$

Answer.

a) $u(x, t) = \dfrac{2\ell}{\pi} \sum\limits_{n=1}^{\infty} \dfrac{(-1)^n}{n} \cdot \sin \dfrac{n\pi x}{\ell} \cdot \exp\left(-\dfrac{a^2 \pi^2 n^2}{\ell^2} t\right) t.$

b) $u(x,t) = \dfrac{4\ell}{\pi^2} \displaystyle\sum_{n=0}^{\infty} \dfrac{(-1)^n}{(2n+1)^2} \cdot \sin \dfrac{(2n+1)\pi x}{\ell} \exp\left(-\dfrac{a^2\pi^2(2n+1)^2}{\ell}t\right).$

Exercise 6.20 *Solve the following problem*

$$\frac{\partial^2 u}{\partial x^2} = \frac{\partial u}{\partial t} + 4u - 20 \qquad (0 < x < \pi, \ t > 0),$$

$$u(0,t) = 5, \quad u(\pi,t) = 5 \qquad (t > 0)$$

$$u(x,0) = 5 + 2x \qquad (0 < x < \pi).$$

Hint. Introduce a new function z as

$$u(x,t) = 5 + z(x,t)e^{-4t}.$$

Then the considered problem becomes the problem appearing in Exercise 6.19 a) for $f(x) = 2x$. The solution is

$$u(x,t) = 5 + 4\sum_{n=1}^{\infty} \frac{(-1)^{n+1}}{n} \sin(nx)\exp(-(n^2+4)t).$$

Example 6.21 *Solve the problem*

$$\frac{\partial u}{\partial t} = a^2\frac{\partial^2 u}{\partial x^2} \qquad (0 < x < \pi, \ t > 0),$$

$$u(0,t) = T_1, \quad u(\pi,t) = T_2, \qquad (t > 0),$$

$$u(x,0) = x(\pi - x) \qquad (0 < x < \pi).$$

Solution. Let us introduce the functions S and v such that

$$u(x,t) = S(x) + v(x,t). \tag{6.13}$$

So, the considered problem can be written as

$$S''(x) + \frac{\partial^2 v(x,t)}{\partial x^2} = a^{-2}\frac{\partial v(x,t)}{\partial t} \qquad (0 < x < \pi, \ t > 0),$$

$$S(0) + v(0,t) = T_1, \quad S(\pi) + v(\pi,t) = T_2 \qquad (t > 0),$$

$$S(x) + v(x,0) = x(\pi - x) \qquad (0 < x < \pi).$$

Let us first solve the steady-state problem

$$S''(x) = 0, \qquad S(0) = T_1, \quad S(\pi) = T_2.$$

The solution of the previous problem is

$$S(x) = T_1 + (T_2 - T_1)\frac{x}{\pi}.$$

Now, we still have to solve the well known problem with homogeneous boundary conditions

$$\frac{\partial^2 v(x,t)}{\partial x^2} = a^{-2}\frac{\partial v(x,t)}{\partial t} \qquad (0 < x < \pi, \quad t > 0),$$

$$v(0,t) = 0, \qquad v(\pi,t) = 0 \qquad (t > 0),$$

$$v(x,0) = x(\pi - x) - S(x) = x(\pi - x) - T_1 - (T_2 - T_1)\frac{x}{\pi}.$$

The solution of this problem is

$$u(x,t) = \sum_{n=1}^{\infty} C_n \exp(-t(na)^2)\sin nx,$$

where

$$C_n = \frac{2}{\pi}\int_0^{\pi}(x(\pi - x) - T_1 - (T_2 - T_1)\frac{x}{\pi})\sin nx \, dx$$

$$= \frac{8}{(2n-1)^3\pi} + \frac{2}{n}((-1)^n T_2 - T_1) \qquad (n \in \mathbf{N}).$$

Therefore the solution is

$$u(x,t) = T_1 + (T_2 - T_1)\frac{x}{\pi}$$

$$+ \sum_{n=1}^{\infty}\left(\frac{8}{(2n-1)^3\pi} + \frac{2}{n}((-1)^n T_2 - T_1)\right)\exp(-t(na)^2)\sin 2nx.$$

Physically, the boundary conditions express that the temperature of the rod at the end $x = 0$ is maintained at value T_1 and the temperature at the end $x = \pi$ is maintained at value T_2.

Exercise 6.22 *Solve the problem*

$$\frac{\partial^2 u}{\partial x^2} = a^{-2} \frac{\partial u}{\partial t} \qquad (0 < x < 1,\ t > 0),$$

$$u(0, t) = 1 \qquad u(1, t) = 2 \qquad (t > 0)$$

$$u(x, 0) = 0 \qquad (0 < x < 1).$$

Answer.

$$u(x, t) = 1 + x + \frac{2}{\pi} \sum_{n=1}^{\infty} \left(\frac{2}{n} ((-1)^n - 1) \right) \exp(-t(na\pi)^2) \sin n\pi x.$$

Example 6.23 *Solve the following nonhomogeneous equation*

$$\frac{\partial^2 u}{\partial x^2} = \frac{\partial u}{\partial t} + u, \qquad (0 < x < 1,\ t > 0),$$

with homogeneous boundary

$$u(0, t) = 0, \qquad u(1, t) = 0 \qquad (t > 0),$$

and initial conditions

$$u(x, 0) = T_0 \qquad (0 < x < \pi),$$

where T_0 is a constant.

Solution. Let us introduce a new function v such that $u(x, t) = z(x, t)e^{-t}$. Then the above problem is equivalent with

$$\frac{\partial^2 z}{\partial x^2} = \frac{\partial z}{\partial t} \qquad (0 < x < 1,\ t > 0),$$

$$z(0, t) = 0, \qquad z(1, y) = 0 \qquad (t > 0),$$

$$z(x, 0) = T_0 \qquad (0 < x < \pi),$$

which solution is of the form

$$z(x, t) = \sum_{n=1}^{\infty} A_n \sin(n\pi x) \exp(-n^2 \pi^2 t),$$

where

$$A_n = \begin{cases} \dfrac{4T_0}{n\pi}, & n = 1, 3, 5, \ldots \\[2ex] 0, & n = 2, 4, 6 \ldots, \end{cases}$$

The solution is

$$u(x, t) = \frac{4T_0}{\pi} \sum_{n=1}^{\infty} \frac{1}{2n - 1} \sin((2n - 1)\pi x) \exp(-(1 + (2n - 1)^2 \pi^2)t).$$

Example 6.24 *Determine the solution of the following problem*

$$\frac{\partial^2 u}{\partial x^2} = a^{-2}\frac{\partial u}{\partial t} \qquad (0 < x < \ell, \ t > 0)$$

$$\frac{\partial u(0,t)}{\partial x} = 0, \qquad \frac{\partial u(\ell,t)}{\partial x} = 0 \qquad (t > 0),$$

$$u(x,0) = x \quad (0 < x < \ell).$$

Solution. Using the method of separation variables we obtain the following problem

$$X'' + \lambda X = 0, \qquad X'(0) = 0, \qquad X'(\ell) = 0,$$

whose eigenfunctions are

$$X_n(x) = \cos\frac{n\pi x}{\ell}.$$

Therefore the solution in of this problem can be written in the form

$$u(x,t) = \frac{A_0}{2} + \sum_{n=1}^{\infty} A_n \exp\left(-t\frac{\pi n a}{\ell^2}\right)\cos\left(\frac{\pi n x}{\ell}\right) \qquad (0 < x < \ell, \ 0 < t < \infty),$$

where the coefficients are obtained from the initial condition as the coefficients of corresponding Fourier series and have the forms

$$A_0 = \ell$$

$$A_n = \frac{2}{\ell}\int_0^\ell x\cos\left(\frac{\pi n x}{\ell}\right)\,dx = \frac{2}{\pi^2 n^2}(\cos n\pi - 1), \quad n = 1,\ldots$$

or

$$A_n = \frac{2}{\pi^2 n^2}((-1)^n - 1) = \begin{cases} -\dfrac{4\ell}{\pi^2 n^2}, & n = 1,3,5,\ldots \\[2ex] 0, & n = 2,4,\ldots. \end{cases}$$

The solution of the considered problem has the form

$$u(x,t) = \frac{\ell}{2} + \frac{4l}{\pi^2}\sum_{n=1}^{\infty}\frac{1}{(2n-1)^2}\exp\left(-t\left(\frac{\pi(2n-1)a}{\ell}\right)^2\right)\cos\left(\frac{\pi(2n-1)x}{\ell}\right),$$

for $0 < x < \ell, \ 0 < t < \infty.$

Exercise 6.25 *Determine the solution of the following problems*

$$\text{a)} \quad \frac{\partial u}{\partial t} = \frac{\partial^2 u}{\partial x^2} \quad (0 < x < 1, t > 0),$$

$$\frac{\partial u(0, t)}{\partial x} = 0, \ u(1, t) = 0 \quad (t > 0),$$

$$u(x, 0) = x^2 - 1 \quad (0 < x < 1).$$

$$\text{b)} \quad \frac{\partial u}{\partial t} = \frac{\partial^2 u}{\partial x^2} - u \quad (0 < x < \ell, t > 0),$$

$$\frac{\partial u(0, t)}{\partial x} = 0, \ u(\ell, t) = 0 \quad (t > 0),$$

$$u(x, 0) = 1 \quad (0 < x < \ell).$$

Answers.

$$\text{a)} \quad \frac{32}{\pi^3} \sum_{n=0}^{\infty} \frac{(-1)^n}{(2n+1)^3} \exp\left(-\left(\frac{(2n+1)\pi}{2}\right)^2 t\right) \cos\frac{2n+1}{2}\pi x.$$

$$\text{b)} \quad \frac{4}{\pi} \sum_{n=0}^{\infty} \frac{1}{2n+1} \exp\left(-\left(\frac{\pi^2(2n+1)^2}{\ell} + 1\right) t\right) \sin\frac{(2n-1)}{\ell}\pi x.$$

Example 6.26 *Solve the problem*

$$a^2 \frac{\partial^2 u}{\partial x^2} = \frac{\partial u}{\partial t} \quad (0 < x < \ell, \ t > 0)$$

$$\frac{\partial u(0, t)}{\partial x} = 0, \quad u(\ell, t) = u_0 \quad (t > 0),$$
$$u(x, 0) = f(x), \quad \lim_{x \to \ell^-} f(x) = u_0 \quad (0 < x < \ell),$$

where u_0 is a given constant.

Solution. Let us write the solution in the form

$$u(x, t) = u_0 + v(x, t),$$

such that v is the solution of the problem

$$a^2 \frac{\partial^2 v}{\partial x^2} = \frac{\partial v}{\partial t}, \quad \frac{\partial v(0, t)}{\partial x} = 0, \quad v(\ell, t) = 0,$$

$$v(x,0) = f(x) - u_0.$$

The solution of the previous problem is

$$v(x,t) \sum_{n=0}^{\infty} A_n \cos\left(\frac{2n+1}{2\ell} x\pi\right) \exp\left(-t\left(\frac{(2n+1)\pi na}{2\ell}\right)^2\right),$$

where

$$A_n = \frac{2}{\ell} \int_0^{\ell} f(x) \cos\left(\frac{2n+1}{2\ell} x\pi\right) dx - \frac{4(-1)^n u_0}{\pi(2n+1)} \quad (n = 0, 1, \ldots).$$

Exercise 6.27 *Solve the following problem*

$$a^2 \frac{\partial^2 u}{\partial x^2} = \frac{\partial u}{\partial t} \quad (0 < x < \ell,\ t > 0),$$

$$\frac{\partial u(0,t)}{\partial x} = q_0, \quad u(\ell,t) = u_0, \quad (t > 0),$$

$$u(x,0) = f(x) \quad (0 < x < \ell),$$

where u_0 and q_0 are arbitrary constants.

Answer. The solution has the form

$$u(x,t) = q_0 x + u_0 + \sum_{n=0}^{\infty} A_n \sin\left(\frac{2n+1}{2\ell} x\pi\right) \exp\left(-t\left(\frac{(2n+1)\pi a}{2\ell}\right)^2\right),$$

where

$$A_n = \frac{2}{\ell} \int_0^{\ell} f(x) \sin\left(\frac{2n+1}{2\ell} x\pi\right) dx - \frac{4}{\pi^2} \frac{(2n+1)\pi u_0 + \ell q_0}{(2n+1)^2}, \quad n = 0, 1, \ldots.$$

Exercise 6.28 *Solve the following problem*

$$a^2 \frac{\partial^2 u}{\partial x^2} = \frac{\partial u}{\partial t} \quad (0 < x < \ell,\ t > 0),$$

$$\frac{\partial u(0,t)}{\partial x} = 0, \quad \frac{\partial u(\ell,t)}{\partial x} = q_0 \quad (t > 0)$$

$$u(x,0) = 0 \quad (0 < x < \ell),$$

where q_0 is an arbitrary constant.

Answer. The solution has the form

$$u(x,t) = q_0 \left(\frac{a^2 t}{\ell} + \frac{3x^2 - \ell^2}{6\ell} + \frac{2\ell}{\pi^2} \sum_{n=0}^{\infty} \frac{(-1)^{n+1}}{n^2} \cos\left(\frac{n\pi x}{\ell}\right) \exp\left(-t \left(\frac{n\pi a}{\ell}\right)^2\right) \right).$$

Example 6.29 *Solve the following problem*

$$a^2 \frac{\partial^2 u}{\partial x^2} = \frac{\partial u}{\partial t} \quad (0 < x < \ell, \ t > 0),$$

$$\frac{\partial u(0,t)}{\partial x} = 0, \quad K_1 \frac{\partial u(\ell,t)}{\partial x} + K_2 u(\ell,t) = 0 \quad (t > 0),$$

$$u(x,0) = f(x) \quad (0 < x < \ell).$$

The values of of K_1 and K_2 are given.

Solution. The boundary conditions for this heat equation express that for time $t = 0$, at the end of the rod $x = 0$ there is no heat flow and the heat is exchanged at the other end with an environment at temperature $0^0 C$.

The method of separation variables bring us to the problem

$$X''(x) + \lambda X(x) = 0 \quad (0 < x < \ell), \qquad X'(0) = 0, \qquad K_1 X'(\ell) + K_2 X(\ell) = 0.$$

and to equation $T' - \lambda a^2 T = 0$ $(t > 0)$.

In order to obtain nontrivial solution of this equation the eigenvalues must be positive i.e., $\lambda = k^2$. Using the boundary conditions we obtain

$$\tan k\ell = \frac{K_2}{K_1 \lambda},$$

wherefrom we obtain the values for the eigenfunctions λ_n. The corresponding eigenfunctions (it may differ only in a constant and we take it to be 1) are of the form

$$X_n = \cos \lambda_n x \quad (n \in \mathbf{N})$$

(they may differ only in a constant and we put it to be 1).

For these eigenvalues the solution of differential equation

$$T' + a^2 \lambda T = 0 \quad (t > 0),$$

is

$$T(t) = C_n \exp(-a\lambda_n t).$$

The final solution of the considered problem is given by

$$u(x,t) = \sum_{n=1}^{\infty} C_n \cos(\lambda_n x) \exp(-a\lambda_n t),$$

where the coefficients are of the form

$$C_n = \int_0^\ell f(x) \cos(\lambda_n x) dx.$$

Exercise 6.30 *Solve the following problem*

$$a^2 \frac{\partial^2 u}{\partial x^2} = \frac{\partial u}{\partial t} \qquad (0 < x < 1,\ t > 0),$$

$$\frac{\partial u(0, t)}{\partial x} = T_1, \qquad \frac{\partial u(1, t)}{\partial x} + u(1, t) = T_2 \qquad (t > 0),$$

$$u(x, 0) = T_1 \qquad (0 < x < 1).$$

Answer. The solution can be written as

$$u(x, t) = T_1 + (T_2 - T_1) \left(\frac{1}{2} x + 2 \sum_{n=1}^{\infty} \frac{\cos k_n \sin k_n x}{k_n(1 + \cos^2 k_n)} \exp\left(-a^2 k_n^2\right) \right),$$

where k_n is the solution of equation

$$\sin k_n + k_n \cos k_n = 0 \qquad (n \in \mathbf{N}).$$

Exercise 6.31 *Determine the solution of equation*

$$\frac{\partial^2 u}{\partial x^2} = \frac{\partial u}{\partial t} + Bu \qquad (0 < x < \ell,\ t > 0),$$

$$\frac{\partial u(0, t)}{\partial x} - hu(0, t) = u(\ell, t) = 0 \qquad (t > 0),$$

$$u(x, 0) = u_0 \qquad (0 < x < \ell).$$

where $h > 0$ is a given constant.

Answer. The function $v(x, t)$, introducing as $u(x, t) = e^{-Bt} v(x, t)$, satisfies the equation from previous example with $a = 1$. The solution has the form

$$u(x, t) = 2u_0 \sum_{n=1}^{\infty} \frac{h - (-1)^n \sqrt{h^2 + \lambda_n^2}}{\lambda_n(\ell(h^2 + \lambda_n^2) + h)} \exp(-ta^2 \lambda_n^2 + Bt) \left(\lambda_n \cos(\lambda_n x) + h \sin(\lambda_n x)\right),$$

where the eigenvalues λ_n are the positive solutions of the equation

$$h \tan \lambda \ell = -\lambda.$$

Example 6.32 *Solve the following mixed problem for the heat equation for $a > 0$ and $\ell > 0$*

$$\frac{\partial u}{\partial t} = a^2 \frac{\partial^2 u}{\partial x^2} \qquad (0 < x < \ell,\ t > 0), \tag{6.14}$$

$$u(x,0) = f(x) \qquad (0 < x < \ell), \tag{6.15}$$

$$\frac{\partial u(0,t)}{\partial x} - h_1(u(0,t) - u_1) = 0 \qquad (t > 0), \tag{6.16}$$

$$\frac{\partial u(\ell,t)}{\partial x} - h_2(u(\ell,t) - u_2) = 0 \qquad (t > 0), \tag{6.17}$$

where $h_i > 0$ and $u_i > 0$ are given constants.

Solution. We try to find the solution of the given mixed problem in the following form

$$u(x,t) = v(x) + w(x,t),$$

where v is a solution of (6.14) which satisfies the boundary conditions (6.16) and (6.17), i.e., v is the solution of the equation $v''(x) = 0$ which is of the form

$$v(x) = C_1 x + C_2,$$

where

$$C_1 = \frac{h_1(u_2 - u_1)}{h_1 + h_2 + h_1 h_2 \ell}, \qquad C_2 = u_1 + \frac{C_1}{h_1}.$$

The function w satisfies the equation (6.14) and the initial condition (6.15), i.e.,

$$w(x,0) = u(x,0) - v(x) = f(x) - v(x) \tag{6.18}$$

and the homogeneous boundary conditions

$$\frac{\partial w(0,t)}{\partial x} - h_1 w(0,t) = 0 \qquad (t > 0), \tag{6.19}$$

$$\frac{\partial w(\ell,t)}{\partial x} - h_2 w(\ell,t) = 0 \qquad (t > 0). \tag{6.20}$$

Applying the Fourier method of separation of variables on the problem (6.14), (6.18), (6.19), (6.20) we obtain

$$w_n(x,t) = C_n e^{-a^2 \mu_n^2 t/\ell^2} \left(\frac{\mu_n}{\ell} \cos \frac{\mu_n}{\ell} x + h_1 \sin \frac{\mu_n}{\ell} x \right),$$

where μ_n are the positive solutions of the following equation

$$\cot \mu = \frac{1}{\ell(h_1 + h_2)} \left(\mu - \frac{h_1 h_2 \ell^2}{\mu} \right).$$

Then the series

$$w(x,t) = \sum_{n=1}^{\infty} C_n e^{-a^2 \mu_n^2 t / \ell^2} \left(\frac{\mu_n}{\ell} \cos \frac{\mu_n}{\ell} x + h_1 \sin \frac{\mu_n}{\ell} x \right)$$

satisfies the equation (6.14) and the boundary conditions (6.16) and (6.17). By the orthogonality of the system of functions on the interval $[0, \ell]$

$$\left\{ \frac{\mu_n}{\ell} \cos \frac{\mu_n}{\ell} x + h_1 \sin \frac{\mu_n}{\ell} x \right\}_{n \in \mathbf{N}}$$

and the initial condition (6.18) we obtain

$$C_n = \left(\int_0^{\ell} \left(\frac{\mu_n}{\ell} \cos \frac{\mu_n}{\ell} x + h_1 \sin \frac{\mu_n}{\ell} x \right)^2 dx \right)^{-1}$$

$$\cdot \int_0^{\ell} (f(x) - v(x)) \left(\frac{\mu_n}{\ell} \cos \frac{\mu_n}{\ell} x + h_1 \sin \frac{\mu_n}{\ell} x \right) dx \quad (n \in \mathbf{N}).$$

Example 6.33 *Solve the nonhomogeneous problem*

$$\frac{\partial^2 u}{\partial x^2} = \frac{\partial u}{\partial t} - 2e^{3x} \quad \left(0 < x < \frac{1}{2}, \ t > 0 \right),$$

$$u(0,t) = 0, \quad u \left(\frac{1}{2}, t \right) = 0 \quad (t > 0), \tag{6.21}$$

$$u(x,0) = \frac{2}{9} \left(1 - e^{3x} \right) \quad \left(0 < x < \frac{1}{2} \right).$$

Solution. Taking $u(x,t) = S(x) + v(x,t)$ the problem (6.21) can be written as

$$S'' + \frac{\partial^2 v}{\partial x^2} = \frac{\partial v}{\partial t} - 2e^{3x}, \quad \left(0 < x < \frac{1}{2}, \ t > 0 \right),$$

$$S(0) + v(0,t) = 0, \quad S \left(\frac{1}{2} \right) + v \left(\frac{1}{2}, 0 \right) = 0 \quad (t > 0),$$

$$S(x) + v(x,0) = \frac{2}{9} \left(1 - e^{3x} \right) \quad \left(0 < x < \frac{1}{2} \right).$$

The solution of the problem

$$S''(x) = -2e^{3x} \quad \left(0 < x < \frac{1}{2} \right), \qquad S(0) = 0, \quad S(1) = 0,$$

is

$$S(x) = \frac{2}{9}\left(1 - e^{3x}\right) - \frac{2}{9}\left(1 - e^3\right)x.$$

The solution of second part of the considered problem, i.e., of the problem

$$\frac{\partial^2 v}{\partial x^2} = \frac{\partial v}{\partial t} \qquad \left(0 < x < \frac{1}{2}, \ t > 0\right),$$

$$v(0,t) = 0, \qquad v\left(\frac{1}{2}, t\right) = 0 \qquad (t > 0), \tag{6.22}$$

$$v(x,0) = \frac{2}{9}\left(1 - e^3\right)x \qquad \left(0 < x < \frac{1}{2}\right),$$

is

$$v(x,t) = \sum_{n=1}^{\infty} C_n \sin(2\pi n x) \exp(-2tn^2\pi^2).$$

Using the boundary conditions we obtain

$$C_n = \frac{4}{9}\left(1 - e^3\right)x \int_{0}^{1/2} x \sin(n\pi x)dx = \frac{4(1 - e^3)(-1)^n}{9n\pi} \qquad (n \in \mathbf{N}).$$

The solution of our problem has the form

$$u(x,t) = \frac{2}{9}\left(1 - e^{3x}\right) - \frac{4}{9}\left(1 - e^3\right)x + \frac{4(1 - e^3)}{9\pi}\sum_{n=1}^{\infty}\frac{(-1)^n \sin(2\pi n x)}{n}\exp(-2tn^2\pi^2).$$

Example 6.34 *Solve the nonhomogeneous problem*

$$\frac{\partial u}{\partial t} - a^2\frac{\partial^2 u}{\partial x^2} = r(x,t) \qquad (0 < x < \ell, \ t > 0),$$

$$u(0,t) = 0, \qquad u(\ell,t) = 0 \qquad (t > 0), \tag{6.23}$$

$$u(x,0) = f(x) \qquad (0 < x < \ell).$$

Solution. Let us suppose that the solution of the problem (6.23) has the form of generalized Fourier series

$$u(x,t) = \sum_{n=1}^{\infty} T_n(t)X_n(x), \tag{6.24}$$

where X_n are eigenfunctions of the Sturm-Liuoville problem $X'' + \lambda X = 0$, with associated homogeneous boundary conditions, with eigenvalues λ_n.

The right-hand side function $r = r(x,t)$ can be expanded also in the generalized Fourier series as

$$r(x,t) = \sum_{n=1}^{\infty} r_n(t) X_n(x),$$

where the coefficients r_n can be written

$$r_n(t) = ||X_n(x)||^{-2} \int_0^{\ell} r(x,t) X_n(x).$$

Assuming that termwise differentiation of the series (6.24) is permitted we obtain

$$\frac{\partial u}{\partial t} = \sum_{n=1}^{\infty} T_n'(t) X_n(t),$$

$$\frac{\partial^2 u}{\partial x^2} = \sum_{n=1}^{\infty} T_n(t) X_n''(x) = -\sum_{n=1}^{\infty} \lambda_n T_n(t) X_n(x).$$

The partial differential equation from (6.23) can be written as

$$\sum_{n=1}^{\infty} \left(T_n'(t) + a^2 \lambda_n T_n(t) \right) X_n(x) = \sum_{n=1}^{\infty} r_n(t) X_n(x).$$

So we obtain the differential equation

$$T_n'(t) + a^2 \lambda_n T_n(t) = r_n(t), \tag{6.25}$$

with the solution

$$T_n(t) = \left(C_n + \int_0^t \exp(a^2 \lambda_n \tau) r_n(\tau) d\tau \right) \exp(-a^2 \lambda_n t).$$

If $t = 0$, then $T_n(0) = C_n$. Taking the initial condition as

$$f(x) = \sum_{n=1}^{\infty} f_n X_n(x),$$

where f_n are the coefficients of the Fourier series i.e.,

$$f_n = ||X_n||^{-2} \int_0^{\ell} f(x) X_n(x) dx,$$

we can say that

$$T_n(0) = f_n = C_n.$$

The solution of equation (6.25), with corresponding condition can be written as

$$T_n(t) = f_n \exp(-\lambda_n a^2 t) + \int_0^t r_n(\tau) \exp(-\lambda_n a^2 (t - \tau)) d\tau.$$

Example 6.35 *Find the solution of the following problem*

$$\frac{\partial u}{\partial t} - \frac{\partial^2 u}{\partial x^2} = \frac{x}{2}\cos t \qquad (0 < x < 1, \ t > 0),$$

$$u(0,t) = 0, \qquad u(1,t) = 0 \qquad (t > 0), \qquad (6.26)$$

$$u(x,0) = 0 \qquad (0 < x < 1).$$

Solution. The eigenvalues for the corresponding Sturm-Liuoville problem are $\lambda_n = n^2\pi^2$ and eigenfunctions are $X_n = \sin n\pi x$ $(n \in \mathbf{N})$. In this case we can write

$$\frac{x}{2}\cos t = \sum_{n=1}^{\infty} T_n(t)\sin n\pi x,$$

where

$$r_n(t) = \|\sin n\pi x\|^{-2} \int_0^1 \frac{x}{2}\cos t \sin n\pi x \, dx = \frac{(-1)^n}{n\pi}\cos t.$$

Let us consider the solution in form

$$u(x,t) = \sum_{n=1}^{\infty} T_n(t)\sin n\pi x.$$

After termwise differentiation we obtain differential equation

$$T_n'(t) + n^2\pi^2 T_n(t) = \frac{(-1)^n}{n\pi}\cos t,$$

whose solution is

$$T_n(t) = \left(C_n + \frac{(-1)^n}{n\pi}\int_0^t \exp(n^2\pi^2\tau)\cos \tau \, d\tau\right)\exp(-n^2\pi^2 t)$$

$$= C_n\exp(-n^2\pi^2 t) + \frac{(-1)^n}{n\pi(1+n^4\pi^4)}(n^2\pi^2(\exp(-n^2\pi^2 t) - \cos t) - \sin t).$$

From $u(x,0) = 0$, it follows that $C_n = 0$ and the solution of the considered problem is

$$u(x,t) = \sum_{n=1}^{\infty} \frac{(-1)^n}{n\pi(1+n^4\pi^4)}(n^2\pi^2(\exp(-n^2\pi^2 t) - \cos t) - \sin t)\sin n\pi x.$$

Example 6.36 *Find the solution*

$$\frac{\partial u}{\partial t} - a^2\frac{\partial^2 u}{\partial x^2} = r(x,t) \qquad (0 < x < \ell,\ t > 0),$$

$$u(0,t) = A(t), \qquad u(\ell,t) = B(t) \qquad (t > 0), \tag{6.27}$$

$$u(x,0) = f(x) \qquad (0 < x < \ell).$$

Solution. Let us introduce the function v as $v(x,t) = u(x,t) - V(x,t)$, where

$$V(x,t) = A(t) + \frac{x}{\ell}(B(t) - A(T)),$$

satisfying the conditions $V(0,t) = A(t), \quad V(\ell,t) = B(t) \quad (t > 0)$.
The function v then satisfies

$$\frac{\partial v}{\partial t} - a^2\frac{\partial^2 v}{\partial x^2} = r(x,t) - \frac{\partial V}{\partial t} \qquad (0 < x < \ell,\ t > 0),$$

$$v(0,t) = 0, \qquad v(\ell,0) = 0 \qquad (t > 0), \tag{6.28}$$

$$u(x,0) = f(x) - V(x,0) \qquad (0 < x < \ell).$$

The function $v(x,t)$ can be found as in the Example 6.34.

Example 6.37 *Find the solution of the following problem*

$$a^2\frac{\partial^2 u}{\partial x^2} = \frac{\partial u}{\partial t} \qquad (0 < x < \ell,\ t > 0),$$

$$u(0,t) = a_0 + a_1 t, \qquad u(\ell,t) = b_0 + b_1 t \qquad (t > 0), \tag{6.29}$$

$$u(x,0) = 0 \qquad (0 < x < \ell).$$

Solution. Taking $v(x,t) = u(x,t) - V(x,t)$, where

$$V(x,t) = a_0 + a_1 t + \frac{x}{\ell}(b_0 - a_0 + (b_1 - a_1)t),$$

we obtain that the function v satisfies the nonhomogeneous equation

$$\frac{\partial v}{\partial t} - a^2\frac{\partial^2 v}{\partial x^2} = -\left(a_1 + \frac{x}{\ell}\right)(b_1 - a_1),$$

with initial condition

$$v(x,0) = -\left(a_0 + \frac{x}{\ell}(b_0 - a_0)\right),$$

and homogeneous boundary conditions

$$v(0,t) = 0, \qquad v(\ell,t) = 0.$$

The solution of previous problem is treated as the generalized Fourier series

$$v(x,t) = \sum_{n=1}^{\infty} v_n(t) \sin \frac{n\pi x}{\ell},$$

where $\sin \dfrac{n\pi x}{\ell}$ are eigenfunctions for eigenvalues $\lambda_n = \dfrac{n^2\pi^2}{\ell}$ for the corresponding Sturm-Liuoville problem.

The right-hand side function $-\left(a_1 + \dfrac{x}{\ell}(b_1 - a_1)\right)$ can be expanded also in the generalized Fourier series as

$$-\left(a_1 + \frac{x}{\ell}(b_1 - a_1)\right) = \sum_{n=1}^{\infty} r_n(t) \sin \frac{n\pi x}{\ell},$$

where the coefficients r_n can be written

$$
\begin{aligned}
r_n(t) &= \ ||\sin \frac{n\pi x}{\ell}||^{-2} \int_0^\ell \left(a_1 + \frac{x}{\ell}(b_1 - a_1)\right) \sin \frac{n\pi x}{\ell} \\[2mm]
&= \ \frac{2}{\pi} \left(\frac{a_1}{n}(1 - (-1)^n) + \frac{b_1 - a_1}{n}(-1)^{n+1}\right).
\end{aligned}
$$

Using differential equation (6.25) we have

$$v_n'(t) + a^2 \frac{n^2\pi^2}{\ell} v_n(t) = r_n(t), \tag{6.30}$$

with the initial conditions

$$v_n(0) = f_n,$$

where f_n are the coefficients of the Fourier series

$$-\left(a_0 + \frac{x}{\ell}(b_0 - a_0)\right) = \sum_{n=1}^{\infty} f_n \sin \frac{n\pi x}{\ell},$$

and have the form

$$
\begin{aligned}
f_n &= \ -||\sin \frac{n\pi x}{\ell}||^{-2} \int_0^\ell \left(a_0 + \frac{x}{\ell}(b_0 - a_0)\right) \sin \frac{n\pi x}{\ell} \, dx, \\[2mm]
&= \ \frac{2}{n\pi} \left(a_0((-1)^n - 1) - (b_0 - a_0)(-1)^{n+1}\right).
\end{aligned}
$$

The solution of equation (6.30), with corresponding condition is

$$v_n(t) = \frac{2}{n\pi}\left(a_0((-1)^n - 1) - (b_0 - a_0)(-1)^{n+1}\right)\exp\left(-\frac{n^2\pi^2}{\ell}a^2 t\right)$$

$$+\int_0^t \frac{2}{\pi}\left(\frac{a_1}{n}(1 - (-1)^n) + \frac{b_1 - a_1}{n}(-1)^{n+1}\right)\exp\left(-\frac{n^2\pi^2}{\ell}a^2(t - \tau)\right)d\tau$$

$$= \frac{1}{\pi n}\left(a_0((-1)^n - 1) - (b_0 - a_0)(-1)^{n+1}n\right)\exp\left(-\frac{n^2\pi^2}{\ell}a^2 t\right)$$

$$+\frac{2}{\pi}\left(\frac{a_1}{n}(1 - (-1)^n) + \frac{b_1 - a_1}{n}(-1)^{n+1}\right)\left(1 - \exp\left(-\frac{n^2\pi^2}{\ell}a^2 t\right)\right)$$

The solution of the problem (6.29) is

$$u(x,t) = a_0 + a_1 t + \frac{x}{\ell}(b_0 - a_0 + (b_1 - a_1)t)$$

$$+\frac{2}{\pi}\sum_{n=1}^{\infty}\frac{1}{n}(a_0((-1)^n - 1) - (b_0 - a_0)(-1)^n)\exp\left(-\frac{n^2\pi^2}{\ell}a^2 t\right)$$

$$+\frac{2}{a^2\pi^2}\sum_{n=1}^{\infty}\frac{1}{n^2}\left(\frac{a_1}{n}(1 - (-1)^n) + \frac{b_1 - a_1}{n}(-1)^{n+1}\right)\left(1 - \exp\left(-\frac{n^2\pi^2}{\ell}a^2 t\right)\right).$$

Example 6.38 *Find the solution of the following problem*

$$\frac{\partial^2 u}{\partial x^2} = \frac{\partial u}{\partial t}, \qquad (0 < x < 1, \quad t > 0),$$

$$u(0,t) = 2t + \sin t, \qquad u(1,t) = 2t \qquad (t > 0), \qquad (6.31)$$

$$u(x,0) = 0 \qquad (0 < x < 1).$$

Solution. Taking $v(x,t) = u(x,t) - V(x,t)$, where $V(x,t) = 2t + (1 - x)\sin t$, we obtain that the function v satisfies the nonhomogeneous equation

$$\frac{\partial v}{\partial t} - \frac{\partial^2 v}{\partial x^2} = -(2 + (1 - x)\cos t) \qquad (0 < x < 1, \ t > 0)$$

$$(6.32)$$

$$v(x,0) = 0 \qquad (0 \le x \le 1), \quad v(0,t) = 0, \qquad v(1,t) = 0 \qquad (t > 0).$$

The solution of this problem is treated as the generalized Fourier series

$$v(x,t) = \sum_{n=1}^{\infty} v_n(t) \sin n\pi x,$$

where $\sin n\pi$ are eigenfunctions for eigenvalues $\lambda_n = n^2\pi^2$, for the corresponding Sturm-Liuoville problem.

The function $-(2 + (1-x)\cos t)$ can be expanded also in the generalized Fourier series as

$$-(2 + (1-x)\cos t) = \sum_{n=1}^{\infty} r_n(t) \sin n\pi x,$$

where the coefficients $r_n(t)$ can be written

$$r_n(t) = -\|\sin n\pi x\|^{-2} \int_0^1 (2 + (1-x)\cos t) \sin n\pi x dx$$

$$= -\frac{2}{n\pi}(\cos t + 2(1-(-1)^n)), \qquad n = 1,2,3,\ldots.$$

From the partial differential equation we have

$$v_n'(t) + n^2\pi^2 v_n(t) = r_n(t),$$

with the initial conditions

$$u_n(0) = 0.$$

Therefore the solution of the problem (6.32) has the form

$$v_n(t) = -\int_0^t \frac{2}{n\pi}(\cos\tau + 2(1-(-1)^n))\exp\left(n^2\pi^2(t-\tau)\right)d\tau$$

$$= \frac{2}{n\pi(1+n^4\pi^4)}\left(n^2\pi^2(\exp\left(-n^2\pi^2 t\right) - \cos) - \sin t\right) - \frac{4(\exp(n^2\pi^2) - 1)}{n^3\pi^3}.$$

The solution of our problem is

$$u(x,t) = \frac{1}{\pi}\sum_{n=1}^{\infty}\left(\frac{2}{n\pi(1+n^4\pi^4)}\left(n^2\pi^2(\exp\left(-n^2\pi^2 t\right) - \cos) - \sin t\right)\right.$$

$$\left. +2t + \sin t - x\sin t - \frac{4(\exp(n^2\pi^2) - 1)}{n^3\pi^3}\right).$$

Example 6.39 *Find the solution*

$$\frac{\partial^2 u}{\partial x^2} = \frac{\partial u}{\partial t} + 1 - 3xt^2 + x \qquad (0 < x < 1,\ t > 0),$$

$$u(0,t) = t, \qquad u(1,t) = t^3 \qquad (t > 0), \tag{6.33}$$

$$u(x,0) = x^2 \qquad (0 < x < 1).$$

Solution. Taking $u(x,t) = V(x,t) + W(x,t)$, we obtain the following problem

$$\frac{\partial^2 V}{\partial x^2} + \frac{\partial^2 W}{\partial x^2} = \frac{\partial V}{\partial t} + \frac{\partial W}{\partial t} + 1 - 3xt^2 + x \qquad (0 < x < 1,\ t > 0),$$

$$V(0,t) + W(0,t) = t, \qquad V(1,t) + W(1,t) = t^3 \qquad (t > 0), \tag{6.34}$$

$$V(x,0) + W(x,0) = x^2 \qquad (0 < x < 1).$$

The function

$$V(x,t) = t + (t^3 - t)x, \tag{6.35}$$

satisfies the conditions $V(0,t) = t$, $V(1,t) = t^3$, $V(x,0) = 0$. Therefore, we have to solve the nonhomogeneous problem

$$\frac{\partial^2 W}{\partial x^2} = \frac{\partial W}{\partial t} + 2 \qquad (0 < x < 1,\ t > 0),$$

$$W(0,t) = 0, \qquad W(1,t) = 0 \qquad (t > 0), \tag{6.36}$$

$$W(x,0) = x^2 \qquad (0 < x < 1).$$

The function $W = W(x,t)$ can be found as in previous two examples. The second way is to introduce two functions S and v such that $W(x,t) = S(x) + v(x,t)$, where $S(x)$ satisfies the following problem

$$S''(x) = 2, \quad S(0) = 0, \quad S(1) = 0. \tag{6.37}$$

The solution of this problem is $S(x) = x^2 - x$.

We still have to find the function v satisfying the equation

$$\frac{\partial^2 v}{\partial x^2} - \frac{\partial v}{\partial t} = 0 \qquad (0 < x < 1,\ t > 0),$$

and the conditions

$$v(0,t) = 0, \qquad v(1,t) = 0 \qquad (t > 0), \qquad v(x,0) = x \qquad (0 < x < 1).$$

The solution of this problem is

$$v(x,t) = \sum_{n=1}^{\infty} A_n \sin(n\pi x) \exp(n^2\pi^2 t),$$

where

$$A_n = 2 \int_0^1 x \sin(n\pi x)\, dx = \frac{2}{n\pi}(-1)^{n+1}.$$

The solution of the considered problem is

$$u(x,t) = t + (t^3 - t)x + x^2 - x + \frac{2}{\pi}\sum_{n=1}^{n} \frac{(-1)^{n+1}}{n} \sin(n\pi x) \exp(n^2\pi^2 t).$$

Example 6.40 *Determine the solution of two dimensional heat conduction problem*

$$\frac{\partial u}{\partial t} = a^2\left(\frac{\partial^2 u}{\partial x^2} + \frac{\partial^2 u}{\partial y^2}\right) \qquad (0 < x < \ell_1,\ 0 < y < \ell_2,\ t > 0),$$

$$u(0,y,t) = 0, \qquad u(\ell_1,y,t) = 0 \qquad (0 < y < \ell_2,\ t > 0),$$

$$u(x,0,t) = 0, \qquad u(x,\ell_2,t) = 0 \qquad (0 < x < \ell_1,\ t > 0),$$

$$u(x,y,0) = f(x,y) \qquad (0 < x < \ell_1,\ 0 < y < \ell_2).$$

Solution. The solution can be found by using the method of separation variables Taking

$$u(x,y,t) = X(x) \cdot Y(y) \cdot T(t),$$

we obtain

$$\frac{X''}{X} = -\frac{Y''}{Y} + \frac{T'}{a^2 T} = \lambda,$$

where λ does not depend on variables x, y, t. Therefore we obtain two Sturm-Liuoville systems $X'' + \lambda^2 X = 0 \quad (0 < x < \ell_1)$, $\quad Y'' + \mu^2 Y = 0 \quad (0 < x < \ell_2)$,

$X(0) = X(\ell_1) = 0, \quad Y(0) = Y(\ell_2) = 0$, with corresponding eigenvalues

$$\lambda_n = \frac{n^2\pi^2}{\ell_1^2}, \qquad \mu_n = \frac{m^2\pi^2}{\ell_2^2},$$

respectively and eigenfunctions

$$X_n(x) = \sin\frac{n\pi x}{\ell_1}, \qquad Y_n(y) = \sin\frac{m\pi y}{\ell_2}.$$

The solution of the first order differential equation

$$T' + a^2(\lambda_n^2 + \mu_n^2)T = 0 \qquad (t > 0),$$

is
$$T(t) = C_{n,m} \exp(-a^2(\lambda_n^2 + \mu_m^2)t).$$

The solution of this heat equation with homogeneous boundary conditions has the form

$$u(x, y, t) = \sum_{n=1}^{\infty} \sum_{m=1}^{\infty} C_{m,n} \exp(-a^2(\lambda_n^2 + \mu_m^2)t) \sin \frac{n\pi x}{\ell_1} \sin \frac{n\pi y}{\ell_2}. \qquad (6.38)$$

For $t = 0$ we obtain the double Fourier series

$$f(x, y) = \sum_{n=1}^{\infty} \sum_{m=1}^{\infty} C_{m,n} \sin \frac{n\pi x}{\ell_1} \sin \frac{n\pi y}{\ell_2}. \qquad (6.39)$$

The coefficients $C_{m,n}$ can be found by

$$C_{m,n} = \frac{4}{\ell_1 \ell_2} \int_0^{\ell_1} \int_0^{\ell_2} f(x, y) \sin \frac{n\pi x}{\ell_1} \sin \frac{n\pi y}{\ell_2}.$$

It is known that if f and f' is are continuous functions in a rectangle $0 < x < \ell_1$, $0 < y < \ell_2$, then the double Fourier series converges to f.

Example 6.41 *Determine the solution of the problem given in Example 6.40 for the initial function $f(x, y) = 2$.*

Solution. The solution of this problem is given by relation (6.38) and the coefficient $C_{n,m}$ can be written as

$$\begin{aligned} C_{n,m} &= \frac{8}{\ell_1 \ell_2} \int_0^{\ell_1} \int_0^{\ell_2} \sin \frac{n\pi x}{\ell_1} \sin \frac{n\pi y}{\ell_2} \\ &= \frac{8}{\ell_1 \ell_2} \frac{(1 - (-1)^n)(1 - (-1)^m)}{mn} = \frac{32}{\ell_1 \ell_2} \frac{1}{(2m+1)(2n+1)}. \end{aligned}$$

Therefore the solution can be written as

$$u(x, y, t) = \frac{32}{\ell_1 \ell_2} \sum_{n=0}^{\infty} \sum_{m=0}^{\infty} \frac{1}{(2m+1)(2n+1)}$$

$$\cdot \exp\left(-a^2 \left(\frac{(2n+1)^2 \pi^2}{\ell_1^2} + \frac{(2m+1)^2 \pi^2}{\ell_2^2}\right) t\right) \sin \frac{(2n+1)\pi x}{\ell_1} \sin \frac{(2m+1)\pi y}{\ell_2}.$$

This series converges together with its partial derivatives $\dfrac{\partial u}{\partial x}, \dfrac{\partial u}{\partial y}, \dfrac{\partial^2 u}{\partial x^2}, \dfrac{\partial^2 u}{\partial y^2}, \dfrac{\partial u}{\partial t}$, for each $t > 0$ and $0 < x < \ell_1$, $0 < y < \ell_2$.

Exercise 6.42 *Show that the solution of two dimensional problem*

$$\frac{\partial u}{\partial t} = a^2 \left(\frac{\partial^2 u}{\partial x^2} + \frac{\partial^2 u}{\partial y^2} \right) \qquad (0 < x < \ell_1, \ 0 < y < \ell_2, \ t > 0),$$

$$-\alpha_1 \frac{\partial u(0,y,t)}{\partial x} + \beta_1 u(0,y,t) = 0 \qquad (0 < y < \ell_2, \ t > 0),$$

$$\alpha_2 \frac{\partial u(\ell_1,y,t)}{\partial x} + \beta_2 u(\ell_1,y,t) = 0 \qquad (0 < y < \ell_2, \ t > 0),$$

$$-\alpha_3 \frac{\partial u(x,0,t)}{\partial y} + \beta_3 u(x,0,t) = 0 \qquad (0 < x < \ell_1, \ t > 0),$$

$$\alpha_4 \frac{\partial u(x,\ell_2,t)}{\partial y} + \beta_4 u(x,\ell_2,t) = 0 \qquad (0 < x < \ell_1, \ t > 0),$$

$$u(x,y,0) = f(x)g(y) \qquad (0 < x < \ell_1, \ 0 < y < \ell_2).$$

is the product of the solutions of the following one dimensional problems

$$\frac{\partial u}{\partial t} = a^2 \frac{\partial^2 u}{\partial x^2} \qquad (0 < x < \ell_1, \ t > 0),$$

$$-\alpha_1 \frac{\partial u(0,t)}{\partial x} + \beta_1 u(0,t) = 0 \qquad (t > 0),$$

$$\alpha_2 \frac{\partial u(\ell_1,t)}{\partial x} + \beta_2 u(\ell_1,t) = 0 \qquad (t > 0),$$

$$u(x,0) = f(x) \qquad 0 < x < \ell_1.$$

$$\frac{\partial u}{\partial t} = a^2 \left(\frac{\partial^2 u}{\partial x^2} + \frac{\partial^2 u}{\partial y^2} \right) \qquad (0 < y < \ell_2, \ t > 0),$$

$$-\alpha_3 \frac{\partial u(0,t)}{\partial y} + \beta_3 u(0,t) = 0 \qquad (t > 0),$$

$$\alpha_4 \frac{\partial u(\ell_2,t)}{\partial y} + \beta_4 u(\ell_2,t) = 0 \qquad (t > 0),$$

$$u(y,0) = g(y) \qquad (0 < y < \ell_2).$$

Exercise 6.43 *Determine the solution of the two dimensional problem*

$$\frac{\partial u}{\partial t} = a^2 \left(\frac{\partial^2 u}{\partial x^2} + \frac{\partial^2 u}{\partial y^2} \right) \qquad (0 < x < \ell_1, \ 0 < y < \ell_2, \ t > 0),$$

$$u(0, y, t) = 0, \qquad u(\ell_1, y, t) = 0 \qquad (0 < y < \ell_2, \ t > 0),$$

$$\frac{\partial u(x, 0, t)}{\partial y} = 0, \qquad \frac{\partial u(x, \ell_2, t)}{\partial y} = 0 \qquad (0 < x < \ell_1, \ t > 0),$$

$$u(x, y, 0) = x \cdot y \qquad (0 < x < \ell_1, \ 0 < y < \ell_2).$$

Answer. The solution follows from Exercises 6.42, Exercises 6.19, Example 6.24 and has the form

$$u(x, t) = \left(\frac{2\ell_1}{\pi} \sum_{n=1}^{\infty} \frac{(-1)^n}{n} \cdot \sin \frac{n\pi x}{\ell_1} \cdot \exp\left(-t \frac{a^2 \pi^2 n^2}{\ell_1^2} \right) \right)$$

$$\cdot \left(\frac{\ell_2}{2} + \frac{4\ell_2}{\pi^2} \sum_{n=1}^{\infty} \frac{1}{(2n-1)^2} \cdot \cos \frac{(2n-1)\pi y}{\ell_2} \cdot \exp\left(-t \frac{a^2 \pi^2 (2n-1)^2}{\ell_2^2} \right) \right).$$

Exercise 6.44

$$\frac{\partial u}{\partial t} = a^2 \left(\frac{\partial^2 u}{\partial x^2} + \frac{\partial^2 u}{\partial y^2} \right) \qquad (0 < x < \ell_1, \ 0 < y < \ell_2, \ t > 0),$$

$$u(0, y, t) = 0, \qquad u(\ell_1, y, t) = 0 \qquad (0 < y < \ell_2, \ t > 0),$$

$$\frac{\partial u(x, 0, t)}{\partial y} = 0, \qquad \frac{\partial u(x, \ell_2, t)}{\partial y} = g(x) \qquad (0 < x < \ell_1, \ t > 0),$$

$$u(x, y, 0) = x \cdot y \qquad (0 < x < \ell_1, \ 0 < y < \ell_2).$$

Hint. Take $u(x, y, t) = S(x, y) + v(x, y, t)$.

Exercise 6.45 *Prove that the three dimensional heat equation*

$$\Delta u = \frac{\partial u}{\partial t} \qquad (x \in \mathbf{R}^3, \ t > 0),$$

can be transformed in a system

$$\Delta W + \lambda W = 0, \qquad T' + \lambda T = 0$$

by using the separation variables method

$$u(x, y, z, t) = W(x, y, z) \cdot T(t).$$

Example 6.46 *Determine the temperature distribution on a very very long rod, (semi-infinite rod), if one end is always kept on the zero temperature while the initial temperature distribution is given by the function f.*

Solution. In this case we consider the heat equation with semi-infinite domain, i.e., the problem

$$\frac{\partial u}{\partial t} = a^2 \frac{\partial^2 u}{\partial x^2} \quad (0 < x < \infty, \ t > 0),$$

$$u(0,t) = 0 \quad (t > 0), \qquad u(x,0) = f(x) \quad (0 < x < \infty).$$

Further on, let us suppose that u and $\dfrac{\partial u}{\partial x}$ are bounded when $x \to \infty$.

Using the separation variables $u(x,t) = X(x)T(t)$ we obtain the problem

$$X''(x) + \lambda X = 0, \quad X(0) = 0,$$

which solution is

$$X(x) = \sin sx, \quad \lambda = s^2.$$

Since the solution must be bounded when $x \to \infty$, we need $\lambda = s^2 > 0$. Let us remark that in the case on bounded intervals the eigenvalues are not discrete, in this case they are arbitrary. The solution of the equation $T' + \lambda a^2 T = 0$ for $\lambda = s^2$, is

$$T(t) = B(s) \exp(a^2 s^2 t).$$

The solution of considered problem can be written as

$$u(x,t) = \int_0^\infty \exp(-a^2 s^2 t) \cdot B(s) \sin(sx) ds. \tag{6.40}$$

In relation (6.40) we have the integral as the suprerposition of separates functions over all values of s and for arbitrary function B. Using the initial condition we have

$$f(x) = \int_0^\infty B(s) \sin(sx) ds \quad (0 < x < \infty).$$

This is the Fourier sine integral representation of f and the function B can be obtained from

$$B(s) = \frac{2}{\pi} \int_0^\infty f(\xi) \sin(s\xi) d\xi. \tag{6.41}$$

Let us replace B form (6.41) into (6.40). So we obtain the following form of solution

$$u(x,t) = \frac{2}{\pi} \int_0^\infty \int_0^\infty f(\xi) \sin(s\xi) \sin(sx) \exp(-a^2 s^2 t) d\xi ds$$

$$= \frac{2}{\pi} \int_0^\infty f(\xi) \left(\int_0^\infty \sin(s\xi) \sin(sx) \exp(-a^2 s^2 t) ds \right) d\xi.$$

This can be transformed as

$$u(x,t) = \frac{2}{\pi} \int_0^\infty f(\xi) \left(\int_0^\infty (\cos s(x-\xi) - \cos s(x+\xi)) \exp(-a^2 s^2 t) ds \right) d\xi$$

$$= \frac{1}{2a\sqrt{\pi t}} \int_0^\infty \left(\exp\left(-\frac{(x-\xi)^2}{4a^2 t}\right) - \exp\left(-\frac{(x+\xi)^2}{4a^2 t}\right) \right) f(\xi) d\xi.$$

Compare with Example 6.5. We used the following rezult

$$\int_0^\infty e^{-bx^2} \cos(cx) = \frac{1}{2}\sqrt{\frac{\pi}{b}} e^{-c^2/4b}.$$

Example 6.47 *Determine the solution of the following problem*

$$\frac{\partial u}{\partial t} = a^2 \frac{\partial^2 u}{\partial x^2} \quad (-\infty < x < \infty, \ t > 0),$$

$$u(x,0) = f(x) \quad (-\infty < x < \infty).$$

Let us also, suppose that u and $\dfrac{\partial u}{\partial x}$ are bounded when $x \to \infty$.

Solution. The solution of considered problem can be written as

$$u(x,t) = \int_0^\infty \exp(-a^2 s^2 t) \cdot (A(s) \cos(sx) + B(s) \sin(sx)) ds. \qquad (6.42)$$

where

$$A(s) = \frac{2}{\pi} \int_{-\infty}^\infty f(\xi) \cos(s\xi) d\xi,$$

$$B(s) = \frac{2}{\pi} \int_{-\infty}^\infty f(\xi) \sin(s\xi) d\xi.$$

Example 6.48 *Determine the solution of problem on an infinite domain*

$$\frac{\partial u}{\partial t} = a^2 \frac{\partial^2 u}{\partial x^2} \quad (-\infty < x < \infty, \ t > 0),$$

$$u(x,0) = e^{-x^2} \quad (-\infty < x < \infty),$$

$$u \quad and \quad \frac{\partial u}{\partial x} \quad are \ bounded \ when \quad x \to \infty.$$

Similarly as in Example 6.47 by using the separation variables $u(x,t) = X(x)T(t)$ we obtain

$$X(x) = A(s)\cos(sx) + B(s)\sin(sx), \quad \lambda = s^2 > 0.$$

So, the solution of considered problem can be written as

$$u(x,t) = \int_0^\infty \exp(-a^2 s^2 t) \cdot (A(s)\cos(sx) + B(s)\sin(sx))ds, \qquad (6.43)$$

Using the initial condition we get

$$e^{-x^2} = \int_{-\infty}^\infty (A(s)\cos(sx) + B(s)\sin(sx))ds,$$

where

$$A(s) = \frac{1}{\pi} \int_{-\infty}^\infty e^{-x^2} \cos sx \, dx \qquad B(s) = \frac{1}{\pi} \int_{-\infty}^\infty e^{-x^2} \sin sx \, dx. \qquad (6.44)$$

In this case, it is allowed to make differentiation under the integral.

$$A'(s) = -\frac{1}{\pi} \int_{-\infty}^\infty x e^{-x^2} \sin sx \, dx$$

$$= \frac{1}{\pi} \frac{e^{-x^2}}{2} \sin(sx)\Big|_{-\infty}^\infty - \frac{s}{2\pi} \int_{-\infty}^\infty e^{-x^2} \cos sx \, dx = -\frac{s}{2} A(s).$$

The solution of the following differential equation

$$A'(s) + \frac{s}{2}A = 0,$$

is $A(s) = A_1 \exp\left(-\frac{s^2}{4}\right)$, and it satisfies the initial condition

$$A(0) = \frac{1}{\pi} \int_{-\infty}^{\infty} e^{-x^2}\, dx = \frac{1}{\sqrt{\pi}}.$$

Therefore

$$A(s) = \frac{1}{\sqrt{\pi}} \exp\left(-\frac{s^2}{4}\right).$$

Since $e^{-x^2}\sin(sx)$ is an odd function, it holds that $B(s) = 0$. So the solution of the considered problem with infinite domain can be written as

$$u(x,t) = \frac{1}{\sqrt{\pi}} \int_{0}^{\infty} \exp\left(-a^2 s^2 t - \frac{s^2}{4}\right) \cdot \cos(sx)\,ds,$$

Exercise 6.49 *Solve the mixed problem for $r > 0$ and $t > 0$*

$$\frac{\partial u}{\partial t} = a^2 \left(\frac{\partial^2 u}{\partial r^2} + \frac{1}{r}\frac{\partial u}{\partial r}\right)$$

with the initial condition

$$u(r,0) = f(r) \qquad r > 0$$

and boundary condition

$$u(R_0, t) = 0 \qquad (t > 0),$$

where the function f and the constant $R_0 > 0$ are given.

Hints. Use the Fourier method of separation of variables in the form

$$u(r,t) = R(r) \cdot T(t).$$

The solution is of the form

$$u(r,t) = \sum_{n=1}^{\infty} a_n J_0\left(\frac{\mu_n r}{R_0}\right) \exp\left(-\left(\frac{a\mu_n}{R_0}\right)^2\right),$$

where μ_n are the positive zeros of the Bessel function J_0, and a_n are determined by the initial condition. The Bessel function J_n $(n \in \mathbf{Z}_+)$ has the following representation by series

$$J_n(x) = \sum_{j=0}^{\infty} \frac{(-1)^j}{\Gamma(j+n+1)\Gamma(j+1)} \left(\frac{x}{2}\right)^{2j+n}$$

6.3   Heat conduction

Our task is to develop the mathematical model of the heat conduction both in the one–dimensional homogeneous thin rod or wire and in the three–dimensional homogeneous body, in which heat can flow freely.

Let us consider a long thin homogeneous rod of length ℓ made of a uniform material. For simplicity, we assume that the road coincides with the segment $[0, \ell]$ on the x–axis. If $u(x, t)$ is the temperature measured at the point x and at the time t, then the rate of heat flow $Q = Q(x, t)$ at the point x and at the time t (the *heat flux*) is, in view of the Fourier's low of heat conduction, equal to

$$Q = -k \frac{\partial u}{\partial x}. \tag{6.45}$$

In (6.45), k is a constant denoting the heat conductivity of the wire. The sign "-" in relation (6.45) shows that the heat flows from hotter to cooler parts.

Let S be the cross section area of the uniform rod. Then, the amount of heat ΔQ_1, entering any cross section S during the time interval $(t, t + \Delta t)$. equals

$$\Delta Q_1 = Q S \Delta t.$$

Thus during the time interval $[t_1, t_2]$ the amount of heat Q_1, written in the ir.egral form is

$$Q_1 = -Sk \int_{t_1}^{t_2} \frac{\partial u}{\partial x} \, dt. \tag{6.46}$$

From the other hand, in order to raise the temperature of the rod for Δu, one needs the amount of heat ΔQ_2 equal to

$$\Delta Q_2 = S c \rho \Delta x \frac{\partial u}{\partial t},$$

where $c > 0$ is the specific heat constant and ρ is the constant mass density.

Thus $[x_1, x_2]$ is a part of the rod, then Q_2, equals to

$$Q_2 = S c \rho \int_{x_1}^{x_2} \frac{\partial u}{\partial t} \, dx. \tag{6.47}$$

Finally, if a heat source $F = F(x, t)$ at the point x and at the time t is present in the rod, then the corresponding amount of heat ΔQ_3 is equal to

$$\Delta Q_3 = S \, F(x, y) \, \Delta x \, \Delta t,$$

or in integral form

$$Q_3 = S \int\limits_{x_1}^{x_2} \int\limits_{t_1}^{t_2} F(x,t)\, dt\, dx. \tag{6.48}$$

The low of conservation of energy implies

$$Q_1 + Q_3 = Q_2,$$

or

$$k \int\limits_{t_1}^{t_2} \left(\frac{\partial u(x_2,t)}{\partial x} - \frac{\partial u(x_1,t)}{\partial x} \right)\, dt + \int\limits_{x_1}^{x_2} \int\limits_{t_1}^{t_2} F(x,t)\, dt\, dx$$

$$= c\rho \int\limits_{x_1}^{x_2} (u(\xi,t_2) - u(\xi,t_1))\, d\xi$$

Applying the mean value theorem, we obtain

$$k\left(\left(\frac{\partial u(x_2,t_3)}{\partial x} - \frac{\partial u(x_1,t_3)}{\partial x} \right) \right) \Delta t + F(x_4,t_4)\,\Delta t\, \Delta x = c\rho(u(x_3,t_2) - u(x_3,t_1))\Delta x,$$

or

$$\frac{\partial}{\partial x}\left(k\frac{\partial u(x_3,t_4)}{\partial x} \right)\Delta t\Delta x + F(x_4,t_4)\Delta t\Delta x = \left(c\rho\frac{\partial u(x_5,t_5)}{\partial t} \right)\Delta t\Delta x,$$

where t_3, t_4, t_5, and x_3, x_4, x_5, are the points from the intervals (t_1,t_2) and (x_1,x_2) respectively. After dividing with $\Delta t\, \Delta x$, we can let $x_1 \to x$, $x_2) \to x$ and $t_1 \to t$, $t_2) \to t$, so we obtain the heat equation

$$\frac{\partial}{\partial x}\left(k\frac{\partial u}{\partial x} \right) + F(x,t) = c\rho\frac{\partial u}{\partial t}.$$

Usually, the last equation is written in the form

$$\frac{\partial u}{\partial t} = a^2\frac{\partial^2 u}{\partial x^2} + f(x,t),$$

where $a^2 = \dfrac{k}{c\rho}$ and $f(x,t) = \dfrac{F(x,t)}{c\rho}$.

We turn now to three dimensional case. Let $P(x,y,z)$ be a point of the three–dimensional body, and $u(P,t) = u(x,y,z,t)$ be the temperature measured at the point P and at the time t. Let $Q = Q(x,y,z,t)$ be the heat flux at the point (x,y,z) and the time t. Now, the Fourier's low of heat conduction implies

$$Q = -k\nabla u. \tag{6.49}$$

Let V be the volume of the body bounded by surface σ. Then, applying the low of conservation of heat energy, similarly as in the one–dimensional case, we have

$$k\rho \int\int\int_V \left(u(P,t_2) - u(P,t_1)\right) dV_p = \int_{t_1}^{t_2} \left(\int\int\int_V F(P,t) dV_p\right) dt$$

$$- \int_{t_1}^{t_2} dt \int\int_\sigma W_n \, d\sigma.$$

In the last multiple integral, $W_n = Q \cdot \mathbf{n}$ is the heat flux in direction $\mathbf{n}$. Applying the divergence theorem, we obtain

$$k\rho \int\int\int_V \left(u(P,t_2) - u(P,t_1)\right) dV_p = \int_{t_1}^{t_2} \int\int\int_V F(P,t) \, dt \, dx$$

$$- c\rho \int_{t_1}^{t_2} dt \int\int\int_V div\mathbf{W} n dV.$$

Similarly as in one-dimensional case we apply the mean value theorem

$$c\rho \left.\frac{\partial u}{\partial t}\right|_{t=t_3, P=P_1} \Delta t \cdot V = - \left. div\mathbf{W}\right|_{t=t_4, P=P_2} \Delta t \cdot V + \left. F\right|_{t=t_5, P=P_2} \Delta t \cdot V,$$

where t_3, t_4, t_5, are the points from the interval (t_1, t_2) and P_1, P_2, P_3, is a point from V.

So we obtain

$$-divW(x,y,z,t) + F(x,y,z,t) = c\rho\frac{\partial u(x,y,z,t)}{\partial t},$$

or

$$\frac{\partial}{\partial x}\left(k\frac{\partial u}{\partial x}\right) + \frac{\partial}{\partial y}\left(k\frac{\partial u}{\partial y}\right) + \frac{\partial}{\partial z}\left(k\frac{\partial u}{\partial z}\right) + F(x,y,z,t) = c\rho\frac{\partial u}{\partial t}.$$

Thus we obtain the three dimensional heat equation. Mostly it is written as

$$\frac{\partial u}{\partial t} = a^2\left(\frac{\partial^2 u}{\partial x^2} + \frac{\partial^2 u}{\partial y^2} + \frac{\partial^2 u}{\partial z^2} + \frac{F}{c\rho}\right), \tag{6.50}$$

or

$$\frac{\partial u}{\partial t} = a^2 \Delta u + F_1,$$

where $a^2 = \dfrac{k}{\rho}$ is positive constant which is called the *diffusivity* of the material of the rod and $F_1 = \dfrac{F}{c\rho}$.

Chapter 7

Numerical Methods

7.0.1 Preliminaries

The most commonly used method for obtaining the approximate solutions of certain partial differential equations is the *finite differences method*. In order to employ this method we replace the continuous independent variables $x, y, z, t, \ldots$, by a finite numbers of discrete variables x_i, y_j, z_m, $t_n, \ldots$, namely we determine suitable mesh points

$$x_i = x_0 + ih, \qquad i = 0, \pm 1, \pm 2, \ldots,$$

$$y_j = y_0 + jk, \qquad j = 0, \pm 1, \pm 2, \ldots,$$

and so on. Replacing each of derivatives by a suitable difference quotient, a difference equation for $i, j = 0, \pm 1, \pm 2, \ldots$, is obtained. It represents a system of algebraic equations, whose solutions can be treated as the approximate solutions of the considered problem at the mesh points.

As usual, we consider the following partial differential equation

$$L(u) = A(x,y)\frac{\partial^2 u(x,y)}{\partial x^2} + B(x,y)\frac{\partial^2 u(x,y)}{\partial y^2} + C(x,y)\frac{\partial u(x,y)}{\partial x}$$

$$+ D(x,y)\frac{\partial u(x,y)}{\partial y} + G(x,y)u(x,y) = F(x,y),$$

(7.1)

where A, B, C, D, G, F are continuous functions on the set $Q \subset \mathbf{R}^2$, and on its boundary ∂Q.

We denote by $u_{i,j}$, $u_{i-1,j}$, $u_{i,j-1}$, $u_{i+1,j}$, $u_{i,j+1}$ the corresponding values for $u(x_i, y_j)$, $u(x_{i-1}, y_j)$, $u(x_i, y_{j-1})$, $u(x_{i+1}, y_j)$, $u(x_i, y_{j+1})$, respectively. The derivatives are replaced by the corresponding difference quotients

- forward difference

$$\frac{\partial u}{\partial x} \approx \frac{u_{i+1,j} - u_{i,j}}{h} \quad \text{and} \quad \frac{\partial u}{\partial y} \approx \frac{u_{i,j+1} - u_{i,j}}{k};$$

227

- backward difference

$$\frac{\partial u}{\partial x} \approx \frac{u_{i,j} - u_{i-1,j}}{h} \quad \text{and} \quad \frac{\partial u}{\partial y} \approx \frac{u_{i,j-1} - u_{i,j}}{k};$$

- centered difference

$$\frac{\partial u}{\partial x} \approx \frac{u_{i+1,j} - u_{i-1,j}}{2h} \quad \text{and} \quad \frac{\partial u}{\partial y} \approx \frac{u_{i,j+1} - u_{i,j-1}}{2k};$$

- centered difference for second derivative

$$\frac{\partial^2 u}{\partial x^2} \approx \frac{u_{i+1,j} - 2u_{i,j} + u_{i-1,j}}{h^2} \quad \text{and} \quad \frac{\partial^2 u}{\partial y^2} \approx \frac{u_{i,j+1} - 2u_{i,j} + u_{i,j-1}}{k^2}.$$

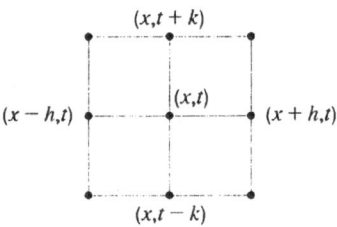

Figure 7.1

If we denote by $A_{i,j}$, $B_{i,j}$, $C_{i,j}$, $D_{i,j}$, $G_{i,j}$, and $F_{i,j}$, for $i = 0, \pm 1, \pm 2, \ldots$, $j = 0, \pm 1, \pm 2, \ldots$, the values of the functions A, B, C, D, G and F, respectively, at the points x_i, y_j, then we obtain the following difference equation

$$L_{h,k} \equiv A_{i,j} \cdot \frac{u_{i+1,j} - 2u_{i,j} + u_{i-1,j}}{h^2} + B_{i,j} \cdot \frac{u_{i,j+1} - 2u_{i,j} + u_{i,j-1}}{k^2}$$

$$+C_{i,j} \cdot \frac{u_{i+1,j} - u_{i-1,j}}{2h} + D_{i,j} \cdot \frac{u_{i,j+1} - u_{i,j-1}}{2k} + G_{i,j}u_{i,j} = f_{i,j}, \tag{7.2}$$

for $i = 0, \pm 1, \pm 2, \ldots$, $j = 0, \pm 1, \pm 2, \ldots$.

The Error of Approximation

The error of approximation for the solution of difference equation (7.2), which can be treated as the approximate solution for differential equation (7.1). can be obtained by using the Taylor formulas. Namely, if we suppose that the solution of a given problem has continuous derivatives up the order four, then there exist $\tilde{x}$, $\tilde{y}$ $\tilde{\tilde{x}}$, $\tilde{\tilde{y}}$ such that

$$x_i - h \le \tilde{x} \le x_i + h, \qquad y_j - k \le \tilde{y} \le x_i - k,$$

$$x_i - h \le \tilde{\tilde{x}} \le x_i + h, \qquad y_j - k \le \tilde{\tilde{y}} \le x_i - k,$$

satisfying

$$\frac{u(x_i + h, y_j) - u(x_i - h, y_j)}{2h} = \frac{\partial u(x_i, y_j)}{\partial x} + \frac{h^2}{6} \cdot \frac{\partial^3 u(\tilde{x}, y_j)}{\partial x^3},$$

$$\frac{u(x_i, y_j + k) - u(x_i, y_j - h)}{2k} = \frac{\partial u(x_i, y_j)}{\partial y} + \frac{k^2}{6} \cdot \frac{\partial^3 u(x_i, \tilde{y})}{\partial y^3},$$

$$\frac{u(x_i + h), y_j) - 2u(x_i, y_j) + u(x_i - h, y_j)}{h^2} = \frac{\partial^2 u(x_i, y_j)}{\partial x^2} + \frac{h^2}{12} \cdot \frac{\partial^4 u(\tilde{\tilde{x}}, y_j)}{\partial x^4},$$

$$\frac{u(x_i, y_j + k) - 2u(x_i, y_j) + u(x_i, y_j - k)}{k^2} = \frac{\partial^2 u(x_i, y_j)}{\partial y^2} + \frac{k^2}{12} \cdot \frac{\partial^4 u(x_i, \tilde{\tilde{y}})}{\partial y^4}.$$

If we take the same denotation as in (7.2), in the neighbourhood of the points (x_i, y_j), it holds

$$
\begin{aligned}
L_{h,k} &= A_{i,j} \frac{\partial^2 u(x_i, y_j)}{\partial x^2} + B_{i,j} \frac{\partial^2 u(x_i, y_j)}{\partial y^2} \\[2mm]
&\quad + C_{i,j} \frac{\partial u(x_i, y_j)}{\partial x} + D_{i,j} \frac{\partial u(x_i, y_j)}{\partial y} + G_{i,j} u(x_i, y_j) \\[2mm]
&\quad \frac{h^2}{12} \left(A_{i,j} \cdot \frac{\partial^4 u(\tilde{\tilde{x}}, y_j)}{\partial x^4} + 2D_{i,j} \cdot \frac{\partial^3 u(\tilde{x}, y_j)}{\partial x^3} \right) \\[2mm]
&\quad + \frac{k^2}{12} \left(B_{i,j} \cdot \frac{\partial^4 u(x_j, \tilde{\tilde{y}})}{\partial y^4} + 2D_{i,j} \cdot \frac{\partial^3 u(x_j, \tilde{y})}{\partial y^3} \right) \\[2mm]
&= [L(u(x_i, y_j)] + R_{i,j},
\end{aligned}
\tag{7.3}
$$

for $i = 0, \pm 1, \pm 2, \ldots$, $j = 0, \pm 1, \pm 2, \ldots$. In (7.3), $R_{i,j}$ is the error of approximation and has the form

$$
\begin{aligned}
R_{i,j} &= \frac{h^2}{12} \left(A_{i,j} \cdot \frac{\partial^4 u(\tilde{\tilde{x}}, y_j)}{\partial x^4} + 2C_{i,j} \cdot \frac{\partial^3 u(\tilde{x}, y_j)}{\partial x^3} \right) \\[2mm]
&\quad + \frac{k^2}{12} \left(B_{i,j} \cdot \frac{\partial^4 u(x_j, \tilde{\tilde{y}})}{\partial y^4} + 2D_{i,j} \cdot \frac{\partial^3 u(x_j, \tilde{y})}{\partial y^3} \right).
\end{aligned}
\tag{7.4}
$$

Taking

$$M_3 = \max_Q \left\{ \left| \frac{\partial^3 u}{\partial x^3} \right|, \left| \frac{\partial^3 u}{\partial y^3} \right| \right\} \qquad M_4 = \max_Q \left\{ \left| \frac{\partial^4 u}{\partial x^4} \right|, \left| \frac{\partial^4 u}{\partial y^4} \right| \right\}, \tag{7.5}$$

we obtain the estimation of the error of approximation as

$$|R_{i,j}| \le \frac{h^2}{12}\left(A_{i,j}M_4 + 2C_{i,j}M_3\right) + \frac{k^2}{12}\left(B_{i,j}M_4 + 2D_{i,j}M_3\right), \tag{7.6}$$

for $i = 0, \pm 1, \pm 2, \ldots$, $j = 0, \pm 1, \pm 2, \ldots$.

7.0.2   Examples and Exercises

Exercise 7.1 *Determine the difference equation for Poisson's equation*

$$\frac{\partial^2 u}{\partial x^2} + \frac{\partial^2 u}{\partial y^2} = F(x,y) \qquad (x,y \in \mathbf{R}),$$

with the condition

$$u|_{\partial Q} = \phi.$$

Solution. Using the finite difference method, we obtain the difference equation

$$\frac{u_{i+1,j} - 2u_{i,j} + u_{i-1,j}}{h^2} + \frac{u_{i,j+1} - 2u_{i,j} + u_{i,j-1}}{k^2} = F_{i,j},$$

and the values at the boundary points satisfy the boundary conditions given by the the function ϕ.

Example 7.2 *Using the finite difference method determine the approximate solution of the Dirichlet problem on a square, which is given by*

$$\frac{\partial^2 u}{\partial x^2} + \frac{\partial^2 u}{\partial y^2} = 0 \qquad (0 < x < 1, \quad 0 < y < 1),$$

$$u(0,y) = 0, \qquad u(1,y) = 10y \qquad (0 < y < 1),$$

$$u(x,0) = 0, \qquad u(x,1) = 10x \qquad (0 < x < 1).$$

Solution. If we choose the same increment $h = k$ on the $x-$ and $y-$ axes, then we obtain the following difference equation corresponding to the Laplace equation

$$4u_{i,j} - (u_{i+1,j} + u_{i-1,j} + u_{i,j+1} + u_{i,j-1}) = 0. \tag{7.7}$$

Taking $n = 5$, i.e., $h = 1/5$, we obtain $i,j = 1,2,3,4,5$. in equation (7.7) and the mesh of the given square is shown in Figure 7.2

The conditions of Dirichlet problem correspond to the following

$$u_{0,j} = 0, \quad u_{5,j} = 2j,$$

$$u_{i,0} = 0, \quad u_{i,5} = 2i, \tag{7.8}$$

for $i = 0, 1, 2, 3, 4, 5$, $j = 0, 1, 2, 3, 4, 5$, The equations (7.7) and (7.8) can be express in terms of grid points in Figure 7.2 as

$$4u_1 - u_2 - u_5 = u_{1,0} + u_{0,1},$$
$$4u_3 - u_4 - u_2 - u_7 = u_{3,0},$$
$$4u_5 - u_6 - u_1 - u_9 = u_{0,2},$$
$$4u_7 - u_3 - u_6 - u_8 + u_{11} = 0,$$
$$4u_9 - u_5 - u_{10} - u_{13} = u_{0,3},$$
$$4u_{11} - u_{10} - u_{12} - u_7 - u_{15} = 0,$$
$$4u_{13} - u_{14} - u_9 = u_{0,4} + u_{1,5},$$
$$4u_{15} - u_{11} - u_{14} - u_{16} = u_{3,5},$$

$$4u_2 - u_3 - u_1 - u_6 = u_{2,0},$$
$$4u_4 - u_3 - u_8 = u_{4,0} + u_{5,1},$$
$$4u_6 - u_2 - u_5 - u_7 - u_{10} = 0,$$
$$4u_8 - u_4 - u_7 - u_{12} = u_{5,2},$$
$$4u_{10} - u_6 - u_9 - u_{11} - u_{14} = 0,$$
$$4u_{12} - u_{11} - u_8 - u_{16} = u_{5,3},$$
$$4u_{14} - u_{13} - u_{10} - u_{15} = u_{2,5},$$
$$4u_{16} - u_{12} - u_{15} = u_{5,4} + u_{4,5},$$

(7.9)

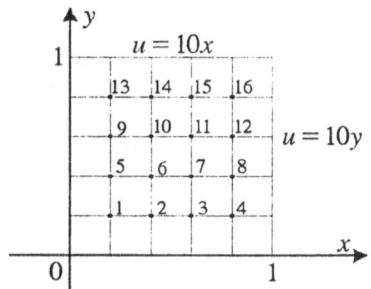

Figure 7.2

Using the boundary conditions we can evaluate the terms on the right-hand side

$$u_{1,0} = u_{2,0} = u_{3,0} = u_{4,0} = u_{5,0} = 0,$$
$$u_{0,1} = u_{0,2} = u_{0,3} = u_{0,4} = u_{0,5} = 0,$$
$$u_{1,5} = 2, \quad u_{2,5} = 4, \quad u_{3,5} = 6, \quad u_{4,5} = 8,$$
$$u_{5,1} = 2, \quad u_{5,2} = 4, \quad u_{5,3} = 6, \quad u_{5,4} = 8.$$

The values $u_1, u_2, \ldots, u_{12}$ are found by solving the system (7.9). We obtain the solution of this system by using the MAPLE subroutine in Scientific WorkPlace 2.5 and they have the values:

$$u_{10} = 1.8269, \quad u_{15} = 4.4637, \quad u_4 = 1.2234, \quad u_5 = .46368,$$
$$u_7 = .22339, \quad u_{12} = 4.3403, \quad u_{13} = 1.4701, \quad u_2 = .34034,$$
$$u_8 = 2.4468, \quad u_{14} = 2.9402, \quad u_6 = .71357, \quad u_{11} = 2.7136,$$
$$u_{16} = 6.201, \quad u_9 = .94016, \quad u_3 = .44678, \quad u_1 = .20101.$$

These are the approximate solutions of the considered problem at the corresponding points.

Example 7.3 *Solve the Laplace equation*

$$\frac{\partial^2 u}{\partial x^2} + \frac{\partial^2 u}{\partial y^2} = 0,$$

in a triangle $0 < x < y < 1$, *with boundary conditions*

$$u(0,y) = 0, \qquad u(x,1) = 5x(1-x), \qquad u(x,x) = 0.$$

Solution. Let us choose the same increment $h = k = \dfrac{1}{5}$ on $x-$ and $y-$ axes, i.e., $n = 5$. From boundary conditions we have

$$u_{0,0} = 0, \quad u_{0,1} = 0, \quad u_{0,2} = 0, \quad u_{0,3} = 0, \quad u_{0,4} = 0, \quad u_{0,5} = 0,$$
$$u_{1,5} = \frac{4}{5}, u_{2,5} = u_{3,5} = \frac{6}{5}, \quad u_{4,5} = \frac{4}{5},$$
$$u_{1,1} = 0, u_{2,2} = 0, \quad u_{3,3} = 0, \quad u_{4,4} = 0, \quad u_{5,5} = 0.$$

The solutions of the system

$$4u_{1,2} - u_{1,3} = u_{0,2} + u_{1,1} + u_{2,2}, \qquad 4u_{1,3} - u_{1,4} - u_{1,2} - u_{2,3} = u_{0,3},$$

$$4u_{2,3} - u_{2,4} - u_{1,3} = u_{3,3} + u_{2,2}, \qquad 4u_{1,4} - u_{1,3} - u_{2,4} = u_{0,4} + u_{1,5},$$

$$4u_{2,4} - u_{1,4} - u_{2,3} - u_{3,4} = u_{2,5}, \qquad 4u_{3,4} - u_{2,4} = u_{3,3} + u_{4,4} + u_{3,5},$$

are obtained from the MAPLE subroutine in Scientific WorkPlace 2.5 and they are

$$u_{2,4} = .54545, \quad u_{1,4} = .37273, \quad u_{1,3} = .14545,$$

$$u_{3,4} = .43636, \quad u_{2,3} = .17273, \quad u_{1,2} = 3.6364 \times 10^{-2}.$$

Example 7.4 *Determine the solution of the Dirichlet problem on the region* Q,

$$\frac{\partial^2 u}{\partial x^2} + \frac{\partial^2 u}{\partial y^2} = 0 \qquad (x,y \in Q)$$

$$u|_{\partial Q} = r(x,y),$$

a) *where* Q *is the triangle with vertices* $(-4,0)$, $(4,0)$, $(0,3)$ *and the function* r *is given by*

$$r(x,y) = \begin{cases} 20 - x^2, & \partial Q \cap \mathbf{R}, \\ 0, & (x,y) \in \partial Q \setminus \mathbf{R}; \end{cases} \qquad (7.10)$$

b) *where* Q *is the ellipsis*

$$\frac{x^2}{16} + \frac{y^2}{9} = 1,$$

and the function r *is given by*

$$r(x,y) = 1 - x^2 + y, \quad x \in \partial Q.$$

Solution.

a) Let us choose the same increment $h = k = 1$, on $x-$ and $y-$ axes, i.e., $n = 8$.

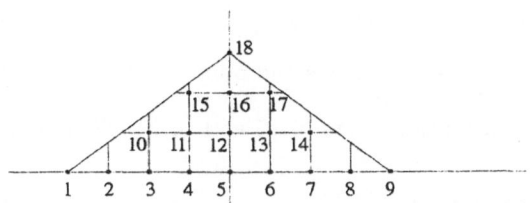

Figure 7.3

The points $(-4, 0)$, $(-3, 0)$, $(-2, 0)$, $(-1, 0)$, $(0, 0)$, $(1, 0)$, $(2, 0)$, $(3, 0)$, $(4, 0)$, and $(0, 3)$ are denoted respectively by 1, 2, 3, 4, 5, 6, 7, 8, 9, 18. The values of the function u at these points are determined directly by the equality $u_i = r(x_i, y_j)$.

The values at these points are

$$u_1 = u_9 = 4, \qquad u_{18} = 0, \qquad u_2 = u_8 = 11,$$
$$u_3 = u_6 = 16, \qquad u_4 = u_7 = 19, \qquad u_5 = 20.$$

The values of the function u at the points $(-1, 1)$, $(0, 1)$ $(1, 1)$, and $(0, 2)$, denoted in Figure 7.3. by 11, 12, 13, 16 respectively can be determined directly, because they together with their neighboring vertices do belong to Q.

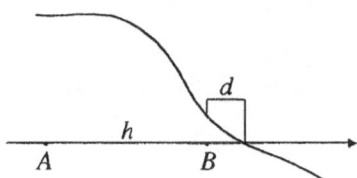

Figure 7.4

The values at the points $(-2,1)$, $(2,1)$, $(-1,2)$, $(1,2)$, denoted by 10, 14, 15, and 17 respectively can not be determined directly because some of their neighboring vertices do not belong to Q. Their values are given by the following formulae (Figure 7.4)

$$u_A = \frac{d \cdot u(x_1 - h, y_1) + h \cdot u(x_1 + d, y_1)}{d + h},$$

where $A(x_1, y_1)$ is one of the mentioned points, $u_B = u(x_1 - h, y_1)$, and d is the distance from the boundary.

The line $y = \frac{3}{4}x + 3$ connects the points $(-4, 0)$ and $(0, 3)$ and the line $y = -\frac{3}{4}x + 3$ connects the points $(4, 0)$ and $(0, 3)$. Therefore

- for the points 10 and 14 the distance from the boundary is $d = \frac{2}{3}$;

- for the points 15 and 17 the distance from the boundary is $d = \frac{1}{3}$.

Therefore we can write

$$u_{10} = \frac{\frac{2}{3}u_{11} + 0}{1 + \frac{2}{3}}, \qquad u_{14} = \frac{\frac{2}{3}u_{13} + 0}{1 + \frac{2}{3}},$$

$$u_{15} = \frac{\frac{1}{3}u_{16} + 0}{1 + \frac{1}{3}}, \qquad u_{17} = \frac{\frac{1}{3}u_{16} + 0}{1 + \frac{1}{3}}. \tag{7.11}$$

Finally we have the following system

$$4u_{11} = u_{12} + u_{15} + u_{10} + u_4, \qquad 4u_{12} = u_{13} + u_{16} + u_{11} + u_5,$$

$$4u_{13} = u_{14} + u_{17} + u_{12} + u_6, \qquad 4u_{16} = u_{17} + u_{18} + u_{15} + u_{12}. \tag{7.12}$$

The solutions of system (7.11) and (7.12) are

$$\begin{array}{cccc}
u_{12} = 9.5293, & u_{11} = 8.1139, & u_{13} = 7.2805, & u_{16} = 2.7226, \\
u_{10} = 3.2455, & u_{14} = 2.9122, & u_{17} = .68066, & u_{15} = .68066.
\end{array}$$

Example 7.5 *The* Poisson *equation*

$$\frac{\partial^2 u}{\partial x^2} + \frac{\partial^2 u}{\partial y^2} = -1 \qquad (0 < x < 1, \quad 0 < y < 1),$$

with boundary conditions

$$u(0,y) = 0, \quad u(1,y) = 0 \quad (0 < y < 1),$$
$$u(x,0) = 0, \quad u(x,1) = 0 \quad (0 < x < 1),$$

characterizes a square membrane under constant tensions on all sides and supporting a uniformly distributed load. Determine the approximate solution by using finite-difference method for $n = 4, \ h = k = \dfrac{1}{4}$.

Solution. Firstly, note that the given equation is nonhomogeneous and on the right-hand side is a constant function. The boundary conditions are equal zero

$$u_{i,0} = u_{i,4} = u_{0,j} = u_{4,j} = 0.$$

Therefore the values of u are symmetric, namely it holds

$$u_{1,1} = u_{3,1}, \quad u_{1,2} = u_{3,2},$$

$$u_{1,3} = u_{3,3}, \quad u_{1,2} = u_{2,1}.$$

This means that it is enough to find the values of u at the points $(1,1)$, $(2,1)$, $(1,2)$ and $(2,2)$.

The difference equation corresponding to the given nonhomogeneous differential equation is

$$u_{i+1,j} + u_{i-1,j} + u_{i,j+1} + u_{i,j-1} - 4u_{i,j} = h^2.$$

So we have the system of equations

$$4u_{1,1} - 2u_{1,2} = -\frac{1}{16}, \quad 4u_{2,1} - 2u_{1,1} - u_{2,2} = -\frac{1}{16}, \quad 4u_{2,2} - 4u_{1,2} = -\frac{1}{16}.$$

The solutions of this system are

$$u_{1,1} = 0,0429, \quad u_{1,2} = 0,0547, \quad u_{2,2} = 0,0703.$$

Example 7.6 *Determine the approximate solution by using the finite-difference method for* $n = 4, \ h = k = \dfrac{1}{4}$, *for PDE*

$$\frac{\partial^2 u}{\partial x^2} + \frac{\partial^2 u}{\partial y^2} = 0 \quad (0 < x < 1, \ 0 < y < 1),$$

with boundary conditions

$$u(0,y) = 0, \quad u(1,y) = 0 \quad (0 < y < 1),$$
$$u(x,0) = 0, \quad u(x,1) = 10 \quad (0 < x < 1).$$

Solution. In this case the boundary conditions imply

$$u_{i,0} = u_{0,j} = u_{4,j} = 0, \qquad u_{i,4} = 10,$$

and the following symmetry

$$u_{1,1} = u_{3,1}, \qquad u_{1,2} = u_{3,2}, \qquad u_{1,3} = u_{3,3}.$$

From the system

$$4u_{1,1} - u_{1,2} - u_{2,1} = 0, \qquad\qquad 4u_{1,2} - u_{1,1} - u_{2,2} - u_{1,3} = 0,$$

$$4u_{2,1} - u_{2,2} - 2u_{1,1} = 0, \qquad\qquad 4u_{1,3} - u_{1,2} - u_{2,3} = 10,$$

$$4u_{2,2} - 2u_{1,2} - u_{2,3} - u_{2,1} = 0, \qquad 4u_{2,3} - 2u_{1,3} - u_{2,2} = 10,$$

we obtain

$$u_{2,3} = 5.2679, \qquad u_{1,3} = 4.2857, \qquad u_{1,2} = 1.875,$$
$$u_{2,2} = 2.5, \qquad u_{2,1} = .98214, \qquad u_{1,1} = .71429.$$

Example 7.7 *Let us consider the mixed type problem for heat equation.*

$$\frac{\partial u}{\partial t} = \frac{\partial^2 u}{\partial x^2} \qquad (t > 0, \quad 0 < x < L),$$

$$u(x,0) = f(x) \qquad (0 < x < L),$$

$$u(0,t) = \phi(t), \qquad u(L,t) = \psi(t) \qquad (t > 0).$$

Using the method of finite differences determine the difference equations by and estimate the error of approximation.

Solution. Taking $x_i = ih$, $t_j = jk$, $i = 0, 1, 2, \ldots,$ $j = 0, 1, 2, \ldots,$ and replacing both the spatial derivative

$$\frac{\partial^2 u}{\partial x^2} \quad \text{and the time derivative} \quad \frac{\partial u}{\partial t}$$

by their difference quotients (taking the forward difference for time derivative and the central one for the second space one) we obtain the following difference equation

$$\frac{u_{i,j+1} - u_{i,j}}{k} = \frac{u_{i+1,j} - 2u_{i,j} + u_{i-1,j}}{h^2}. \tag{7.13}$$

The difference equation

$$\frac{u_{i,j} - u_{i,j-1}}{k} = \frac{u_{i+1,j} - 2u_{i,j} + u_{i-1,j}}{h^2}, \tag{7.14}$$

is obtained for the backward difference for time derivative. Taking $\sigma = \dfrac{k}{h^2}$, the equation (7.13) can be written in the form

$$u_{i,j+1} = (1 - 2\sigma)u_{i,j} + \sigma(u_{i-1,j} + u_{i+1,j}), \qquad (7.15)$$

while equation (7.14) becomes

$$(1 + 2\sigma)u_{i,j} - \sigma(u_{i+1,j} + \sigma u_{i-1,j}) = u_{i,j-1}. \qquad (7.16)$$

One can prove that

- the difference equation (7.15) is stable for $0 < \sigma \le \dfrac{1}{2}$;

- the difference equation (7.16) is stable for every σ.

If $\sigma = \dfrac{1}{2}$, then the difference equation (7.15) has the form

$$u_{i,j+1} = \frac{u_{i-1,j} + u_{i+1,j}}{2}. \qquad (7.17)$$

If $\sigma = \dfrac{1}{6}$, then the difference equation (7.15) has the form

$$u_{i,j+1} = \frac{u_{i-1,j} + 4u_{i,j} + u_{i+1,j}}{6}. \qquad (7.18)$$

The error of approximation for the approximate solution obtained by using equations (7.17), (7.18) and (7.16) are respectively

$$|u - \tilde{u}| \le \frac{T}{3}M_1 h^2,$$

$$|u - \tilde{u}| \le \frac{T}{135}M_2 h^4, \qquad (7.19)$$

$$|u - \tilde{u}| \le \frac{T}{12}\left(6k + h^2\right)M_1,$$

for $0 \le x \le L$, $0 \le t \le T$, where $\tilde{u}$ is the exact solution of the considered problem and u is the approximate one and

$$M_1 = \max_{0 \le x \le L,\ 0 \le t \le T}\{|f^{(4)}(x)|, |\phi^{(2)}(t)|, |\psi^{(2)}(t)|\},$$

$$M_2 = \max_{0 \le x \le L,\ 0 \le t \le T}\{|f^{(6)}(x)|, |\phi^{(4)}(t)|, |\psi^{(4)}(t)|\}.$$

Example 7.8 *Determine the difference equations corresponding to the nonhomogeneous parabolic equation*

$$\frac{\partial u}{\partial t} = \frac{\partial^2 u}{\partial x^2} + F(x,t) \qquad (t > 0, \quad 0 < x < L),$$

for $\sigma = \dfrac{1}{2}$ and $\sigma = \dfrac{1}{6}$. Estimate the error of approximation for $0 \le t \le T$.

Solution. Taking the forward difference for the time derivative and central one for the second space derivative we obtain the following difference equation

$$u_{i,j+1} = (1 - 2\sigma)u_{i,j} - \sigma(u_{i+1,j} + u_{i-1,j}) + kF_{i,j}. \tag{7.20}$$

If $\sigma = \dfrac{1}{2}$ then the difference equation (7.15) has the form

$$u_{i,j+1} = \frac{u_{i-1,j} + u_{i+1,j}}{2} + kF_{i,j}. \tag{7.21}$$

If $\sigma = \dfrac{1}{6}$ then the difference equation (7.15) has the form

$$u_{i,j+1} = \frac{u_{i-1,j} + 4u_{i,j} + u_{i+1,j}}{6} + kF_{i,j}. \tag{7.22}$$

The error of approximations for the approximate solutions obtained by using equations (7.21) and (7.22) are

$$|u - \tilde{u}| \le \frac{T}{4}\left(M_2 + \frac{M_4}{3}\right)h^2,$$

$$\tag{7.23}$$

$$|u - \tilde{u}| \le \frac{T}{72}\left(\frac{M_3}{3} + \frac{M_6}{5}\right)h^4,$$

where $\tilde{u}$ is the exact solution of the considered problem and u is the approximate one and

$$M_i = \max_{0 \le x \le L, \ 0 \le t \le T}\left\{\left|\frac{\partial^i u}{\partial x^i}\right|\right\}, \qquad i = 2, 4,$$

$$M_i = \max_{0 \le x \le L, \ 0 \le t \le T}\left\{\left|\frac{\partial^i u}{\partial t^i}\right|\right\}, \qquad i = 3, 6.$$

Example 7.9 *Determine the numerical solution for the following problem*

$$\frac{\partial u}{\partial t} = \frac{\partial^2 u}{\partial x^2} \qquad (t > 0, \quad 0 < x < 1),$$

$$u(0,t) = u(1,t) = 0, \qquad (t > 0),$$
$$u(x,0) = x, \qquad (0 < x < 1),$$

by using the finite difference method for the

a) *mesh $h = \dfrac{1}{5}$, and $k = \dfrac{1}{50}$;*

b) *mesh $h = k = \dfrac{1}{5}$.*

Solution.

a) Taking $x_i = ih$, $t_j = jk$, $h = \dfrac{1}{5}$, $k = \dfrac{1}{50}$, we have that $\sigma = \dfrac{1}{2}$ and therefore the difference equation corresponding to this problem is

$$u_{i,j+1} = \frac{u_{i-1,j} + u_{i+1,j}}{2}.$$

From the conditions we get

$$u_{0,j} = x_{5,j} = 0, \quad j = 0,1,2,\ldots, \qquad u_{i,0} = \frac{i}{5}, \ i = 0,1.2,3,4,5.$$

So we get

$$u_{1,1} = \frac{u_{0,0} + u_{2,0}}{2} = \frac{1}{5}, \quad u_{2,1} = \frac{u_{1,0} + u_{3,0}}{2} = \frac{2}{5}, \quad u_{3,1} = \frac{u_{2,0} + u_{4,0}}{2} = \frac{3}{5},$$

$$u_{4,1} = \frac{u_{3,0} + u_{5,0}}{2} = \frac{4}{5}, \quad u_{1,2} = \frac{u_{0,1} + u_{2,1}}{2} = \frac{1}{5}, \quad u_{2,2} = \frac{u_{1,1} + u_{3,1}}{2} = \frac{2}{5},$$

$$u_{3,2} = \frac{u_{2,1} + u_{4,1}}{2} = \frac{3}{5}, \quad u_{4,2} = \frac{u_{3,1} + u_{5,1}}{2} = \frac{3}{10}, \quad u_{1,3} = \frac{u_{0,2} + u_{2,2}}{2} = \frac{1}{5},$$

$$u_{2,3} = \frac{u_{1,2} + u_{3,2}}{2} = \frac{2}{5}, \quad u_{3,3} = \frac{u_{2,2} + u_{4,2}}{2} = \frac{7}{20}, \quad u_{4,3} = \frac{u_{3,2} + u_{5,2}}{2} = \frac{3}{10}.$$

b) In this case we have $\sigma = 5$, and the difference equation has the form

$$u_{i,j+1} = -9u_{i,j} + 5(u_{i-1,j} + u_{i+1,j}). \tag{7.24}$$

Show that this system is not stable.

Example 7.10 *Determine the numerical solution for the following problem*

$$\frac{\partial u}{\partial t} = \frac{\partial^2 u}{\partial x^2} \quad (t > 0, \ 0 < x < 1),$$

$$u(x,0) = x \quad (0 < x < 1),$$

$$\frac{\partial u(0,t)}{\partial t} = 0, \quad \frac{\partial u(1,t)}{\partial t} = 0 \quad (t > 0).$$

Solution. If $h = \dfrac{1}{5}$, $k = \dfrac{1}{50}$, then $\sigma = \dfrac{1}{2}$ and therefore we have

$$u_{i,j+1} = \frac{u_{i-1,j} + u_{i+1,j}}{2}. \tag{7.25}$$

From the given conditions we get

$$\frac{u_{0,1} - u_{1,j}}{h} = 0, \qquad \frac{u_{4,1} - u_{5,j}}{h} = 0,$$

wherefrom

$$u_{0,j} = u_{1,j}, \ u_{4,j} = u_{5,j}, \quad j = 0,1,2,\ldots, \qquad u_{i,0} = \frac{i}{5}, \ i = 0,1,2,3,4,5.$$

So we obtain

$$u_{1,1} = \frac{u_{0,0} + u_{2,0}}{2} = u_{0,1} = \frac{1}{5}, \qquad u_{2,1} = \frac{u_{1,0} + u_{3,0}}{2} = \frac{2}{5},$$

$$u_{3,1} = \frac{u_{2,0} + u_{4,0}}{2} = \frac{3}{5}, \qquad u_{4,1} = \frac{u_{3,0} + u_{5,0}}{2} = u_{5,1} = \frac{4}{5},$$

$$u_{0,2} = u_{1,2} = \frac{u_{0,1} + u_{2,1}}{2} = \frac{3}{10}, \qquad u_{2,2} = \frac{u_{1,1} + u_{3,1}}{2} = \frac{2}{5},$$

$$u_{3,2} = \frac{u_{2,1} + u_{4,1}}{2} = \frac{3}{5}, \qquad u_{4,2} = u_{5,2} = \frac{u_{3,1} + u_{5,1}}{2} = \frac{7}{10},$$

$$u_{0,3} = u_{1,3} = \frac{u_{0,2} + u_{2,2}}{2} = \frac{7}{20}, \qquad u_{2,3} = \frac{u_{1,2} + u_{3,2}}{2} = \frac{9}{20},$$

$$u_{3,3} = \frac{u_{2,2} + u_{4,2}}{2} = \frac{11}{20}, \qquad u_{4,3} = u_{5,3} = \frac{u_{3,2} + u_{5,2}}{2} = \frac{13}{20}.$$

Example 7.11 *Determine the numerical solution for the following problem*

$$\frac{\partial u}{\partial t} = \frac{\partial^2 u}{\partial x^2} + \frac{\partial^2 u}{\partial y^2} \qquad (t > 0, \quad 0 < x < y < 1),$$

with boundary conditions

$$u(0,y,t) = u(x,0,t) = u(x,x,t) = 0,$$

and initial condition $u(x,y,0) = x(1-y)$.

Solution. The given PDE is the two dimensional heat equation. Let us take

$$x_i = ih, \qquad y_j - kj, \qquad h = k,$$

for $i = 0, 1, 2, 3, 4, 5$, $j = 0, 1, 2, 3, 4, 5$. Then after replacing the second derivatives with corresponding difference quotients we obtain

$$\frac{\partial^2 u}{\partial x^2} + \frac{\partial^2 u}{\partial y^2} \approx \frac{u_{i+1,j} + u_{i-1,j} + u_{i,j+1} + u_{i,j-1} - 4u_{i,j}}{h^2}.$$

Let us replace the time derivative $\dfrac{\partial u}{\partial t}$ with

$$\frac{u_{i,j}(t + \Delta t) - u_{i,j}(t)}{\Delta t}.$$

Then the difference equation corresponding to the two dimensional heat equation is

$$\frac{u_{i,j}(t + \Delta t) - u_{i,j}(t)}{\Delta t} = \frac{u_{i+1,j} + u_{i-1,j} + u_{i,j+1} + u_{i,j-1} - 4u_{i,j}}{h^2},$$

or

$$u_{i,j}(t + \Delta t) = \frac{\Delta t}{h^2} \left(u_{i+1,j}(t) + u_{i-1,j}(t) + u_{i,j+1}(t) + u_{i,j-1}(t) \right) + \left(1 - 4\frac{\Delta t}{h^2} \right) u_{i,j}(t).$$

Let us use the mesh $h = \dfrac{1}{4}$, $\Delta t = \dfrac{1}{64}$. Then $\dfrac{\Delta t}{h^2} = \dfrac{1}{4}$, and the corresponding difference equation has the form

$$u_{i,j}(t + \Delta t) = \frac{\Delta t}{h^2} \left(u_{i+1,j}(t) + u_{i-1,j}(t) + u_{i,j+1}(t) + u_{i,j-1}(t) \right).$$

Using the initial and boundary conditions together with the previous equation we obtain

$$u_{0,j} = u_{i,4} = 0, \quad u_{i,i} = \frac{i}{4} \cdot \frac{4 - i}{4} \quad \text{for} \quad t = 0,$$

$$u_{0,j} = u_{i,1} = u_{i,i} = 0, \quad i = 0, 1, 2, 3, 4, \quad j = 0, 1, 2, 3, 4,$$

for each value $t = \dfrac{1}{64}$, $t = \dfrac{1}{32}$, $t = \dfrac{3}{64}$, $t = \dfrac{1}{16}$.

We have

$$u_{1,2}\left(\frac{1}{64}\right) = \frac{1}{4} \left(u_{2,2}(0) + u_{1,3}(0) + u_{1,1}(0) \right) = \frac{1}{8};$$

$$u_{1,3}\left(\frac{1}{64}\right) = \frac{1}{4} \left(u_{2,3}(0) + u_{1,2}(0) \right) = \frac{1}{16};$$

$$u_{2,3}\left(\frac{1}{64}\right) = \frac{1}{4} \left(u_{3,3}(0) + u_{1,3}(0) + u_{3,2}(0) \right) = \frac{1}{8};$$

$$u_{1,2}\left(\frac{1}{32}\right) = u_{2,3}\left(\frac{1}{32}\right) = \frac{1}{4} \cdot u_{1,3}\left(\frac{1}{64}\right) = \frac{1}{64};$$

$$u_{1,3}\left(\frac{1}{32}\right) = \frac{1}{4}\left(u_{2,3}\left(\frac{1}{64}\right) + u_{1,2}\left(\frac{1}{64}\right)\right) = \frac{1}{16};$$

$$u_{1,2}\left(\frac{3}{64}\right) = u_{2,3}\left(\frac{3}{64}\right) = \frac{1}{4}\cdot u_{1,3}\left(\frac{1}{32}\right) = \frac{1}{64};$$

$$u_{1,3}\left(\frac{3}{64}\right) = \frac{1}{4}\left(u_{2,3}\left(\frac{1}{32}\right) + u_{1,2}\left(\frac{1}{32}\right)\right) = \frac{1}{128};$$

$$u_{1,2}\left(\frac{1}{16}\right) = u_{2,3}\left(\frac{1}{16}\right) = \frac{1}{4}\cdot u_{1,3}\left(\frac{3}{64}\right) = \frac{1}{512};$$

$$u_{1,3}\left(t + \frac{1}{64}\right) = \frac{1}{4}\left(u_{2,3}\left(\frac{3}{64}\right) + u_{1,2}\left(\frac{3}{64}\right)\right) = \frac{1}{128};$$

$$u_{1,2}\left(\frac{1}{16} + \frac{1}{64}\right) = u_{2,3}\left(\frac{1}{16} + \frac{1}{64}\right) = \frac{1}{4}\cdot u_{1,3}\left(\frac{1}{16}\right) = \frac{1}{512};$$

$$u_{1,3}\left(\frac{1}{16} + \frac{1}{64}\right) = \frac{1}{4}\left(u_{2,3}\left(\frac{1}{16}\right) + u_{1,2}\left(\frac{1}{16}\right)\right) = \frac{1}{1024}.$$

Exercise 7.12 *Determine the numerical solution for the following problem*

$$\frac{\partial u}{\partial t} = \frac{\partial^2 u}{\partial x^2} + \frac{\partial^2 u}{\partial y^2} \qquad (t > 0, \quad 0 < x < 1, \quad 0 < y < 1),$$

with boundary conditions

$$u(0, y, t) = u(x, 0, t) = u(1, y, t) = u(x, 1, t) = 0,$$

and initial condition $u(x, y, 0) = xy$. *Use the mesh* $h = k = \dfrac{1}{5}$, $\Delta t = \dfrac{1}{125}$.

Exercise 7.13 *Determine the difference equation for three dimensional Laplace equation*

$$\frac{\partial^2 u}{\partial x^2} + \frac{\partial^2 u}{\partial y^2} + \frac{\partial^2 u}{\partial z^2} = 0 \qquad (t > 0, \quad 0 < x < 1, \quad 0 < y < 1, \quad 0 < z < 1),$$

by using the finite differences method.

Exercise 7.14 *Determine the difference equation for three dimensional heat equation*

$$\frac{\partial u}{\partial t} = \frac{\partial^2 u}{\partial x^2} + \frac{\partial^2 u}{\partial y^2} + \frac{\partial^2 u}{\partial z^2} \qquad (t > 0, \quad 0 < x < 1, \quad 0 < y < 1, \quad 0 < z < 1),$$

by using the finite differences method.

Example 7.15 *Determine the difference equation and the error of approximation corresponding to the mixed type problem for the hyperbolic equation*

$$\frac{\partial^2 u}{\partial t^2} = \frac{\partial^2 u}{\partial x^2} \qquad (t > 0, \quad 0 < x < L),$$

$$u(x,0) = f(x), \qquad \frac{\partial u(x,0)}{\partial t} = g(x) \qquad (0 < x < L),$$

$$u(0,t) = \phi(t), \qquad u(L,t) = \psi(t) \qquad (t > 0).$$

Estimate the error of approximation for $0 \le t \le T$.

Solution. Taking $x_i = ih$, $t_j = jk$, $i = 0,1,2,\dots$, $j = 0,1,2,\dots$, and replacing both spatial derivative

$$\frac{\partial^2 u}{\partial x^2} \quad \text{and the time derivative} \quad \frac{\partial^2 u}{\partial t^2}$$

by their difference quotients we get

$$\frac{u_{i,j+1} - 2u_{i,j} + u_{i,j-1}}{k^2} = \frac{u_{i+1,j} - 2u_{i,j} + u_{i-1,j}}{h^2}, \tag{7.26}$$

for $i = 0,1,2,\dots$, $j = 0,1,2,\dots$. Denoting by $\alpha = \dfrac{k}{h}$, we obtain

$$u_{i,j+1} = 2u_{i,j} - u_{i,j-1} + \alpha(u_{i,j+1} - 2u_{i,j} + u_{i,j+1}). \tag{7.27}$$

One can prove that the difference equation (7.27) is stable for $0 < \alpha \le 1$.

The error of approximation for the approximate solution obtained by using equations (7.27)

$$|u - \tilde{u}| \le \frac{T}{12}\left((M_4 h + 2M_3)T + T^2 M_4\right), \tag{7.28}$$

where $\tilde{u}$ is the exact solution and u is the exact solution of the considered problem and

$$M_i = \max_{0 \le x \le L, \ 0 \le t \le T}\left\{\left|\frac{\partial^i u}{\partial x^i}\right|, \left|\frac{\partial^i u}{\partial t^i}\right|\right\}, \qquad i = 3,4.$$

If we put

$$\frac{\partial u(x,0)}{\partial t} = g(x),$$

where g is from initial conditions

$$\frac{u_{i,1} - u_{i,0}}{k} = g(x_i) = g_i,$$

then from the relations

$$u_{i,0} = f_i, \qquad u_{i,1} = f_i + kg_i,$$

we obtain the values of the function u at the points (i,j), $i = 0, 1, \ldots,$ $j = 0, 1$. The error of approximation can be determined as

$$|u_{i,1} - \tilde{u}_{i,1}| \leq \frac{\alpha h}{2} M_2. \tag{7.29}$$

where

$$M_i = \max_{0 \leq x \leq L, \; 0 \leq t \leq T} \left\{ \left| \frac{\partial^2 u}{\partial x^2} \right|, \left| \frac{\partial^2 u}{\partial t^2} \right| \right\}$$

Exercise 7.16 *Determine the approximate solution of the problem*

$$\frac{\partial^2 u}{\partial x^2} = \frac{\partial^2 u}{\partial t^2} \qquad (0 < x < 1, \quad t > 0)$$

$$u(0,t) = u(1,t) = 0 \qquad (t > 0),$$

$$u(x,1) = x(1-x), \qquad \frac{\partial u(x,0)}{\partial t} = 0 \qquad (0 < x < 1).$$

Answer. Let us choose the same increment $h = k = 0,1$ on the $x-$ and the $t-$ axes, ($\alpha = 1$.) From the initial conditions we have

$$u_{i,0} = \frac{i}{10}\left(1 - \frac{i}{10}\right), \qquad i = 0, 1, \ldots, 10$$

$$u_{i,1} = u_{i,0} = \frac{i}{10}\left(1 - \frac{i}{10}\right), \qquad i = 0, 1, \ldots, 10.$$

We still have to solve the following system of equations

$$u_{i,j+1} = u_{i-1,j} + u_{i,j} - u_{i,j-1}, \qquad i = 1, 2, \ldots, 10, \quad j = 1, 2, \ldots 10.$$

Example 7.17 *Determine the numerical solution of the problem*

$$\frac{\partial^2 u}{\partial t^2} = \frac{\partial^2 u}{\partial x^2} \qquad (0 < x < 1, \quad t > 0),$$

$$u(x,0) = 0, \qquad \frac{\partial u(x,0)}{\partial t} = 1 \qquad (0 < x < 1),$$

$$u(0,t) = 0, \qquad u(1,t) = 0 \qquad (t > 0).$$

Estimate the error of approximation for $0 \leq t \leq 1$.

Solution. Taking $h = k = \dfrac{1}{10}$, we obtain the difference equation

$$u_{i,j+1} = u_{i+1,j} + u_{i-1,j} - u_{i,j-1}. \qquad (7.30)$$

If we replace $\dfrac{\partial u(x,0)}{\partial t}$ with $\dfrac{u_{i,1} - u_{i,0}}{k}$, then from the initial conditions we get

$$u_{i,0} = 0, \qquad u_{i,1} = \frac{1}{10} \cdot 1 \quad i = 0, 1, 2, \ldots, 10.$$

Also from the boundary conditions we have

$$u_{0,j} = u_{10,j} = 0, \quad j = 0, 1, \ldots 10.$$

From the previous equations we get

$$u_{1,2} = u_{2,1} + u_{0,1} - u_{1,0} = \frac{1}{10}, \qquad u_{2,2} = u_{3,1} + u_{1,1} - u_{2,0} = \frac{2}{10},$$

$$u_{3,2} = u_{4,1} + u_{2,1} - u_{3,0} = \frac{2}{10}, \qquad u_{4,2} = u_{5,1} + u_{3,1} - u_{4,0} = \frac{2}{10}.$$

Similarly we obtain

$$u_{5,2} = u_{6,2} = u_{7,2} = u_{8,2} = u_{9,2} = \frac{2}{10},$$

$$u_{1,3} = u_{2,2} + u_{0,2} - u_{1,1} = \frac{1}{10}, \qquad u_{2,3} = u_{3,2} + u_{1,2} - u_{2,1} = \frac{2}{10},$$

$$u_{3,3} = u_{4,2} + u_{2,2} - u_{3,1} = \frac{3}{10}, \qquad u_{4,3} = u_{5,2} + u_{3,2} - u_{3,1} = \frac{3}{10},$$

$$u_{5,3} = u_{6,3} = u_{7,3} = u_{8,3} = u_{9,3} = \frac{3}{10},$$

$$u_{1,4} = u_{2,3} + u_{0,3} - u_{1,2} = \frac{1}{10}, \qquad u_{2,4} = u_{3,3} + u_{2,3} - u_{2,2} = \frac{3}{10},$$

$$u_{3,4} = u_{4,3} + u_{2,3} - u_{3,2} = \frac{4}{10}, \qquad u_{4,4} = u_{5,3} + u_{3,3} - u_{3,2} = \frac{4}{10},$$

$$u_{5,4} = u_{6,4} = u_{7,4} = u_{8,4} = u_{9,4} = \frac{2}{5},$$

$$u_{1,5} = \frac{3}{10}, \quad u_{2,5} = u_{3,5} = u_{4,5} = u_{5,5} = u_{6,5} = u_{7,5} = u_{8,5} = u_{9,5} = \frac{1}{2},$$

$$u_{1,6} = \frac{3}{10}, \quad u_{2,6} = u_{3,6} = u_{4,6} = u_{5,6} = u_{6,6} = u_{7,6} = u_{8,6} = u_{9,6} = \frac{6}{10},$$

$$u_{1,7} = \frac{3}{10}, \quad u_{2,7} = u_{3,7} = u_{4,7} = u_{5,7} = u_{6,7} = u_{7,7} = u_{8,7} = u_{9,7} = \frac{7}{10},$$

$$u_{1,8} = \frac{4}{10}, \quad u_{2,8} = u_{3,8} = u_{4,8} = u_{5,8} = u_{6,8} = u_{7,8} = u_{8,8} = u_{9,8} = \frac{8}{10},$$

$$u_{1,9} = \frac{5}{10}, \quad u_{2,9} = u_{3,9} = u_{4,8} = u_{5,9} = u_{6,9} = u_{7,9} = u_{8,9} = u_{9,9} = \frac{9}{10}.$$

Exercise 7.18 *Determine the numerical solution of the problem*

$$\frac{\partial^2 u}{\partial t^2} - \frac{\partial^2 u}{\partial x^2} = x + t \qquad (t > 0, \quad 0 < x < 1),$$

$$u(x,0) = x^2 \qquad \frac{\partial u(x,0)}{\partial t} = x, \qquad (0 < x < 1),$$

$$u(0,t) = 0, \qquad u(1,t) = 0, \qquad (t > 0).$$

Answer. Taking $h = k = \dfrac{1}{10}$, we obtain the difference equation

$$u_{i,j+1} = u_{i+1,j} + u_{i-1,j} - u_{i,j-1} + x_j + t_i = u_{i+1,j} + u_{i-1,j} - u_{i,j-1} + \frac{i}{10} + \frac{j}{10}.$$

Example 7.19 (von Neumann's criteria for stability) *Examine the behavior of the special exponential solution*

$$u_{i,j} = e^{\imath i \theta} e^{\imath j \theta}, \tag{7.31}$$

where $\theta \in \mathbf{R}$, $\imath = \sqrt{-1}$ *and* $\lambda = \lambda_1 + \imath \lambda_2$, *of the difference equation (correspoding to the wave equation)*

$$\frac{u_{i,j+1} - 2u_{i,j} + u_{i,j-1}}{k^2} - \frac{u_{i+1,j} - 2u_{i,j} + u_{i-1,j}}{h^2} = 0, \tag{7.32}$$

with initial conditions

$$u_{i,0} = f_i \tag{7.33}$$

and

$$u_{i,1} = f_i + g_i k + \frac{f_i''}{2} k^2, \tag{7.34}$$

where $f_i = f(ih), g_i = g(ih)$ *and* $f_i'' = f''(ih)$ *for*

a) $k/h > 1$;

b) $k/h \leq 1$.

Solution. Putting the special form of the solution (7.31) in (7.32) we obtain the equation

$$\sin^2 \frac{\lambda}{2} = \frac{k^2}{h^2} \sin^2 \frac{\theta}{2}. \tag{7.35}$$

For $i = 0$ we obtain from (7.31) the initial condition $u_{i,0} = e^{\iota i\theta}$.

The solution (7.31) is continuously dependent from the initial condition if for every real θ the solution from (7.31) remains bounded as $j \to \infty$. This will be satisfied if the zeros λ of of the equation (7.35) have nonnegative imaginary part for every θ.

a) If $k/h > 1$ we can always find θ such that the right part of the equation (7.35) is greater of one. Then for such θ the zeros λ of the equation (7.35) are complex and conjugate to each other. Therefore one of them is with negative imaginary part and by the general statement from the beginning it is not stable as $h \to 0$ and $k \to 0$.

b) If $k/h \leq 1$, then all zeros of the equation (7.35) are real. Therefore $u_{i,j}$ from (7.31) are bounded as $i \to \infty$. Hence the solution $u_{i,j}$ of (7.32), (7.33) and (7.34) will converges to the solution u of the corresponding Cauchy problem for one-dimensional wave equation.

Exercise 7.20 *Let A, B, C, D be the corners of the parallelogram whose sides are the characteristics of the homogeneneous one-dimensional wave equation*

$$\frac{\partial^2 u}{\partial x^2} - \frac{\partial^2 u}{\partial t^2} = 0. \tag{7.36}$$

Prove that a function $u \in C^2(\mathbf{R}^2)$ is a solution of the equation (7.36) if and only if the function u satisfies the functional equation

$$u(A) + u(C) = u(B) + u(D) \tag{7.37}$$

for all parallelograms $ABCD$ obtained by the given construction.

Exercise 7.21 *Apply the preceding result for the construction of the solution of the mixed type problem for the one- dimensional wave equation on the strip $S = \{(x,y)|0 < x < \ell, t > 0\}$ with the usual initial conditions*

$$u(x,0) = f(x), \quad \frac{\partial u(x,0)}{\partial t} = g(x) \quad (0 < x < \ell),$$

and boundary conditions

$$u(0,t) = m_1(t), \quad u(\ell,t) = m_2(t) \quad (t > 0),$$

where $f, m_1, m_2 \in C^2, g \in C^1$ and $m_1(0) = f(0), m_1'(0) = g(0), m_1''(0) = f''(0)$, $m_2(0) = f(\ell), m_2'(0) = g(\ell), m_2''(0) = f''(\ell)$.

Hints. Divide the strip S with characteristics of the one-dimensional wave equation in the following way. First take the characteristics at the endpoints $(0,0)$ and $(\ell,0)$ of the interval $[0,\ell]$ till the borders $x = 0$, $x = \ell$. Then at the points of the intersections with the borders take characteristics at these intersection points, and continue this procedure. At the first part, which is a triangle on the interval $[0,\ell]$, we can find all values of the solution by the D'Alambert formula. Then in the next two neighbourhood triangles we can find all values of the solution by the functional equation (7.37). Continuing this procedure we can reach every point since it belongs to some part of the constructed partition of the strip.

Chapter 8

Lebesgue's Integral and the Fourier Transform

8.1 Lebesgue's Integral and the $L_2(Q)$ Space

8.1.1 Preliminaries

Let $A \subset \mathbb{R}^n$ be a (Lebesgue) measurable set. For a measurable function $f : A \to [0, \infty]$ its Lebesgue integral on a measurable subset E of A is given by

$$\int_E f(x)\,dx = \sup \int_E s(x)\,dx,$$

where the supremum is taken over all simple functions s, $s = \sum_{i=1}^{n} a_i \chi_{E_i}$, which satisfy the inequality $0 \le s \le f$ (we are using the convention $0 \cdot \infty = 0$) and χ_{E_i} are the characteristic functions of sets $E_i \in \Sigma$ $(i = 1, \dots, n)$ and $\cup E_i = E$, where Σ is the σ-algebra of measurable sets.

Theorem 8.1 (Beppo–Levi) *Let $\{f_n\}_{n \in \mathbb{N}}$ be a sequence of measurable functions on the set A such that*
1) $0 \le f_1(x) \le f_2(x) \le \dots < \infty$ for every $x \in A$,
2) $\lim_{n \to \infty} f_n(x) = f(x)$ for every $x \in A$.
Then f is measurable and $\lim_{n \to \infty} \int_A f_n(x)\,dx = \int_A f(x)\,dx$.

Theorem 8.2 (Fatou lemma) *If $\{f_n\}_{n \in \mathbb{N}}$ is a sequence of measurable functions defined on the set A and with values in $[0, \infty]$, then*

$$\int_A \liminf_{n \to \infty} f_n(x)\,dx \le \liminf_{n \to \infty} \int_A f_n(x)\,dx.$$

Let $L_1(A)$ denote the set of all measurable complex functions defined on the measurable set A with the property $\int_A |f(x)|\,dx < \infty$.

249

Theorem 8.3 (Lebesgue theorem on convergence) *Let $\{f_n\}_{n \in \mathbb{N}}$ be a sequence of measurable functions on the set A such that $f(x) = \lim_{n \to \infty} f_n(x)$ exists almost everywhere on A. If g is an integrable function such that $|f_n(x)| \leq |g(x)|$ a.e., $n = 1, 2, \ldots$, then f is integrable and*

$$\lim_{n \to \infty} \int_A |f_n(x) - f(x)| \, dx = 0.$$

Consider now the Lebesgue measure on the real line. Let function f be integrable on the interval $[a, b]$ and define

$$F(x) = \int_a^x f(t) \, dt \qquad (a \leq x \leq b).$$

Then F has a finite derivative almost everywhere on the interval $[a, b]$ and $F' = f$ almost everywhere on $[a, b]$ and F is *absolute continuous* on the interval, i.e., for every $\varepsilon > 0$ there exists $\delta > 0$ such that

$$\sum_{k=1}^{n-1} |F(x_{k+1}) - F(x_k)| < \varepsilon$$

for

$$\sum_{k=1}^{n-1} |x_{k+1} - x_k| < \delta \text{ and } a = x_1 \leq x_2 \leq \ldots \leq x_n = b.$$

Theorem 8.4 (Fubini) *If f is an integrable function on a measurable set $A = A_1 \times A_2$ from $\mathbb{R}^n$, where A_1 and A_2 are measurable subsets of $\mathbb{R}^m$ $(m \leq n)$ and $\mathbb{R}^{n-m}$, respectively, then the functions h and g defined by*

$$h(x_1, \ldots, x_m) = \int_{A_2} f(x_1, \ldots, x_n) \, dx_{m+1} \ldots dx_n$$

and

$$g(x_{m+1}, \ldots, x_n) = \int_{A_1} f(x_1, \ldots, x_n) \, dx_1 \ldots dx_m$$

are integrable and

$$\int_A f(x_1, \ldots, x_n) \, dx_1 \ldots dx_n = \int_{A_1} (\int_{A_2} f(x_1, \ldots, x_n) \, dx_{m+1} \ldots dx_n) dx_1 \ldots dx_m$$

$$= \int_{A_1} (\int_{A_2} f(x_1, \ldots, x_n) \, dx_{m+1} \ldots dx_n) \, dx_1 \ldots dx_m.$$

Let X be a vector space. A function $\| \cdot \| : X \to [0, \infty)$ is a *norm* on X if for every $x, y \in X$ and every scalar α it satisfies the following conditions:

(i) $||x|| = 0$ iff $x = 0$;

(ii) $||\alpha x|| = |\alpha| \cdot ||x||$;

(iii) $||x + y|| \leq ||x|| + ||y||$.

Then the pair $(X, || \cdot ||)$ is called *normed space*.

Let p be a fixed real number greater or equal one. Then we say that a function defined on a measurable subset Q of $\mathbf{R}^n$ belongs to $L_p(Q)$ if $\int_Q |f(x)|^p \, dx < \infty$, where functions are equal if they are equal almost everywhere, i.e., they differ only on a set of a measure zero.

A complete normed space is called Banach space. The set $L_p(Q)$ endowed with the following norm

$$||f||_{L_p(Q)} = (\int_Q |f(x)|^p \, dx)^{\frac{1}{p}}$$

is a Banach space.

Let $f, g \in L_p(Q)$ and $1/p + 1/q = 1$. Then we have the Hölder inequality

$$|\int_Q f(x)g(x) \, dx| \leq ||f||_{L_p(Q)}||g||_{L_q}$$

and the Minkowski inequality

$$||f + g||_{L_p(Q)} \leq ||f||_{L_p(Q)} + ||g||_{L_p(Q)}.$$

If $Q = \mathbf{R}^n$, then we denote $L_p = L_p(\mathbf{R}^n)$. In the special case $p = 2$ the space $L_2(Q)$ is a Hilbert space, i.e., a Banach space whose norm is induced by the following scalar product (see more in Chapter 10.)

$$(f|g)_{L_2(Q)} = \int_Q f(x)\overline{g(x)} \, dx,$$

where $\overline{g}$ is the conjugate of the complex valued function g.

Theorem 8.5 (Continuity of the integral) *Every function f from $L_2(Q)$ is continuous in mean, i.e., for every $\varepsilon > 0$ there exists $\delta > 0$ such that*

$$||f(x + v) - f(x)||_{L_2(Q)} < \varepsilon \text{ for } |v| < \delta, \, x + v \in Q.$$

We denote by $L_{loc}^p(Q)$ the set of all functions which belong to $L_1(K)$ for some compact subset K of Q.

Change of variables

Let $Q \subset \mathbf{R}^n$ be a bounded open set with a boundary ∂Q of measure zero which is mapped with a function $\Phi : \overline{Q} \to \mathbf{R}^n$, $(\Phi_1, \ldots, \Phi_n)$ of class C^1 as a bijection on a bounded open set Ω with a boundary $\partial \Omega$ of measure zero with the Jacobian

$$J = \frac{\partial(\Phi_1, \ldots, \Phi_n)}{\partial(x_1, \ldots, x_n)} \neq 0.$$

Then the following holds.

1) The functions Φ and Φ^{-1} map every set of measure zero on a set of a measure zero;

2) The function $f : \Omega \to \mathbf{R}$ is (Lebesgue) integrable on the set Ω if and only if the function $(f \circ \Phi)|J|$ is (Lebesgue) integrable on the set Q. Then the following equality holds

$$\int\limits_{\Omega} f(y) \, dy = \int\limits_{Q} (f \circ \Phi)(x)|J| \, dx.$$

8.1.2   Examples and Exercises

Example 8.1 *Prove that the Dirichlet function defined by*

$$\mathcal{D}(x) = \begin{cases} 1 & for\ x\ rational\,, \\ 0 & for\ x\ irrational. \end{cases}$$

is Lebesgue integrable on the interval $[0, 1]$ but not Riemann integrable.

Solution. The function $\mathcal{D}$ is equal almost everywhere zero, which implies that its Lebesgue integral is zero. On the other hand, for this function all upper Darboux sums are equal one and all lower Darboux sums are equal zero, which imply the non-existence of the Riemann integral.

Exercise 8.2 *Prove that the following functions are Lebesgue integrable and find the corresponding Lebesgue integrals:*

a)

$$f(x) = \begin{cases} x^2 & for\ x\ irrational\ and\ x > \frac{1}{3}, \\ x^3 & for\ x\ irrational\ and\ x < \frac{1}{3}, \\ 0 & for\ x\ rational. \end{cases}$$

b)

$$\mathcal{R}(x) = \begin{cases} \frac{1}{n} & for\ x = \frac{m}{n}, m\ and\ n\ have\ no\ common\ divisors\,, \\ 0 & for\ x\ irrational \end{cases}$$

(the Riemann function).

c) $g(x) = sgn(\sin \frac{\pi}{x})$.

Answers. a) $\frac{35}{108}$, b) 0, c) $1 - 2\ln 2$.

Exercise 8.3 *For which real number a the function*

$$f(x) = \frac{1}{|x|^a}$$

is integrable on $B(0,1) = \{x \mid |x| < 1\}$?

Answer. $a < n$.

Exercise 8.4 *Show that for the integral*

$$\int\limits_0^\infty \int\limits_0^\infty e^{-xy} \sin x \sin y \, dx dy$$

there exist the iterated integrals but not the double integral.

Exercise 8.5 *Show that for the integral*

$$\int\limits_0^1 \int\limits_0^1 \frac{x^2 - y^2}{(x^2 + y^2)^2} \, dx dy$$

there exist the iterated integrals but they are different.

Example 8.6 *Show that for the integral*

$$\int\limits_{-1}^1 \int\limits_{-1}^1 \frac{xy}{(x^2 + y^2)^2} \, dx dy$$

there exist the iterated integrals and they are equal, but the double integral does not exist.

Solution. Both iterated integrals are zero. If we suppose that there exists the double integral then it would exist also on the square $\{(x,y) \mid 0 \leq x \leq 1, \, 0 \leq y \leq 1\}$ and it would be possible to apply the Fubini theorem, but for $x = 0$ the function

$$\int\limits_0^1 \frac{xy}{(x^2 + y^2)^2} \, dy = \frac{1}{2x} - \frac{x}{2(x^2 + 1)}$$

is not integrable.

Exercise 8.7 *Let f be a bounded measurable function on the region Q. Prove that the function*

$$g(x) = \int_Q \frac{f(z)}{|x - z|^a}\, dz$$

(Hilbert transform) belongs to $C^k(\mathbf{R}^n)$ if $k < n - [a]$, where $[a]$ is the integer part of a.

Exercise 8.8 *Find which functions belong to the space $C^k(Q)$ or $L_p(Q)$:*
  a) $f(x) = \dfrac{\sin x}{x^{3/4}}$ *on $Q = (0,1)$;*

  b)

$$g(x) = \begin{cases} x^{-1/3}\cos x & \text{for } x \text{ irrational,} \\ x^{-1/3} & \text{for } x \text{ rational and } x \neq 0, \\ 0 & \text{for } x = 0, \end{cases}$$

on $Q = [0,1]$;
  c)

$$h(x) = \begin{cases} \dfrac{\sin(xy)}{x^2 + y^2} & \text{for } x^2 + y^2 \neq 0, \\ 0 & \text{for } (x,y) = (0,0). \end{cases}$$

Exercise 8.9 *Find the real number a such that the function $|x|^{-a}$ belongs $L_2(Q)$ if*
a) $Q = \{x \mid |x| < 1\}$,   b) $Q = \{x \mid |x| > 1\}$,   c) $Q = \mathbf{R}^n$.

Answers. a) $a < \frac{n}{2}$, b) $a > \frac{n}{2}$, c) there is no a.

Exercise 8.10 *Let σ be a continuous and positive function defined on a region Q. Denote by $L_{2,\sigma}(Q)$ the space of functions such that $\sigma |f|^2 \in L_1(Q)$. Prove that*

$$(f|g) = \int_Q \sigma(x) f(x) \overline{g(x)}\, dx$$

is a scalar product (see more details in Chapter 10.) on $L_{2,\sigma}(Q)$.

Exercise 8.11 *If σ is a bounded function on Q (see Exercise 8.10) then*

$$L_2(Q) \subset L_{2,\sigma}(Q).$$

Example 8.12 *Show that the following systems of functions are orthogonal in the corresponding spaces of functions.*
a) $1, \sin x, \cos x, \sin 2x, \cos 2x, \ldots$ in $L_2(0, 2\pi)$;

b) $P_n(x) = \dfrac{1}{2^n n!}\sqrt{\dfrac{2n+1}{2}}\dfrac{d^n}{dx^n}(x^2 - 1)$, $n = 0, 1, 2, \ldots$ in $L_2(-1,1)$ (Legendre

polynomials);
c) $H_n(x) = (-1)^n e^{x^2} \frac{d^n}{dx^n} e^{-x^2}$, $\quad n = 0, 1, 2, \ldots \quad$ *in* $L_{2, e^{-x^2}}(-\infty, +\infty)$ *(Hermite polynomials).*
d)

$$
\varphi_{m,n}(x) = \begin{cases} 2^{m/2} & for \ \dfrac{n-1}{2^m} \leq x < \dfrac{n - \frac{1}{2}}{2^m}, \\[2mm] -2^{m/2} & for \ \dfrac{n - \frac{1}{2}}{2^m} \leq x < \dfrac{n}{2^m}, \\[2mm] 0 & for \ x \notin [\dfrac{n-1}{2^m}, \dfrac{n}{2^m}] \end{cases}
$$

for $m \in \mathbf{N}, 1 \leq n \leq 2^m$ *in* $L_2(0,1)$ *(Haar's system).*

Exercise 8.13 *Let* $A \subset \mathbf{R}^n$ *be a (Lebesgue) measurable set and* f *a function* $f : A \times (a,b) \to \mathbf{C}$ *such that for almost every (fixed)* $x \in A$ *the function* $t \mapsto f(x,t)$ *is continuous on the interval* (a,b), *and for* $t \in (a,b)$ *the function* $x \mapsto f(x,t)$ *is measurable, and there exists an integrable function* h *on* A *such that for every* $t \in (a,b)$ *we have* $|f(x,t)| \leq h(x)$ *for every* $x \in A$. *Prove that the function* F *defined by*

$$F(t) = \int_A f(x,t)dx$$

is continuous on the interval (a,b).

Hints. Apply the Lebesgue theorem on convergence on the sequence of functions $\{f(x, t_i)\}_{i \in \mathbf{N}}$, where $\{t_i\}_{i \in \mathbf{N}}$ is a convergent sequence from the interval (a,b).

Exercise 8.14 *Let* $A \subset \mathbf{R}^n$ *be a (Lebesgue) measurable set and* f *a function* $f : A \times (a,b) \to \mathbf{C}$ *such that for almost every (fixed)* $x \in A$ *the functions* $t \mapsto f(x,t)$ *and* $t \mapsto D_t f(x,t)$ *are continuous on the interval* (a,b), *and for* $t \in (a,b)$ *the functions* $x \mapsto f(x,t)$ *and* $x \mapsto D_t f(x,t)$ *are measurable and there exist integrable functions* h_1 *and* h_2 *on* A *such that for every* $t \in (a,b)$ *we have* $|f(x,t)| \leq h_1(x)$ *and* $|D_t f(x,t)| \leq h_2(x)$ *for almost every* $x \in A$. *Prove that the function* F *defined by*

$$F(t) = \int_A f(x,t)\,dx$$

belongs to $C^1(a,b)$ *and*

$$F'(t) = \int_A D_t f(x,t)\,dx \qquad (t \in (a,b)).$$

Hints. Apply the Lagrange Mean value theorem and Lebesgue theorem on convergence.

8.2 Delta Nets

8.2.1 Preliminaries

A continuous function φ has a *compact support* in region Q if $\varphi(x) = 0$ outside of some compact subset of Q. We denote

$$\text{supp } \varphi = \overline{\{x \mid \varphi(x) \neq 0\}}$$

and by $C_0^s(Q)$ the set of all functions which have continuous derivative till the order s on Q and they have compact support contained in Q.

We introduce a function δ_1 defined by

$$\delta_1(x) = \begin{cases} ce^{-\dfrac{1}{1-t^2}} & \text{for } -1 < t < 1, \\ 0 & \text{for } |t| \geq 1. \end{cases}$$

δ_1 has the following properties:
a) it belongs to $C^\infty(Q)$;
b) $\delta_1(t) \geq 0$, $\delta_1(t) = 0$ for $|t| \geq 1$.
We choose the constant c such that the following condition is satisfied
c) $\int_{\mathbf{R}} \delta_1(t)dt = 1$.
In the following considerations we can take instead the function δ_1 any other function which satisfies the conditions a), b) and c).

For $h > 0$ the function δ_h defined by

$$\delta_h(x) = \frac{1}{h^n}\delta_1\left(\frac{|x|}{h}\right) \qquad (x \in \mathbf{R}^n)$$

has the following properties
1) it belongs to $C^\infty(Q)$;
2) $\delta_h(t) \geq 0$, $\delta_h(t) = 0$ for $|t| \geq h$;
3) $\int_{\mathbf{R}^n} \delta_h(t)dt = 1$;
4) For every $\alpha \in \mathbf{Z}_+^n$ and every $x \in \mathbf{R}^n$ we have

$$|D^\alpha \delta_h(|x|)| \leq \frac{c_\alpha}{h^{n+|\alpha|}},$$

where $c_\alpha > 0$ is a constant. We shall call $\{\delta_h\}_{h>0}$ *delta net*.

If $f \in L_2(Q)$ (or $L_1(Q)$) then

$$\|f_h - f\|_{L_2(Q)} \to 0 \text{ (or } \|f_h - f\|_{L_1(Q)} \to 0) \text{ as } h \to 0,$$

where f_h is defined by

$$f_h(x) = \int_Q f(y)\delta_h(|x - y|)\,dy \qquad (x \in Q)$$

and f_h is called *mollifier*. If $f \in C(\overline{Q})$, then $|f_h(x) - f(x)| \to 0$ as $h \to 0$ uniformly on each compact subset of Q.

8.2.2 Examples and Exercises

Example 8.15 *Let*

$$f_h(x) = \int_Q f(y)\delta_h(|x-y|)\, dy.$$

Prove

a) *If $f \in L_2(Q)$, then*

$$\lim_{h\to 0} \|f_h - f\|_{L_2(Q)} = 0;$$

b) *If $f \in L_1(Q)$, then*

$$\lim_{h\to 0} \|f_h - f\|_{L_1(Q)} = 0;$$

c) *If $f \in C(\overline{Q})$, then*

$$\lim_{h\to 0} |f_h(x) - f(x)| = 0$$

   uniformly on every compact subset of Q.

Solution.

a) Using the properties 2) and 3) of δ_h and the Cauchy–Schwartz inequality we have

$$|f_h(x) - f(x)|^2 = \left| \int_{|x-y|<h} f(y)\delta_h(|x-y|)\, dy - f(x) \int_{|x-y|<h} \delta_h(|x-y|)\, dy \right|^2$$

$$\leq \int_{|x-y|<h} \delta_h^2(|x-y|)\, dy \int_{|x-y|<h} |f(y) - f(x)|^2\, dy$$

$$\leq \frac{M}{h^n} \int_{|v|<h} |f(x+v) - f(x)|^2\, dv$$

for $M > 0$. Applying the Fubini theorem 8.4 we obtain further

$$\|f_h - f\|_{L_2(Q)}^2 = \frac{M}{h^n} \int_Q dx \int_{|v|<h} |f(x+v) - f(x)|^2\, dv$$

$$= \frac{M}{h^n} \int_{|v|<h} dv \int_Q |f(x+v) - f(x)|^2\, dv.$$

Using now the continuity of the integral we obtain the desired conclusion.

b) The proof is analogous to a).

Example 8.16 *Prove that the set $C_0^\infty(Q)$ is dense in the spaces $L_1(Q)$ and $L_2(Q)$.*

Solution. We shall prove only the case $L_2(Q)$ since for $L_1(Q)$ the proof is analogous. For an arbitrary $\varepsilon > 0$ by the absolute continuity of the Lebesgue integral there exists $\delta > 0$ such that

$$\int\limits_{Q \setminus Q_\delta} |f(x)|^2 \, dx < \frac{\varepsilon^2}{4},$$

where Q_δ is a subset of Q which contains all points whose distance till the boundary ∂Q is greater than δ. We introduce a function $g \in L_2(Q)$ in the following way

$$g(x) = \begin{cases} f(x) & \text{for } x \in Q_\delta \\ 0 & \text{for } x \in Q \setminus Q_\delta. \end{cases}$$

It is obvious that

$$\|g - f\|_{L_2(Q)} \le \frac{\varepsilon}{2}.$$

By Example 8.15 there exists $h_0 > 0$ such that $\|g_h - g\|_{L_2(Q)} < \frac{\varepsilon}{2}$ for $0 < h \le h_0$. For enough small h we obtain $g_h \in C_0^\infty(Q)$, and we have

$$\|f - g_h\|_{L_2(Q)} \le \|f - g\|_{L_2(Q)} + \|g - g_h\|_{L_2(Q)} < \varepsilon.$$

Hence $C_0^\infty(Q)$ is dense in $L_2(Q)$.

Example 8.17 *Let $f \in L_1$ be a function which is almost everywhere zero outside of some compact subset K of Q. Prove that the function*

$$f_h(x) = \int\limits_Q f(y)\delta_h(|x - y|)dy$$

belongs to $C^\infty(Q)$ for $h > 0$ which is smaller than the distance of the set K to the border ∂Q (for $Q = \mathbf{R}^n$ we can take for h any positive number).

Solution. Since $\delta_h \in C^\infty(Q)$ we obtain by the property of Lebesgue integral on exchange with the derivative that $f_h \in C^\infty(Q)$.

If $h > 0$ is smaller than the distance d of the set K to the border ∂Q, then we have for every x from Q whose distance from K is greater than d

$$f_h(x) = \int\limits_K f(y)\delta_h(|x - y|) \, dy$$

since for $y \in K$ and $|x - y| \ge d$ always $\delta_h(|x - y|) = 0$ holds.

Example 8.18 *Let $Q \subset \mathbf{R}^n$ be an open set and $K \subset Q$ an compact set. Then there exists a function φ from $C_0^\infty(Q)$ such that $\varphi(x) = 0$ for every x in some neighbourhood of the set K and $0 \le \varphi \le 1$.*

Solution. Let $d > 0$ be the distance between the sets K and ∂Q. We introduce the set

$$K_{d/2} = \{x \mid d(x, K) \le \frac{d}{2}\}$$

and a function g by

$$g(x) = \begin{cases} 1 & \text{for } x \in K_{d/2}, \\ 0 & \text{for } x \in Q \setminus K_{d/2}. \end{cases}$$

Using Example 8.17, since the distance between $K_{d/2}$ and ∂Q is greater or equal with $d/2$, we have for $h = d/2$ that $g_h \in C_0^\infty(Q)$. We have for $x \in K_{d/2}$

$$g_h(x) = \int_Q g(y)\delta_h(|x - y|)\, dy = \int_{K_{d/2}} \delta_h(|x - y|)\, dy = \int_{|x-y|\le d/4} \delta_h(|x - y|)\, dy = 1.$$

We take $\varphi = g_h$.

Example 8.19 (Partition of the unity) *Let $K \subset \mathbf{R}^n$ be a compact set covered by open sets $O_1, \ldots, O_s$, i.e., $K \subset \cup_1^s O_i$. Then*
a) There exist open sets $O_1', \ldots, O_s'$ such that $\overline{O_i'}$ are compact, $\overline{O_i'} \subset O_i$ $(i = 1, \ldots, s)$ and $K \subset \cup_{i=1}^s O_i'$.
b) There exist functions $\psi_i \in C_0^\infty(\mathbf{R}^n)$ $(i = 1, \ldots, s)$ such that $\operatorname{supp}\psi_i \subset Q$, $0 \le \psi_i \le 1$ and $\sum_{i=1}^s \psi_i = 1$ in some neighbourhood of the set K.

Solution. a) We choose O_1' such that $O_1' = \{x \mid d(x, K') < d/2\}$ where $K' = K \setminus (\cup_{i=2}^s O_i)$ and $d = d(K', \partial O_1)$. We construct O_2' in an analogous way starting by $O_1', O_2, \ldots, O_s$.
b) Applying Example 8.18 on the compact set $K_i = \overline{O_i'}$ and open set O_i $(i = 1, \ldots, s)$ we obtain that there exist functions $\varphi_i \in C_0^\infty(O_i)$ with the properties $\varphi_i = 1$ in some neighbourhood of the set K and $0 \le \varphi_i \le 1$. We define the functions φ_i equal zero outside of O_i, respectively. We define

$$\psi_1 = \varphi_1, \quad \psi_i = \varphi_i(1 - \varphi_1)\cdots(1 - \varphi_{i-1}) \ (i = 2, \ldots, s).$$

It is easy to check that ψ_i satisfy the prescribed properties. Namely, by the construction $\psi_i \in C_0^\infty(\mathbf{R}^n)$, $\operatorname{supp}\psi_i \subset O_i$ and $0 \le \psi_i \le 1$. We have to prove only that $\sum_{i=1}^s \psi_i = 1$ in some neighbourhood of the set K. We can represent the functions ψ_i in the following form

$$\psi_1 = 1 - (1 - \varphi_1)$$

$$\psi_i = (1 - \varphi_1)\ldots(1 - \varphi_{i-1}) - (1 - \varphi_1)\cdots(1 - \varphi_{i-1})(1 - \varphi_i)$$

for $i = 2, \ldots, s$. This implies

$$\sum_{i=1}^s \psi_i = 1 - (1 - \varphi_1)(1 - \varphi_2)\cdots(1 - \varphi_s).$$

Then by the fact that $\varphi_i = 1$ in some neighbourhood of the set K_i we obtain the desired equality.

Exercise 8.20 *Let $O \subset \mathbf{R}^n$ be an open set. Prove that the set of functions $C_0^\infty(O)$ is a normed space with the norm*

$$\|f\| = \sum_{|\alpha| \le k} \sup_{x \in O} |D^\alpha f(x)| \qquad (f \in C_0^\infty(O)),$$

but it is not a complete space!

Exercise 8.21 *Let $Q \subset \mathbf{R}^n$ be a bounded region . Prove that the set $C_0^\infty(\overline{Q})$ endowed with the norm from Exercise 8.20 which takes the following form*

$$\|f\| = \sum_{|\alpha| \le k} \max_{x \in \overline{Q}} |D^\alpha f(x)| \qquad (f \in C_0^\infty(\overline{Q}))$$

is a Banach space.

8.3   The Surface Integrals

8.3.1   Preliminaries

Let $U, V \subset \mathbf{R}^n$ be bounded regions and Φ a bijection of the class C^k from $\overline{U}$ onto $\overline{V}$ such that Φ^{-1} is also of the class C^k. Such function Φ is called *dipheomorphism* between $\overline{U}$ and $\overline{V}$.

Let $Q \subset \mathbf{R}^n$ $(n \ge 2)$ be a bounded region Q is *locally quadratic* if for every point x_0 on the boundary ∂Q there exists region U which contains the point x_0 and a C^1 dipheomorphism Φ from $\overline{U}$ onto $\overline{V}$ for some region $V \subset \mathbf{R}^n$ such that it maps the set $U \cap Q$ on a n–dimensional parallelepiped $P = (a_1, b_1) \times \cdots \times (a_n, b_n)$ and the set $\overline{U} \cap \partial Q$ maps on a side or union of sides of P.

A compact subset S of $\mathbf{R}^n$ $(n \ge 2)$ is a *locally quadratic $(n-1)$–dimensional part of surface* if for every point $x_0 \in S$ there exists a region U which contains the point x_0 and there is a C^1 -dipheomorphism Φ from $\overline{U}$ onto $\overline{V}$, where V is a region, such that it maps the set $S \cap \overline{U}$ on some side or union of sides of P.

A compact subset S of $\mathbf{R}^n$, $n \ge 2$ is *locally quadratic $(n-1)$–dimensional part of surface* if for each point x_0 from S there exist a region U which contains the point x_0 and a C^1–dipheomorphism Φ from $\overline{U}$ onto $\overline{V}$, where V is a region and Φ maps $S \cap \overline{U}$ on some side or union of sides of some generalized parallelepiped $P = (a_1, b_1) \times \cdots \times (a_n, b_n)$. By compactness of S it can be covered by finite number of regions $U_1, \ldots, U_s$ and there are dipheomorphisms $\Phi^{(i)}$ which map $\overline{U}_i$ onto $\overline{V}_i$, for some regions V_i $(i = 1, \ldots, s)$ such that $S \cap \overline{U}_i$ is mapped on some side or union of sides of some generalized parallelepiped $P \subset U_i$. For $U_1, \ldots, U_s$ we take the corresponding partition of the unity (see Example 8.19) $\varphi_i \in C_0^\infty(U_i)$ with $\sum_{i=1}^{s} \varphi_i(x) = 1$ in some neighbourhood of S. A function $f : A \to \mathbf{C}$ is *Lebesgue integrable on the surface S* if for every $i, i = 1, \ldots, s$, the function $\varphi_i f \circ (\Phi^{(i)})^{-1}$

is Lebesgue integrable on the sides $\Phi^{(i)}(S \cap \overline{U}_i)$. Then the corresponding *Lebesgue surface integral* of f on S is given by

$$\int\limits_S f \, dS = \sum_{i=1}^{s} \int\limits_{\Phi^{(i)}(S \cap \overline{U}_i)} \varphi_i f \circ (\Phi^{(i)})^{-1} |J_i| \, dx, \qquad (8.1)$$

where the integration goes through parallelepipeds from $\mathbf{R}^{n-1}$, $\Phi^{(i)}(S \cap \overline{U}_i)$ is a part of some

$$I_1 \times \cdots \times I_{m-1} \times \{\alpha_m\} \times I_{m+1} \times \cdots \times I_n$$

for $m = m_i$ and J_i is a vector valued determinant of the following form

$$J_i = \begin{vmatrix} e_1 & e_2 & \cdots & e_n \\ D_1((\Phi^{(i)})^{-1})_1 & D_1((\Phi^{(i)})^{-1})_2 & \cdots & D_1((\Phi^{(i)})^{-1})_n \\ \vdots & \vdots & & \vdots \\ D_{m-1}((\Phi^{(i)})^{-1})_1 & D_{m-1}((\Phi^{(i)})^{-1})_2 & \cdots & D_{m-1}((\Phi^{(i)})^{-1})_n \\ D_{m+1}((\Phi^{(i)})^{-1})_1 & D_{m+1}((\Phi^{(i)})^{-1})_2 & \cdots & D_{m+1}((\Phi^{(i)})^{-1})_n \\ \vdots & \vdots & & \vdots \\ D_n((\Phi^{(i)})^{-1})_1 & D_n((\Phi^{(i)})^{-1})_2 & \cdots & D_n((\Phi^{(i)})^{-1})_n \end{vmatrix},$$

where

$$e_1 = (1,0,\ldots,0), e_2 = (0,1,0,\ldots,0),\ldots, e_n = (0,0,\ldots,1)$$

and

$$(\Phi^{(i)})^{-1} = \left(((\Phi^{(i)})^{-1})_1, \ldots, ((\Phi^{(i)})^{-1})_n \right).$$

$L_p(S)$, $1 \le p < \infty$, be the set of all functions $f : S \to \mathbf{C}$ such that for all $i, i = 1, \ldots, s$, we have

$$\varphi^{1/p} f(\Phi^{(i)})^{-1} \in L_p(\Phi^{(i)}(S \cap \overline{U}_i)).$$

8.3.2 Examples and Exercises

Example 8.22 *Let Q be a bounded region of $\mathbf{R}^n$. Prove that if $\partial Q \in C^k, k \ge 1$, is a $(n-1)$-dimensional surface (manifold), then Q is a generalized quadratic region.*

Solution. Let

$$\partial Q \cap P = \{(x_1, \ldots, x_n) | x_n = \Phi(x_1, \ldots, x_{n-1}), (x_1, \ldots, x_n) \in P_n\}$$

for $\Phi \in C^k(\overline{P}_n)$ and $P_n = (a_1, b_1) \times \ldots \times (a_{n-1}, b_{n-1})$. Taking enough small P_n we can obtain that

$$a'_n = \inf_{\overline{P}_n} \Phi > a_n \quad \text{and} \quad b'_n = \sup_{\overline{P}_n} \Phi < b_n.$$

Let

$$Q \cap P = \{(x_1, \ldots, x_n) | x_n < \Phi(x_1, \ldots, x_{n-1})\}.$$

Then the C^k–dipheomorphism $\Phi = (\Phi_1, \ldots, \Phi_n)$ defined by

$$\Phi_1(x_1, \ldots, x_n) = x_1, \ldots, \Phi_{n-1}(x_1, \ldots, x_n), \; \Phi_n(x_1, \ldots, x_n) = x_n - \Phi(x_1, \ldots, x_{n-1})$$

maps, for $a''_n = a'_n - a_n$ and $b''_n = b_n - b'_n$, the region

$$U = \Big\{ (x_1, \ldots, x_n) | \; \Phi(x_1, \ldots, x_{n-1}) - a''_n < x_n < \Phi(x_1, \ldots, x_{n-1}) + b''_n,$$

$$(x_1, \ldots, x_n) \in P_n \Big\} \subset P$$

on the closure of the region

$$V = (a_1, b_1) \times \ldots (a_{n-1}, b_{n-1}) \times (-a''_n, b''_n),$$

such that $U \cap Q$ is mapped on the generalized parallelepiped

$$(a_1, b_1) \times \ldots (a_{n-1}, b_{n-1}) \times (-a''_n, 0),$$

and $\overline{U} \cap \partial Q$ is mapped on the side $\overline{P}_n \times \{0\}$.

Exercise 8.23 *Let $Q \subset \mathbf{R}^n$ be a locally quadratic bounded region. Prove that $(a, b) \times Q$, for $a, b \in \mathbf{R}$, $a < b$, is a locally quadratic region in $\mathbf{R}^{n+1}$.*

Exercise 8.24 *Let $Q \subset \mathbf{R}^n$ be a locally quadratic bounded region. Prove that ∂Q is a locally quadratic part of surface.*

Example 8.25 *Prove that the cylinder*

$$V = \{(x, y, z) | \; x^2 + y^2 < R^2, 0 < z < H\}.$$

for $R > 0$ and $H > 0$, is a locally quadratic region and ∂Q is a locally quadratic part of the surface.

Solution. We introduce the cylindric coordinate system $x = r\sin\varphi$, $y = r\sin\varphi$ and $z = z$. Then we can cover the border by

$$U_1 = \{(x,y,z)|\; \frac{R}{3} < r < \frac{3R}{2}, -\frac{3\pi}{4} < \varphi < \frac{3\pi}{4}, -1 < z < H+1\},$$

$$U_2 = \{(x,y,z)|\; \frac{R}{3} < r < \frac{3R}{2}, \frac{\pi}{4} < \varphi < \frac{7\pi}{4}, -1 < z < H+1\},$$

$$U_3 = \{(x,y,z)|\; |x| < \frac{R}{2}, |y| < \frac{R}{2}, |z| < \frac{H}{2}\},$$

$$U_4 = \{(x,y,z)|\; |x| < \frac{R}{2}, |y| < \frac{R}{2}, \frac{H}{2} < z < \frac{3H}{2}\}.$$

We introduce for $i = 1,2$ the functions $\alpha^{(i)} = (\alpha_1^{(i)}, \alpha_2^{(i)}, \alpha_3^{(i)})$, $\alpha^{(i)} : V_i \to U_i$, in the following way

$$\alpha_1^{(i)} = r\cos\varphi, \quad \alpha_2^{(i)} = r\sin\varphi, \quad \alpha_3^{(i)} = z, \tag{8.2}$$

where

$$V_1 = \{(r,\varphi,z)|\; \frac{R}{3} < r < \frac{3R}{2}, -\frac{3\pi}{4} < \varphi < \frac{3\pi}{4}, -1 < z < H+1\},$$

$$V_2 = \{(r,\varphi,z)|\; \frac{R}{3} < r < \frac{3R}{2}, \frac{\pi}{4} < \varphi < \frac{7\pi}{4}, -1 < z < H+1\}.$$

In this way we obtain a C^1–dipheomorphism from $\overline{U}_i$ onto $\overline{V}_i$ ($i = 1,2$), such that $\Phi^{(1)}$ maps the set $U_1 \cap V$ onto the parallelepiped

$$P_1 = \{(r,\varphi,z)|\; \frac{R}{3} < r < R, -\frac{3\pi}{4} < \varphi < \frac{3\pi}{4}, 0 < z < H\},$$

and $\Phi^{(2)}$ maps the set $U_2 \cap V$ onto the parallelepiped

$$P_2 = \{(r,\varphi,z)|\; \frac{R}{3} < r < R, \frac{\pi}{4} < \varphi < \frac{7\pi}{4}, 0 < z < H\}.$$

On the borders $\Phi^{(1)}$ maps $\overline{U}_1 \cap \partial V$ onto $P_{11} \cup P_{12} \cup P_{13}$, where

$$P_{11} = \{(r,\varphi,z)|\; \frac{R}{3} \leq r \leq R, -\frac{3\pi}{4} \leq \varphi \leq \frac{3\pi}{4}, z = 0\},$$

$$P_{12} = \{(r,\varphi,z)|\; \frac{R}{3} \leq r \leq R, \frac{-3\pi}{4} \leq \varphi < \frac{3\pi}{4}, z = H\},$$

$$P_{13} = \{(r,\varphi,z)|\; r = R, \frac{-3\pi}{4} \leq \varphi \leq \frac{3\pi}{4}, 0 \leq z \leq H\},$$

and $\Phi^{(2)}$ maps $\overline{U}_2 \cap \partial V$ onto $P_{21} \cup P_{22} \cup P_{23}$, where

$$P_{21} = \{(r,\varphi,z)|\; \frac{R}{3} \leq r \leq R, \frac{\pi}{4} \leq \varphi \leq \frac{7\pi}{4}, z = 0\},$$

$$P_{22} = \{(r, \varphi, z) | \frac{R}{3} \leq r \leq R, \frac{\pi}{4} \leq \varphi \leq \frac{7\pi}{4}, z = H\},$$

$$P_{23} = \{(r, \varphi, z) | r = R, \frac{\pi}{4} \leq \varphi \leq \frac{7\pi}{4}, 0 \leq z \leq H\}.$$

We choose for C^1-dipheomorphisms $\Phi^{(3)}$ and $\Phi^{(4)}$ the identical mappings. Then $\Phi^{(3)}$ maps $\overline{U}_3 \cap \partial V$ on the side

$$P_{31} = \{(x, y, z) | \ |x| < \frac{R}{2}, |y| < \frac{R}{2}, z = 0\},$$

and $\Phi^{(4)}$ maps the set $\overline{U}_4 \cap \partial V$ on the side

$$P_{41} = \{(x, y, z) | \ |x| < \frac{R}{2}, |y| < \frac{R}{2}, z = H\}.$$

Since V is a locally quadratic region we have by the preceding Exercise that ∂V is a locally quadratic part of a surface.

Example 8.26 *Find the surface integral $\int_{\partial V} x ds$, where V is the cylinder from Example 8.25.*

Solution. By the Example 8.25 we have

$$\bigcup_{i=1}^{4} U_i \supset \partial V.$$

We take the corresponding partition of the unity $\varphi_i \in C_0^\infty(U_i)$ $(i = 1, 2, 3, 4)$,

$$\sum_{i=1}^{n} \varphi_i(x) = 1 \qquad (x \in \partial V). \tag{8.3}$$

We shall find the surface integral by (8.1). For that purpose we shall first find the absolute value of the determinant J_i by on sides $P_{11}, P_{12}, P_{13}, P_{21}, P_{22}, P_{23}, P_{31}, P_{41}$. We have on sides P_{ij} $(i, j = 1, 2)$

$$J_i = \begin{vmatrix} e_1 & e_2 & e_3 \\ D_1((\Phi^{(i)})^{-1})_1 & D_1((\Phi^{(i)})^{-1})_2 & D_1((\Phi^{(i)})^{-1})_3 \\ D_2((\Phi^{(i)})^{-1})_1 & D_2((\Phi^{(i)})^{-1})_2 & D_2((\Phi^{(i)})^{-1})_3 \end{vmatrix}$$

$$= \begin{vmatrix} e_1 & e_2 & e_3 \\ \cos\varphi & \sin\varphi & 0 \\ -r\sin\varphi & r\cos\varphi & 0 \end{vmatrix}$$

$$= re_3.$$

Hence $|J_i| = r$. We have on sides P_{i3} $(i = 1, 2)$

$$
J_i = \begin{vmatrix} e_1 & e_2 & e_3 \\ D_2((\Phi^{(i)})^{-1})_1 & D_2((\Phi^{(i)})^{-1})_2 & D_2((\Phi^{(i)})^{-1})_3 \\ D_3((\Phi^{(i)})^{-1})_1 & D_3((\Phi^{(i)})^{-1})_2 & D_3((\Phi^{(i)})^{-1})_3 \end{vmatrix}
$$

$$
= \begin{vmatrix} e_1 & e_2 & e_3 \\ -R\sin\varphi & R\cos\varphi & 0 \\ 0 & 0 & 1 \end{vmatrix}
$$

$$
= R\cos\varphi \cdot e_1 + R\sin\varphi \cdot e_2.
$$

Hence $|J_i| = \sqrt{R^2\cos^2\varphi + R^2\sin^2\varphi} = R$. It is obvious that on sides P_{31} and P_{41} we have $|J_i| = 1$.

Using (8.1) we obtain

$$
\int_{\partial V} f \, ds = \sum_{i=1}^{2}\sum_{j=1}^{3} \int_{P_{ij}} (\varphi_i f \circ (\Phi^{(i)})^{-1})|J_i| \, d\varphi dr + \sum_{i=3}^{4} \int_{P_{i1}} (\varphi_i f \circ (\Phi^{(i)})^{-1})|J_i|. \qquad (8.4)
$$

By (8.3) we have

$$
\int_{P_{11}} (\varphi_1 f \circ (\Phi^{(1)})^{-1})|J_1| + \int_{P_{21}} (\varphi_2 f \circ (\Phi^{(2)})^{-1})|J_2| + \int_{P_{31}} (\varphi_3 f \circ (\Phi^{(3)})^{-1})|J_3|
$$

$$
= \int_{\frac{R}{3}}^{R} (\int_{-\frac{3\pi}{4}}^{\frac{3\pi}{4}} (\varphi_1 f)(r\cos\varphi, r\sin\varphi, 0) r \, d\varphi) dr
$$

$$
+ \int_{\frac{R}{3}}^{R} (\int_{-\frac{\pi}{4}}^{\frac{7\pi}{4}} (\varphi_2 f)(r\cos\varphi, r\sin\varphi, 0) r \, d\varphi) dr
$$

$$
+ \int_{-\frac{R}{2}}^{\frac{R}{2}} (\int_{-\frac{R}{2}}^{\frac{R}{2}} (\varphi_3 f)(r\cos\varphi, r\sin\varphi, 0) r \, dy) dx
$$

$$
= \int_{0}^{R} (\int_{0}^{2\pi} (\varphi_1 f)(r\cos\varphi, r\sin\varphi, 0) r \, d\varphi) dr
$$

$$
+ \int_{0}^{R} (\int_{0}^{2\pi} (\varphi_2 f)(r\cos\varphi, r\sin\varphi, 0) r \, d\varphi) dr
$$

$$+ \int_0^R (\int_0^{2\pi} (\varphi_3 f)(r\cos\varphi, r\sin\varphi, 0)r \, d\varphi)dr.$$

Hence

$$\int_{P_{11}} (\varphi_1 f \circ (\Phi^{(1)})^{-1})|J_1| + \int_{P_{21}} (\varphi_2 f \circ (\Phi^{(2)})^{-1})|J_2| + \int_{P_{31}} (\varphi_3 f \circ (\Phi^{(3)})^{-1})|J_3|$$

$$= \int_0^R (r \int_0^{2\pi} f(r\cos\varphi, r\sin\varphi, 0) \, d\varphi)dr. \tag{8.5}$$

We obtain in an analogous way

$$\int_{P_{12}} (\varphi_1 f \circ (\Phi^{(1)})^{-1})|J_1| + \int_{P_{22}} (\varphi_2 f \circ (\Phi^{(2)})^{-1})|J_2| + \int_{P_{41}} (\varphi_4 f \circ (\Phi^{(4)})^{-1})|J_4|$$

$$= \int_0^R (r \int_0^{2\pi} f(r\cos\varphi, r\sin\varphi, H) \, d\varphi)dr. \tag{8.6}$$

Finally, we have

$$\int_{P_{13}} (\varphi_1 f \circ (\Phi^{(1)})^{-1})|J_1| + \int_{P_{23}} (\varphi_2 f \circ (\Phi^{(2)})^{-1})|J_2|$$

$$= R \int_0^{2\pi} (\int_0^R f(R\cos\varphi, R\sin\varphi, z) \, dz)d\varphi. \tag{8.7}$$

Putting (8.5), (8.6) and (8.7) in (8.4) we obtain

$$\int_{\partial V} f \, ds = \int_0^R (r \int_0^{2\pi} f(r\cos\varphi, r\sin\varphi, 0) \, d\varphi)dr + \int_0^R (r \int_0^{2\pi} f(r\cos\varphi, r\sin\varphi, H) \, d\varphi)dr$$

$$+ R \int_0^{2\pi} (\int_0^H f(R\cos\varphi, R\sin\varphi, z) \, dz)d\varphi. \tag{8.8}$$

Hence

$$\int_{\partial V} x \, dS = 2 \int_0^R (r \int_0^{2\pi} r\cos\varphi \, d\varphi)dr + R \int_0^{2\pi} (\int_0^H R\cos\varphi \, dz)d\varphi.$$

Exercise 8.27 *Find the surface integral*

$$\int_{\partial V} (x^2 + y^2) \, dS,$$

where V is the cylinder from Example 8.25.

Hint. Use the equality (8.8) from Example 8.26.

Exercise 8.28 *Prove that the space $L_p(S)$ is independent of the cover $U_1, \ldots, U_s$, C^1–dipheomorphisms $\Phi^{(i)}$ and partition of the unity φ_i ($i = 1, \ldots, s$).*

Hint. Use the facts that a C^1–dipheomorphism maps a measurable function on a measurable function and it maps an integrable function on an integrable function.

Exercise 8.29 *Prove that the space $L_p(S)$, $1 \leq p < \infty$ is a normed space equipped with the following norm*

$$\|f\|_{L_p(S)} = \left(\int_S |f|^p \, ds \right)^{\frac{1}{p}} = \sum_{i=1}^{s} \int_{\Phi^{(i)}(S \cap \overline{U}_i)} (\varphi_i |f|^p \circ (\Phi^{(i)})^{-1}) |J_i|)^{\frac{1}{p}}.$$

Remark 8.29.1. The space $L_p(S)$ is a Banach space with respect to the introduced norm $\|f\|_{L_p(S)}$.

8.4   The Fourier Transform

8.4.1   Preliminaries

The Fourier transform of a function $f \in L_1(\mathbf{R}^n)$ is defined by

$$(\mathcal{F}f)(x) = \hat{f}(x) = \frac{1}{(2\pi)^{\frac{n}{2}}} \int_{\mathbf{R}^n} e^{-\imath z x} f(z) \, dz \quad (x \in \mathbf{R}^n),$$

where $z \cdot x = \sum_{i=1}^{n} z_i x_i$.

Theorem 8.6 *Let $f \in L_1(\mathbf{R}^n)$. Then:*

1) $\hat{f} \in C(\mathbf{R}^n)$;

2) if additionally $z_k \cdot f(z)$ is also a function from $L_1(\mathbf{R}^n)$, then there exists $D_k \hat{f}(x)$ and

$$D_k \hat{f}(x) = -\imath \widehat{z_k f}(x);$$

more general, if additionally $z_k^{\alpha_k} \cdot f(z)$ for $k = 1, \ldots, n$ are also functions from $L_1(\mathbf{R}^n)$, then there exists $D^\alpha \hat{f}(x)$ and

$$D_\alpha \hat{f}(x) = \widehat{(-\imath z)^\alpha f}(x)$$

for $\alpha = (\alpha_1, \ldots, \alpha_n) \in \mathbf{Z}_+^n$;

3) if additionally $D_k f \in L_1(\mathbf{R}^n) \cup C(\mathbf{R}^n)$, then

$$\widehat{D_k f}(x) = \imath x_k \hat{f}(x);$$

more general, if additionally $D_k^{\alpha_k} f \in L_1(\mathbf{R}^n) \cup C(\mathbf{R}^n)$ for $k = 1, \ldots, n$, then

$$\widehat{D^\alpha f}(x) = (-\imath x)^\alpha \hat{f}(x)$$

for $\alpha = (\alpha_1, \ldots, \alpha_n) \in \mathbf{Z}_+^n$.

Theorem 8.7 *For a linear differential operator*

$$L(D) = \sum_{|\alpha| \leq m} a_\alpha D^\alpha$$

and a function $f \in C^m(\mathbf{R}^n)$ with $D^\alpha f \in L_1(\mathbf{R}^n)$ for $|\alpha| \leq m$ the following equality holds

$$L(\widehat{D})f(x) = L(\imath x)\hat{f}(x).$$

Definition 8.8 *The convolution of two functions $f, g \in L_1(\mathbf{R}^n)$ is defined by*

$$(f * g)(x) = \int_{\mathbf{R}^n} f(y)g(x - y)\, dy.$$

The following *exchange formula* allows us to transfer the convolution in the usual product.

Theorem 8.9 *If $f, g \in L_1(\mathbf{R}^n)$, then*

$$\widehat{f * g}(x) = (2\pi)^{\frac{n}{2}} \hat{f}(x) \cdot \hat{g}(x).$$

Definition 8.10 *The inverse Fourier transform $\mathcal{F}^{-1}$ of a function $f \in L_1(\mathbf{R}^n)$ is defined by*

$$\mathcal{F}^{-1}(f)(x) = \hat{f}(-x).$$

Theorem 8.11 *If $f \in L_1(\mathbf{R}^n), \hat{f} \in L_1(\mathbf{R}^n)$ and $f(z)$ is a continuous function at the point $z = z^0$, then $f(z^0) = \mathcal{F}^{-1}(\hat{f})(z^0)$, i.e.,*

$$f(z^0) = (2\pi)^{-\frac{n}{2}} \int_{\mathbf{R}^n} e^{-\imath z^0 x} \hat{f}(y)\, dy.$$

Definition 8.12 *The equation*

$$L(D, D_t)u = \sum_{|\alpha| \leq m} a_\alpha D^\alpha u(x, t) = 0 \tag{8.9}$$

is hyperbolic *(with respect to the manifold $t = 0$) if the Cauchy problem (8.9) with the conditions*

$$\frac{\partial^j u(x,0)}{\partial t^j} = \begin{cases} 0 & \text{for } 0 \leq j < m - 1, \\ f(x) & \text{for } j = m \end{cases} \tag{8.10}$$

always has a solution $u(x, t) \in C^m(\mathbf{R}^{n+1})$ for every $f \in C_0^s(\mathbf{R}^n)$ for enough big s.

8.4.2   Examples and Exercises

Example 8.30 *Find the Fourier transform of the following functions*

a) $e(x) = (2\pi)^{-\frac{n}{2}} \exp(-\frac{|x|^2}{2})$;

b) $f(x) = (2t)^{-\frac{n}{2}} \exp(-\frac{|x|^2}{4t})$.

Solution.

a) We have

$$
\begin{aligned}
(\mathcal{F}e)(x) &= (2\pi)^{-\frac{n}{2}} \prod_{k=1}^{n} \left((2\pi)^{-\frac{1}{2}} \exp(-\frac{x_k^2}{2}) \int_{-\infty}^{+\infty} \exp(-\frac{(z_k + ix_k)^2}{2}) \, dz_k \right) \\
&= (2\pi)^{-\frac{n}{2}} \prod_{k=1}^{n} \left((2\pi)^{-\frac{1}{2}} \exp(-\frac{x_k^2}{2}) \int_{-\infty}^{+\infty} \exp(-\frac{z^2}{2}) \, dz \right) \\
&= (2\pi)^{-\frac{n}{2}} \exp(-\frac{|x|^2}{2}) = e(x),
\end{aligned}
$$

where we have used that $\displaystyle\int_{-\infty}^{+\infty} \exp(-z^2/2) \, dz = \sqrt{2\pi}$.

b) We obtain by a)
$$(\mathcal{F}f)(x) = e^{-|x|^2 t}.$$

Example 8.31 *Solve the Cauchy problem for the heat equation*

$$\frac{\partial u}{\partial t} = \Delta u \qquad (x \in \mathbf{R}^n, t > 0),$$

$$u(x, 0) = f(x) \qquad (x \in \mathbf{R}^n),$$

using the Fourier transform.

Solution. Applying the Fourier transform $\mathcal{F}$ on the both equation we obtain

$$\frac{\partial \hat{u}(z, t)}{\partial t} + |z|^2 \hat{u}(z, t) = 0,$$

$$\hat{u}(z, 0) = \hat{f}(z),$$

supposing that we can exchange the order of the derivative with respect to t and the Fourier transform. The solution of this ordinary differential equation is given by

$$\hat{u}(z, t) = \hat{f}(z) e^{-|z|^2 t}.$$

Applying now the inverse Fourier transform $\mathcal{F}^{-1}$ on both sides we obtain

$$u(x,t) = (2\pi)^{-\frac{n}{2}} \int_{\mathbf{R}^n} \hat{f}(z) e^{-|z|^2 t} e^{\imath z x} \, dz.$$

using the exchange formula and Example 8.30 we have

$$
\begin{aligned}
u(x,t) &= (2\pi)^{-\frac{n}{2}} \int_{\mathbf{R}^n} \hat{f}(z)(2t)^{-n/2} \mathcal{F}(\exp(-\frac{|z|^2}{4t})) e^{\imath z x} \, dz \\
&= (2t)^{-n/2} \mathcal{F}^{-1}(\mathcal{F}(f) \cdot \mathcal{F}(\exp(-\frac{|z|^2}{4t}))) \\
&= (2t)^{n/2}(2\pi)^{-\frac{n}{2}} \mathcal{F}^{-1}(\mathcal{F}(f * \exp(-\frac{|z|^2}{4t}))) \\
&= (4\pi t)^{-\frac{n}{2}} f(x) * \exp(-\frac{|x|^2}{4t}) \\
&= (4\pi t)^{-\frac{n}{2}} \int_{\mathbf{R}^n} f(y) \exp(-\frac{|x-y|^2}{4t}) \, dy.
\end{aligned}
$$

Exercise 8.32 *Prove that if $Q_1 \subset \mathbf{R}^n$ and $Q_2 \subset \mathbf{R}^m$ are bounded regions, $\{\varphi_i\}_{i \in \mathbf{N}}$ an orthonormal complete system of functions in $L_2(Q_1)$ and $\{\psi_j\}_{j \in \mathbf{N}}$ an orthonormal complete system of functions in $L_2(Q_2)$, then the system of functions $\{\varphi_i \cdot \psi_j\}_{i,j \in \mathbf{N}}$ defined by*

$$(\varphi_i \cdot \psi_j)(x,y) = \varphi_i(x) \cdot \psi_j(y) \qquad (i,j \in \mathbf{N}, x \in Q_1, y \in Q_2)$$

orthonormal and complete in the space $L_2(Q_1 \times Q_2)$.

Hints. For the proof of the completeness of the system of functions $\{\varphi_i \cdot \psi_j\}_{i,j \in \mathbf{N}}$ use the denseness of $C(\overline{Q}_1 \times \overline{Q}_2)$ in $L_2(Q_1 \times Q_2)$ and the fact that a system of functions $\{h_i\}_{i \in \mathbf{N}}$ is complete in the space $L_2(Q)$ if for every function f the Parseval's identity holds

$$\sum_{i=1}^{\infty} |(f|h_i)_{L_2(Q)}|^2 = \int_Q |f|^2 \, dx,$$

and use the Fubini theorem.

Exercise 8.33 *Prove that for $f,g,h \in L_1(\mathbf{R}^n)$ a) $f * g = g * f$; b) $(f * g) * h = f * (g * h)$; c) $\|f * g\|_{L_1} \le \|f\|_{L_1} \cdot \|g\|_{L_1}$.*

Example 8.34 *A function $f \in L_2(Q)$ is continuous in mean (quadratic) , i.e., for every $\varepsilon > 0$ there exists $\delta > 0$ such that*

$$\|f(x+v) - f(x)\|_{L_2(Q)} < \varepsilon$$

for each v which satisfies $|v| < \delta$ $(x + v \in Q)$.

Solution. Let $f \in L_2(Q)$ and $a > 0$ such that $Q \subset\subset B(0, a)$. We define the function F in the following way

$$F(x) = \begin{cases} f(x) & \text{for } x \in Q \\ 0 & \text{for } x \in B(0, 3a) \setminus Q. \end{cases}$$

F belongs to $L_2(B(0, 3a))$. Since the set $C(\overline{Q})$ is dense in $L_2(\overline{B(0, 3a)})$, for $\varepsilon > 0$ there exists a function $\tilde{F} \in C(\overline{B(0, 3a)})$ such that

$$\|F(x) - \tilde{F}(x)\|_{L_2(B(0,3a))} < \frac{\varepsilon}{3}. \tag{8.11}$$

We can suppose that $\tilde{F}(x) = 0$ for $x \in B(0, 3a) \setminus B(0, a)$. Namely, we can always multiply $\tilde{F}$ with a "cutoff" function on $B(0, a)$. We have for $|z| \leq a$

$$\|F(x + z) - \tilde{F}(x + z)\|_{L_2(B(0,3a))} = \|F(x) - \tilde{F}(x)\|_{L_2(B(0,a))} \leq \frac{\varepsilon}{3}. \tag{8.12}$$

Since the function $\tilde{F}$ is uniformly continuous on $\overline{B(0, 2a)}$, there exists $\delta > 0$ ($\delta < a$) such that

$$\|\tilde{F}(x + z) - \tilde{F}(x)\|_{L_2(B(0,2a))} \leq \frac{\varepsilon}{3}, \ |z| < \delta. \tag{8.13}$$

Then we have by (8.12) and (8.13) for $|z| < \delta$

$$\|f(x + z) - f(x)\|_{L_2(Q)} = \|F(x + z) - F(x)\|_{L_2(B(0,2a))}$$

$$\leq \|F(x + z) - \tilde{F}(x + z)\|_{L_2(B(0,2a))} + \|\tilde{F}(x + z) - \tilde{F}(x)\|_{L_2(B(0,2a))}$$

$$+ \|\tilde{F}(x) - F(x)\|_{L_2(B(0,2a))} \leq \frac{\varepsilon}{3} + \frac{\varepsilon}{3} + \frac{\varepsilon}{3} = \varepsilon.$$

Exercise 8.35 (Gårding hyperbolicity condition) *Prove that if*

$$L(\imath x, \imath \lambda) \neq 0 \qquad (x \in \mathbf{R}^n)$$

for all complex numbers such that $\Im \lambda \leq -c$, then the equation (8.9) is hyperbolic.

Remark 8.35.1 The given condition is also necessary for the hyperbolicity of the equation (8.9).

Hints. Apply the Fourier transform on (8.9) and (8.10) to obtain the formal solution

$$u(x, t) = (2\pi)^{-\frac{n}{2}} \int_{\mathbf{R}^n} e^{\imath x y} K(y, t) \hat{f}(y) \, dy,$$

where the function K satisfies with respect to the variable t the following ordinary differential equation

$$L(\imath y, D_t) K(y, t) = 0$$

and conditions

$$D_t^j K(y,t) = \begin{cases} 0 & \text{for } 0 \le j < m-1 \\ 1 & \text{for } j = m-1. \end{cases}$$

Prove that the function K has the following representation

$$K(y,t) = \frac{1}{2\pi i} \int_C \frac{e^{zt}\,dz}{L(iy,z)} = \frac{1}{2\pi} \int_C \frac{e^{i\lambda t}}{L(iy,i\lambda)}\,d\lambda,$$

where C is a closed countor which goes one times around all zeros of the polynomial $L(iy,i\lambda) = 0$. Use this result to prove the absolute convergence of the integrals

$$\int_{\mathbf{R}^n} e^{ixy} D_t^j y^\alpha K(y,t) \hat{f}(y)\,dy$$

for $|\alpha| + j \le m$ which will imply that the given formal solution is the solution of the problem.

Exercise 8.36 *If all zeroes λ_k of $L(iy,i\lambda) = 0$ are different, then $K(y,t)$ given by*

$$K(y,t) = \frac{1}{2\pi} \int_C \frac{e^{i\lambda t}}{L(iy,i\lambda)}\,d\lambda,$$

where C is a closed path which goes only one times around each zero of $L(iy,i\lambda) = 0$, can be represented in the following form

$$K(y,t) = \sum_{j=1}^m \frac{e^{|\lambda|t}}{D_t L(iy,i\lambda)}.$$

Hint. Use theorem on residiums (see [21]).

Exercise 8.37 *Prove that the wave equation*

$$D_t u - \sum_{j=1}^n D_j^2 u = 0$$

with initial conditions

$$u(x,0) = 0 \text{ and } D_t u(x,0) = g(x) \text{ for } g \in C_0^{n+3}(\mathbf{R}^n),$$

is hyperbolic and that the solution is of the following form

$$u(x,t) = (2\pi)^{-\frac{n}{2}} \int_{\mathbf{R}^n} e^{ixy} \frac{\sin(|y|t)}{|y|} \hat{g}(y)\,dy.$$

Hints. *Use the Gårding condition and Exercise 8.36.*

Example 8.38 *Let* $\varphi \in C_0^\infty(\mathbf{R}^n)$ *such that* $\operatorname{supp}\varphi \subset \overline{B(0,R)}$. *Then its Fourier transform* $\hat{\varphi} = \mathcal{F}\varphi$ *can be extended from* $\mathbf{R}^n$ *onto* $\mathbf{C}^n$ *such that it will be analytical function on* $\mathbf{C}^n$ *with the property that for every* $s \in \mathbf{N}$ *there exists* $C_s > 0$ *such that*

$$|\hat{\varphi}(z)| \leq C_s (1 + |z|^2)^{-s} e^{R\Im z} \quad (z \in \mathbf{C}^n),$$

where $z = (z_1, \ldots, z_n)$.

Solution. For every $z \in \mathbf{C}^n$ the following integral is well- defined

$$\hat{\varphi}(z) = (2\pi)^{-\frac{n}{2}} \int_{\mathbf{R}^n} e^{-\imath z x} \varphi(x)\, dx$$

and it is an analytical function (since we can differentiate under the integral). Since $\operatorname{supp}\varphi \subset \overline{B(0,R)}$ we have

$$\hat{\varphi}(z) = (2\pi)^{-\frac{n}{2}} \int_{B(0,R)} e^{-\imath z x} \varphi(x)\, dx.$$

Applying the partial integration we obtain

$$\begin{aligned}
z^\alpha \hat{\varphi}(z) &= (2\pi)^{-\frac{n}{2}} \int_{B(0,R)} \frac{D_x^\alpha (e^{-\imath z x})}{(-\imath)^{|\alpha|}} \varphi(x)\, dx \\
&= \frac{(-\imath)^{|\alpha|}}{(2\pi)^{\frac{n}{2}}} \int_{B(0,R)} e^{-\imath z x} D^\alpha \varphi(x)\, dx.
\end{aligned}$$

The last equality give us the following estimation

$$\begin{aligned}
|z_1|^{\alpha_1} \cdots |z_n|^{\alpha_n} |\hat{f}(z)| &= |z^\alpha| \cdot |\hat{\varphi}(z)| \\
&\leq (2\pi)^{-\frac{n}{2}} \int_{B(0,R)} |e^{-\imath z x}| \cdot |D^\alpha \varphi(x)|\, dx. \\
&\leq (2\pi)^{-\frac{n}{2}} \int_{B(0,R)} e^{\Im(z \cdot x)} \cdot |D^\alpha \varphi(x)|\, dx \\
&= (2\pi)^{-\frac{n}{2}} \int_{B(0,R)} e^{x \cdot \Im z} \cdot |D^\alpha \varphi(x)|\, dx \\
&\leq (2\pi)^{-\frac{n}{2}} \sup_{x \in B(0,R)} e^{x \cdot \Im z} \int_{B(0,R)} |D^\alpha \varphi(x)|\, dx \\
&= (2\pi)^{-\frac{n}{2}} e^{R \cdot |\Im z|} \int_{B(0,R)} |D^\alpha \varphi(x)|\, dx \\
&= C_\alpha e^{R \cdot |\Im z|}.
\end{aligned}$$

Since for $s \in \mathbf{N}$ the expression

$$(1 + |z|^2)^s = (1 + |z_1|^2 + \cdots + |z_n|^2)^s$$

consists of the finite sum of the expressions $|z_1|^{\alpha_1} \cdots |z_n|^{\alpha_n}$, we obtain that by the preceding estimation there exists $C_s > 0$ such that

$$(1 + |z|^2)^2 |\hat{\varphi}(z)| \le C_s e^{R \cdot |\Im z|}.$$

Remark 8.38. 1. The opposite statement is also true. Therefore the following theorem is true.

Theorem 8.13 (Paley–Wiener) *Entire analytical function g is a Fourier transform of a function from the space $C_0^\infty(\mathbf{R}^n)$ with a support in $B(0, R)$ if and only if for every $s \in \mathbf{N}$ there exists a constant $C_s > 0$ such that for every $z \in \mathbf{C}^n$ we have*

$$|g(z)| \le C_s(1 + |z|^2)^{-s} e^{R\Im z} \qquad (z \in \mathbf{C}^n).$$

Example 8.39 *Let $\mathcal{S}$ be the set of all rapidly decreasing functions, i.e., $f \in \mathcal{S}$ if f has derivative of any order on $\mathbf{R}$ and for every pair $(k, l) \in \mathbf{N}_0$ we have*

$$\sup_{x \in \mathbf{R}} (1 + |x|)^k \cdot |f^{(l)}(x)| \le C_k. \tag{8.14}$$

a) Prove that the function $\tilde{f}$ defined by

$$\tilde{f}(y) = \int_{-\infty}^{\infty} e^{\imath z x} f(x) \, dx \quad (y \in \mathbf{R}) \tag{8.15}$$

belongs to $\mathcal{S}$ if $f \in \mathcal{S}$.

b) Prove that the function $u = u(x, t)$ defined by

$$u(x, t) = \frac{1}{2\pi} \int_{-\infty}^{\infty} dy \int_{-\infty}^{\infty} f(z) \exp(\imath z y - \imath t y^2 - \imath x y) \, dz \tag{8.16}$$

for every $t \in \mathbf{R}$ belongs $\mathcal{S}$.

c) Prove that the function u given by (8.16) is the solution of the Cauchy problem for the Schrödinger equation

$$\frac{\partial u}{\partial t} = \imath \frac{\partial^2 u}{\partial x^2} \quad ((x, t) \in \mathbf{R} \times (0, \infty)), \qquad u(x, 0) = f(x) \quad (x \in \mathbf{R}), \tag{8.17}$$

where $f \in \mathcal{S}$.

d) Prove that if the function u is given by (8.16) then the function $v(x, t) = u(x, T - t)$ is a solution of the following problem

$$\frac{\partial v}{\partial t} = -\imath \frac{\partial^2 v}{\partial x^2}, \quad (x \in \mathbf{R}, t \in [0, T]) \text{ and } v(x, T) = g(x) \quad (x \in \mathbf{R}), \tag{8.18}$$

where g is a given function such that $f = g$.

e) *Let ω_R, $R > 1$, be a function defined on $\mathbf{R}$ which has derivative of arbitrary order such that $\omega_R(x) = 1$ for $|x| < R-1$, $\omega_R(x) = 0$ for $|x| > R$ and $|\omega_R^{(k)}(x)| \leq C_k$, $k \in \mathbf{N}$, $C_k > 0$ independently of R. Starting from the equality*

$$\int_0^T dt \int_{-R}^R v(x,t)\omega_R(x) \cdot \left(\frac{\partial u}{\partial t}(x,t) - \imath \frac{\partial^2 u}{\partial x^2}(x,t) \right) dx = 0, \qquad (8.19)$$

prove that if u is a solution of the problem (8.16) and v a solution of the problem (8.14), then the following is true

$$\int_{-\infty}^{\infty} f(x)v(x,0)\, dx = \int_{-\infty}^{\infty} u(x,T)g(x)\, dx. \qquad (8.20)$$

f) *Prove that there is a unique solution of the problem (8.17) in the set of slowly increasing functions u, i.e., there exist $C > 0$ and $k \in \mathbf{N}$ such that for $t > 0$*

$$|u(x,t)| \leq C(1 + |x|)^k.$$

Solution.

a) The condition (8.14) implies

$$|\tilde{f}(y)| \leq C_{k,0} \int_{-\infty}^{\infty} \frac{dx}{(1 + |x|)^2} < \infty$$

and also absolute and uniform convergence of the following integral

$$F_l(y) = \int_{-\infty}^{\infty} (\imath x)^2 f(x) e^{\imath x y}\, dx \quad (l \in \mathbf{N}).$$

Then by the property of Fourier transform we have $\tilde{f}^{(l)}(y) = F_l(y)$ for every $l \in \mathbf{N}$. Hence the function $\tilde{f}$ given by (8.15) has derivative of arbitrary order. Further, we have for $y \neq 0$ and $k \in \mathbf{N}$

$$\int_{-\infty}^{\infty} e^{-\imath y x} f(x)\, dx = \left(\frac{\imath}{y}\right)^k \int_{-\infty}^{\infty} e^{-\imath y x} f^{(k)}(x)\, dx.$$

From this easily follows $f \in \mathcal{S}$.

Remark 8.39. 1. a) implies that the inverse Fourier transform is an inner operation on $\mathcal{S}$. It can be proved that it as well the direct Fourier transform are isomorphism of the space $\mathcal{S}$ on itself.

b) Since we have

$$u(x,t) = \frac{1}{2\pi} \int_{-\infty}^{\infty} \tilde{f}(y) \exp(-\imath t y^2 - \imath x y)\, dy,$$

we can prove the desired equality in an analogous way as for a).

c), d) Can be checked directly using a) and b).

e) Using the partial integration and the properties of the function ω_R we obtain

$$\int_{-R}^{R} \int_0^T v(x,t)\omega_R(x)\frac{\partial u}{\partial t}(x,t)\,dtdx$$

$$= \int_{-R}^{R} \omega_R(x)(u(x,T)v(x,T) - u(x,0)v(x,0) - \int_0^T u(x,t)\frac{\partial v}{\partial t}(x,t)\,dt)\,dx,$$

and

$$(-\imath)\int_0^T \int_{-R}^{R} v(x,t)\omega_R(x)\frac{\partial^2 u}{\partial x^2}(x,t)\,dxdt$$

$$= (-\imath)\int_0^T \int_{-R}^{R} u(x,t)\frac{\partial^2 (v(x,t)\omega_R(x))}{\partial x^2}\,dxdt.$$

Adding the last two equalities we obtain by (8.14)

$$\int_{-R}^{R} \omega_R(x)(u(x,t)v(x,t) - u(x,0)v(x,0))\,dx$$

$$= \int_{-R}^{R} \int_0^T u(x,t)\cdot(\frac{\partial v}{\partial t}(x,t)+\imath\cdot(\frac{\partial^2 v}{\partial x^2}(x,t)\omega_R(x)+2\frac{\partial v}{\partial x}(x,t)\omega_R(x)+\omega''_R(x)))\,dtdx.$$

$$(8.21)$$

Since

$$\lim_{R\to\infty} \omega_R(x) = 0, \quad \lim_{R\to\infty} \omega'_R(x) = 0, \quad \lim_{R\to\infty} \omega''_R(x) = 0,$$

uniformly on every compact set and v is a solution of the equation (8.18) the left part in (8.21) converges as $R \to \infty$ to

$$\int_{-\infty}^{\infty} u(x,T)v(x,T)\,dx - \int_{-\infty}^{\infty} u(x,0)v(x,0)\,dx$$

$$= \int_{-\infty}^{\infty} u(x,T)g(x)\,dx - \int_{-\infty}^{\infty} f(x)v(x,0)\,dx.$$

Since the right part in (8.21) converges to 0 we obtain the equality (8.20).

f) For a given $f \in S$ the function u given by the integral (8.16) is a solution of the problem (8.17). Suppose that an other slowly increasing function u_1 is also a solution of the problem (8.17). Then the function z defined by $z(x,t) = u(x,t) - u_1(x,t)$ is a solution of the problem

$$\frac{\partial z}{\partial t} = \imath\frac{\partial^2 z}{\partial x^2} \quad (x \in \mathbf{R}, t > 0), \quad z(x,0) = 0.$$

If v is a solution of the problem (8.18), then we obtain by (8.20)

$$\int_{-\infty}^{\infty} z(x,T)g(x)\,dx = 0 \text{ for all } T > 0,$$

$$(8.22)$$

and all functions $g \in S$. The integral in (8.22) exists for all $T > 0$ since
the function z is slowly increasing and g rapidly decreasing, which implies
that there product is an integrable function. The equality implies $z(x,t) \equiv 0$ for all $t > 0$. Hence $u = u_1$.

Example 8.40 *Prove that the solution of the integral equation*

$$u(x) = f(x) + \int_{-\infty}^{\infty} k(x - y)u(y) \, dy \qquad (8.23)$$

for given functions f and k, has the following representation (if the expressions are defined)

$$u(x) = f(x) + \int_{0}^{x} s(x - y)f'(t) \, dt, \qquad (8.24)$$

where

$$s(x) = \mathcal{F}(\frac{\mathcal{F}^{-1}(k)(z)}{1 - \sqrt{2\pi}\mathcal{F}^{-1}(k)(z)}). \qquad (8.25)$$

Solution. We apply the inverse Fourier transform $\mathcal{F}^{-1}$ on (8.23)

$$
\begin{aligned}
\mathcal{F}^{-1}(u)(z) &= \frac{1}{\sqrt{2\pi}} \int_{-\infty}^{\infty} (f(x) + \int_{-\infty}^{\infty} k(x-y)u(y)\,dy)e^{\imath zx}\,dx \\
&= \mathcal{F}^{-1}(f)(z) + \frac{1}{\sqrt{2\pi}} \int_{-\infty}^{\infty} u(y)\,dy \int_{-\infty}^{\infty} k(x-y)e^{-\imath zx}\,dx \\
&= \mathcal{F}^{-1}(f)(z) + \frac{1}{\sqrt{2\pi}} \int_{-\infty}^{\infty} u(y)\,dy \int_{-\infty}^{\infty} k(t)e^{\imath(y+t)z}\,dt \\
&= \mathcal{F}^{-1}(f)(z) + \sqrt{2\pi}\mathcal{F}^{-1}(u)\mathcal{F}^{-1}(k).
\end{aligned}
$$

This implies

$$\mathcal{F}^{-1}(u)(z) = \frac{\mathcal{F}^{-1}(f)(z)}{1 - \sqrt{2\pi}\mathcal{F}^{-1}(f)(z)}.$$

Applying on the last equality the Fourier transform we obtain

$$u(x) = \frac{1}{\sqrt{2\pi}} \int_{-\infty}^{\infty} \frac{\mathcal{F}^{-1}(f)(z)}{1 - \sqrt{2\pi}\mathcal{F}^{-1}(f)(z)} e^{-\imath zz}\,dz.$$

Using that $f = \mathcal{F}(\mathcal{F}^{-1}(f))$ we obtain by the preceding equality

$$u(x) - f(x) = \frac{1}{\sqrt{2\pi}} \int_{-\infty}^{\infty} (\frac{\mathcal{F}^{-1}(f)(z)}{1 - \sqrt{2\pi}\mathcal{F}^{-1}(f)(z)} - \mathcal{F}^{-1}(f)(z))e^{-\imath zz}\,dz$$

$$= \int_{-\infty}^{\infty} \mathcal{F}^{-1}(f)(z)\frac{\mathcal{F}^{-1}(f)(z)}{1 - \sqrt{2\pi}\mathcal{F}^{-1}(f)(z)} e^{-\imath zz}\,dz. \qquad (8.26)$$

On the other hand, from the equality $\mathcal{F}^{-1}(u * v) = \sqrt{2\pi}\mathcal{F}^{-1}(u)\mathcal{F}^{-1}(v)$ we obtain $u * v = \mathcal{F}(\sqrt{2\pi}\mathcal{F}^{-1}(u) \cdot \mathcal{F}^{-1}(v))$. Applying the last equality on (8.26) we obtain

$$u(x) = f(x) + \int_{-\infty}^{\infty} s(x - t)f(t)\, dt,$$

where the function s is given by (8.25).

Exercise 8.41 *Let $f \in L_2(\mathbf{R})$ and $k \in L_1(\mathbf{R})$. Prove that there exists a solution in the space $L_2(\mathbf{R})$ of the integral equation*

$$f(x) = \int_{-\infty}^{\infty} k(x - y)u(y)\, dy$$

if and only if

$$\frac{\mathcal{F}^{-1}(f)(z)}{\mathcal{F}^{-1}(k)(z)} \in L_2(\mathbf{R}).$$

Hint. Using the procedure from Example 8.40 prove that

$$u(x) = \sqrt{2\pi} \int_{-\infty}^{\infty} \frac{\mathcal{F}^{-1}(f)(z)}{\mathcal{F}^{-1}(k)(z)} e^{-\imath x z}\, dz.$$

Chapter 9

Generalized Derivative and Sobolev Spaces

9.1 Generalized Derivative

9.1.1 Preliminaries

Let $\alpha = (\alpha_1, \cdots, \alpha_n) \in \mathbf{Z}_+^n$ and $Q \subset \mathbf{R}^n$ is a region.

Definition 9.1 *A function $f^{(\alpha)} \in L_2(Q)$ is α–generalized derivative on the region Q of the function $f \in L_2(Q)$ if for every function $\varphi \in C_0^{|\alpha|}(Q)$ we have*

$$\int_Q f(x) D^\alpha \overline{\varphi}(x) \, dx = (-1)^{|\alpha|} \int_Q f^{(\alpha)}(x) \overline{\varphi}(x) \, dx.$$

For more general case, we can take the space $L_{loc}^2(Q)$ instead of $L_2(Q)$. Instead of the notation $f^{(\alpha)}$ we are often using also the usual notations for the classical derivative, such as $D^\alpha f$ and $\dfrac{\partial^{|\alpha|} f}{\partial x_1^{\alpha_1} \cdots \partial x_n^{\alpha_n}}$.

If $f \in C^{|\alpha|}(Q)$ then there exists the generalized derivative $f^{(\alpha)}$ and it is equal to the classical derivative $D^{(\alpha)} f$.

9.1.2 Examples and Exercises

Example 9.1 *Prove that the generalized derivative $f^{(-1)}$ of the function f given by*

$$f(x) = \begin{cases} 0 & \text{for } x < 0, \\ x & \text{for } x \geq 0, \end{cases}$$

on $\mathbf{R}$ is the function g given by

$$g(x) = \begin{cases} 0 & \text{for } x < 0, \\ 1 & \text{for } x \geq 0. \end{cases}$$

Solution. For an arbitrary but fixed $\varphi \in C_0^1(\mathbf{R})$ there exists $R > 0$ such that $\varphi(x) = 0$ and $\frac{d\varphi(x)}{dx} = 0$ for every $x \geq R$. Therefore

$$\int_{-\infty}^{+\infty} f(x)\overline{\frac{d\varphi(x)}{dx}}\, dx = \int_0^R x \overline{\frac{d\varphi(x)}{dx}}\, dx = x\overline{\varphi(x)}\Big|_0^R - \int_0^R 1 \cdot \overline{\varphi(x)}\, dx = -\int_0^R \overline{\varphi(x)}\, dx.$$

On the other side we have

$$\int_{-\infty}^{+\infty} g(x)\overline{\varphi(x)}\, dx = \int_0^R \overline{\varphi(x)}\, dx.$$

The preceding two equalities imply

$$\int_{-\infty}^{+\infty} \left(f(x)\overline{\frac{d\varphi(x)}{dx}} + g(x)\overline{\varphi(x)} \right) dx = 0$$

for every $\varphi \in C_0^1(\mathbf{R})$. This implies that the generalized derivative $f^{(1)}$ of f is equal g on $\mathbf{R}$, although the classical derivative does not exists on $\mathbf{R}$ (it exists on $\mathbf{R} \setminus \{0\}$).

Example 9.2 *Prove that if for a function $f \in L_2(Q)$ there exists the $\alpha-$ generalized derivative $f^{(\alpha)}$ then it is unique and it is independent of the order of the differentiation.*

Solution. Suppose the contrary, i.e, that for $f \in L_2(Q)$ there are two α–derivatives $f_1^{(\alpha)}$ and $f_2^{(\alpha)}$. By Definition 9.1 we have for an arbitrary but fixed $\varphi \in C_0^{|\alpha|}(Q)$

$$\int_Q (f_1^{(\alpha)}(x) - f_2^{(\alpha)}(x)) \cdot \overline{\varphi(x)}\, dx = 0.$$

Since $f_1^{(\alpha)} - f_2^{(\alpha)} \in L_2(K)$ for every compact subset of Q, we obtain by Example 8.16

$$f_1^{(\alpha)} - f_2^{(\alpha)} = 0$$

almost everywhere on K. Since K was an arbitrary compact subset of Q it follows $f_1^{(\alpha)} - f_2^{(\alpha)} = 0$ almost everywhere on Q.

The second part of example follows by the same property of functions from $C_0^{|\alpha|}(Q)$ and Definition 9.1.

Example 9.3 *Prove that if a function $f \in L_2(Q)$ has the generalized derivative $D^\alpha f \in L_2(Q)$, then for every $y \in Q$ and for enough small $h > 0$ we have*

$$(D^\alpha f)_h(y) = D^\alpha f_h(y),$$

where

$$f_h(x) = \int_Q f(y)\delta_h(|x - y|)\, dy.$$

Moreover, for a compact subset K of Q we have

$$\lim_{h \to 0} \|D^\alpha f_h - D^\alpha f\|_{L_2(K)} = 0.$$

Solution. We have

$$
\begin{aligned}
(D^\alpha f)_h(y) &= \int_Q D^\alpha f(x) \delta_h(|x - y|)\, dx \\
&= (-1)^{|\alpha|} \int_Q f(x) D_x^\alpha \delta_h(|x - y|)\, dx \\
&= \int_Q f(x) D_y^\alpha \delta_h(|x - y|)\, dx.
\end{aligned}
$$

We shall need the following consequence of Lebesgue theorem on convergence 8.3 (which we can easily prove).

Theorem 9.2 *Let Q_1 be a region of $\mathbf{R}^n$ and Q_2 a region of $\mathbf{R}^m$. If for some $k \geq 0$ the function F defined on $Q_1 \times \overline{Q_2}$ belongs for almost every $x \in Q_1$ to $C^k(\overline{Q_2})$ and for some integrable function g defined on Q_1 for every $y \in \overline{Q_2}$*

$$
|D_y^\alpha F(x, y)| \leq g(x) \quad (|\alpha| \leq k),
$$

almost everywhere on Q_1, then

$$
\int_Q F(x, y)\, dx \in C^k(\overline{Q_2}).
$$

By this theorem and the property 4) (section 8.2) of the function δ_h we obtain

$$
\int_Q f(x) D_y^\alpha \delta_h(|x - y|)\, dx = D_y^\alpha f_h(y).
$$

We shall prove now the second part of this exercise. Let K be a compact subset of the region Q. Then there exists $h_0 > 0$ such that for every $h \leq h_0$ we have

$$
(D^\alpha f)_h(y) = D^\alpha f_h(y) \quad \text{for every } y \in K.
$$

By Example 8.15 we obtain

$$
\lim_{h \to 0} \|D^\alpha f_h - D^\alpha f\|_{L_2(K)} = \lim_{h \to 0} \|(D^\alpha f)_h - D^\alpha f\|_{L_2(K)} = 0.
$$

Remark 9.3.1.

(i) The preceding procedure can be applied also on an arbitrary subregion $Q' \subset\subset Q$ (instead of K) such that we obtain

$$
\lim_{h \to 0} \|D^\alpha f_h - D^\alpha f\|_{L_2(Q')} = 0.
$$

(ii) By the preceding remark (i) and the fact that every bounded subset of a Hilbert space is weakly compact (see Example 10.1) it can be proved that the generalized derivative $D^\alpha f$ of a function $f \in L_2(Q)$ exists if and only if for every subregion $Q' \subset\subset Q$ there exist constants $c = c(Q') > 0$ and $h_0 = h_0(Q') > 0$ such that $\|D^\alpha f_h\|_{L_2(Q')} \leq c$ for every $h < h_0$.

Example 9.4 *Prove that if all derivatives of the first order of a function f are zero, then this function f is constant almost everywhere.*

Solution. We shall use Exercises 9.3. We have for every subregion $Q' \subset\subset Q$ and enough small h

$$(D_i f)_h = 0 \quad (i = 1, \cdots, n).$$

By Example 9.3 we have

$$D_i f_h = (D_i f)_h = 0 \quad (i = 1, \cdots, n).$$

Hence f_h is equal to a constant $C = C(h)$. We have

$$\lim_{h \to 0} \|f_h - f\|_{L_2(Q')} = \lim_{h \to 0} \|C(h) - f\|_{L_2(Q')} = 0. \tag{9.1}$$

On the other hand the following inequality is true

$$
\begin{aligned}
\sqrt{|Q'|} \cdot |C(h_1) - C(h_2)| &= \|C(h_1) - C(h_2)\|_{L_2(Q')} \\
&\leq \|C(h_1) - f\|_{L_2(Q')} + \|f - C(h_2)\|_{L_2(Q')},
\end{aligned}
$$

where $|Q'| = \int_{Q'} dx$. By (9.1) the right part of the preceding inequality converges to zero as $h_1, h_2 \to 0$. Hence $C(h)$ converges uniformly as $h \to 0$ on $\overline{Q'}$ to some constant. This means that f is equal to a constant on Q' and since Q' was arbitrary it follows that f is equal to a constant on the whole Q.

Example 9.5 *The function $f(x) = \operatorname{sgn} x_1$ has no generalized derivative $\dfrac{\partial f}{\partial x_1}$ on the unit ball $B(0,1) = \{x \mid |x| < 1\}$.*

Solution. Suppose the contrary, i.e, that there exists a function $g \in L_2(B(0,1))$ such that for every $\varphi \in C_0^1(B(0,1))$

$$\int_{B(0,1)} g(x)\overline{\varphi(x)}\, dx = -\int_{B(0,1)} \operatorname{sgn} x_1 \frac{\overline{\partial \varphi(x)}}{\partial x_1}\, dx.$$

Hence

$$
\begin{aligned}
\int_{B(0,1)} g(x)\overline{\varphi(x)}\, dx &= -\int_{B(0,1) \cap \{x \mid x_1 > 0\}} \frac{\overline{\partial \varphi(x)}}{\partial x_1}\, dx + \int_{B(0,1) \cap \{x \mid x_1 < 0\}} \frac{\overline{\partial \varphi(x)}}{\partial x_1}\, dx \\
&= 2 \int_{B(0,1) \cap \{x \mid x_1 = 0\}} \overline{\varphi(x)}\, dx_2 \ldots dx_n.
\end{aligned}
$$

Since the preceding equality holds for every $\varphi \in C_0^1(B(0,1))$, taking in a special function φ which is zero on $B(0,1) \cap \{x \mid x_1 < 0\}$ we obtain

$$\int_{B(0,1) \cap \{x \mid x_1 > 0\}} g(x)\overline{\varphi(x)}\, dx = 0,$$

which implies $g = 0$ almost everywhere on $B(0,1) \cap \{x \mid x_1 > 0\}$. In an analogous way we obtain that $g = 0$ almost everywhere on $B(0,1) \cap \{x \mid x_1 < 0\}$. Therefore $g = 0$ almost everywhere on $B(0,1)$. Then we have

$$\int_{B(0,1)\cap\{x \mid x_1=0\}} \overline{\varphi(x)} \, dx_2 \dots dx_n = 0$$

for every $\varphi \in C_0^1(B(0,1))$, which is impossible.

Example 9.6 *Prove that the function $f : B(0,1) \to \mathbf{R}$ defined by*

$$f(x) = sgn \, x_1 + sgn \, x_2$$

has not the derivatives $\dfrac{\partial f}{\partial x_1}$ and $\dfrac{\partial f}{\partial x_2}$ on $B(0,1)$, although there exists $\dfrac{\partial^2 f}{\partial x_1 \partial x_2}$ on $B(0,1)$.

Solution. The non-existence of the derivatives $\dfrac{\partial f}{\partial x_1}$ and $\dfrac{\partial f}{\partial x_1}$ follows by Example 9.5.

We shall prove that there exists the generalized derivative $\dfrac{\partial^2 f}{\partial x_1 \partial x_2}$ on $B(0,1)$. We have that for every $\varphi \in C_0^2(B(0,1))$

$$\int_{B(0,1)} f(x) \frac{\partial^2 \overline{\varphi(x)}}{\partial x_1 \partial x_2} \, dx = \int_{B(0,1)} sgn \, x_1 \frac{\partial^2 \overline{\varphi(x)}}{\partial x_1 \partial x_2} \, dx + \int_{B(0,1)} sgn \, x_2 \frac{\partial^2 \overline{\varphi(x)}}{\partial x_1 \partial x_2} \, dx$$

$$= - \int_{B(0,1)\cap\{x \mid x_1<0\}} \frac{\partial^2 \overline{\varphi(x)}}{\partial x_1 \partial x_2} \, dx + \int_{B(0,1)\cap\{x \mid x_1>0\}} \frac{\partial^2 \overline{\varphi(x)}}{\partial x_1 \partial x_2} \, dx$$

$$- \int_{B(0,1)\cap\{x \mid x_2<0\}} \frac{\partial^2 \overline{\varphi(x)}}{\partial x_1 \partial x_2} \, dx + \int_{B(0,1)\cap\{x \mid x_2<0\}} \frac{\partial^2 \overline{\varphi(x)}}{\partial x_1 \partial x_2} \, dx$$

$$= 0 = \int_{B(0,1)} 0 \cdot \overline{\varphi(x)} \, dx.$$

Hence there exists the generalized derivative

$$\frac{\partial^2 f}{\partial x_1 \partial x_2} = 0 \text{ on } B(0,1).$$

This example shows the difference with respect the classical derivative for which it can not happen that there exists the derivative of higher order although it does not exist the derivatives of lower order.

Example 9.7 *Prove that a function $f \in L_2(-1,1)$ has a generalized derivative if and only if it is absolutely continuous and $f' \in L_2(-1,1)$.*

Solution. Suppose that a function $f \in L_2(-1,1)$ has a generalized derivative $f' \in L_2(-1,1)$, i.e.,

$$\int_{-1}^{1} f(x)\overline{\varphi'(x)}\, dx = -\int_{-1}^{1} f'(x)\overline{\varphi(x)}\, dx \text{ for all } \varphi \in C_0^1[-1,1].$$

Applying the partial integration we get

$$\int_{-1}^{1} f(x)\overline{\varphi'(x)}\, dx = \int_{-1}^{1} \int_{-1}^{t} f'(x)\, dx\, \overline{\varphi'(t)}\, dt.$$

Introducing a function $u = u(t)$ in the following way

$$u(t) = f(t) - \int_{-1}^{1} f'(x)\, dx \qquad (9.2)$$

we obtain $u \in L_2(-1,1)$ and

$$\int_{-1}^{1} u(t)\overline{\varphi'(t)}\, dt = 0 \text{ for all } \varphi \in C_0^1[-1,1].$$

If we denote by F the class of all functions $\varphi \in C_0^1[-1,1]$ with the property $\int_{-1}^{1} \varphi(x)\, dx = 0$, then such functions $\varphi \in F$ satisfy

$$\int_{-1}^{1} u(t)\varphi(x)\, dx = 0.$$

F is dense in the subspace of $L_2(-1,1)$ consisting of the functions $g \in L_2(-1,1)$ such that $\int_{-1}^{1} g(t)\, dt = 0$. Hence u is almost everywhere constant. Therefore by (9.2) the function f is equal to the function $t \mapsto \int_{-1}^{1} f'(x)\, dx$ almost everywhere up to a constant. Hence f is absolutely continuous function, since f is absolutely continuous if and only if

$$f(t) - f(-1) = \int_{-1}^{t} f'(x)\, dx.$$

Using now the preceding procedure in the opposite direction it can be easily proved that each absolutely continuous function f with $f' \in L_2(-1,1)$ has a generalized derivative (Starting from (9.2) we have that u is absolutely continuous and $u(t) = 0$ almost everywhere on $[-1,1]$).

Remark 9.7.1. It is obvious that in the preceding example we can take any finite interval $[a,b]$ instead of the interval $[-1,1]$. The supposition of the existence of the derivative f' does not imply the absolutely continuity of the function f (we have the both conditions in Example 9.7). For example the function f given by

$$f(x) = \begin{cases} x^2 \sin x^{-2} & \text{for } 0 < x < 1, \\ 0 & \text{for } x = 0 \end{cases}$$

has the derivative on the interval $[0,1]$ but $f' \notin L_1[0,1]$ and so $f' \notin L_2[0,1]$. Even if for some function f we have $f' \in L_1[0,1]$ this does not imply always the absolutely continuity of f.

Exercise 9.8 *Prove that for the function*

$$f(x,y) = \begin{cases} a & \text{for } x^2 + y^2 < 1, y > 0 \\ b & \text{for } x^2 + y^2 < 1, y < 0, \end{cases}$$

there exist the generalized derivatives of the first order on both semidiscs, for $y > 0$ and $y < 0$, but for $a \neq b$ there is no generalized derivative on the whole disc with respect to y.

9.2 Sobolev Spaces

9.2.1 Preliminaries

Definition 9.3 *Let k be a non-negative integer. Sobolev space $W^k(Q)$ consists of all functions $f \in L_2(Q)$ which have all generalized derivatives $D^\alpha f$ for $0 \leq |\alpha| \leq k$ and they belong to $L_2(Q)$ equipped with a scalar product (see Section 10.1)*

$$(f|g)_{W^k(Q)} = \sum_{|\alpha| \leq k} \int_Q D^\alpha f \, D^\alpha \bar{g} \, dx.$$

For $k = 0$ is $W^0(Q) = L_2(Q)$.

The scalar product $(f|g)_{W^k(Q)}$ induces the norm

$$\|f\|_{W^k(Q)} = \left(\sum_{|\alpha| \leq k} \int_Q |D^\alpha f|^2 \, dx \right)^{\frac{1}{2}}.$$

Theorem 9.4 *If Q is a bounded region, then*

$$\overline{C^k(\overline{Q})} = W^k(Q) \qquad (k = 0, 1, 2, \cdots),$$

where the closure of $C^k(\overline{Q})$ is taken with respect to the norm $\| \cdot \|_{W^k(Q)}$.

Definition 9.5 *Sobolev space $\overset{o}{W}{}^k(Q)$ is defined by*

$$\overset{o}{W}{}^k(Q) = \overline{C_0^k(Q)} \quad (k = 0, 1, 2, \ldots),$$

where the closure of $C^k(Q)$ is taken with respect to the norm $\| \cdot \|_{W^k(Q)}$.

Theorem 9.6 (Sobolev lemma) *Let Q be a bounded region with $\partial Q \in C^k$. If $k > s + \frac{n}{2}, s \in \mathbf{N} \cup \{0\}$, and $f \in \overset{o}{W}{}^k(Q)$, then $f \in C^s(\overline{Q})$.*

Theorem 9.7 (Rellich) *If Q is a bounded region of $\mathbf{R}^n$ and $m > k$ $(m, k \in \mathbf{Z}_+)$, then the embedding map of the space $\overset{o}{W}{}^m(Q)$ in the space $\overset{o}{W}{}^k(Q)$ is a compact operator.*

9.2.2   Examples and Exercises

Exercise 9.9 *Find for which α the given function f belongs to the given Sobolev space*

a) $f(x) = |x|^{-\alpha} \sin |x|$, *space* $W^2(B(0,1))$, *where* $B(0,1)$ *is the unit disc in* $\mathbf{R}^2$ *with center in* $(0,0)$ *and radius* 1;

b) $f(x) = r^\alpha \varphi$, *the space* $W^2(B(0,1))$ *for* $x = (x_1, x_2) = (r\cos\varphi, r\sin\varphi)$ *for* $0 \le \varphi < 2\pi$

Answer. a) $\alpha < 0$.

Exercise 9.10 *A function $f \in L_2(-\pi, \pi)$ belongs to the space $W^1(-\pi, \pi)$ if and only if the series*

$$\sum_{n=0}^{\infty} n^2(a_n^2 + b_n^2)$$

converges, where

$$a_n = \frac{1}{\pi}\int_{-\pi}^{\pi} f(x)\cos nx \, dx, \quad b_n = \frac{1}{\pi}\int_{-\pi}^{\pi} f(x)\sin nx \, dx \text{ for } n = 0,1,2,\cdots.$$

Exercise 9.11 *For $f \in W^1(-\pi, \pi)$ we have*

$$\|f\|_{W^1(-\pi,\pi)} = \int_{-\pi}^{\pi} (f^2(x) + f'^2(x)) \, dx = \pi \sum_{n=0}^{\infty}(n^2+1)(a_n^2 + b_n^2) + \pi\left(\frac{a_0}{2}\right)^2.$$

Example 9.12 *Prove that the Sobolev space $W^k(Q), k \ge 0$, is a Hilbert space (see Section 10.1).*

Solution. It is easy to check the properties of the scalar product for the bilinear functional

$$(f|g)_{W^k(Q)} = \sum_{|\alpha|\le k}\int_Q D^\alpha f \, D^\alpha \bar{g} \, dx.$$

We shall prove the completeness of the space $W^k(Q)$ endowed with the induced norm

$$\|f\|_{W^k(Q)} = \left(\sum_{|\alpha|\le k}\int_Q |D^\alpha f(x)|^2 \, dx\right)^{\frac{1}{2}}.$$

Let $\{f_m\}_{m\in\mathbf{N}}$ be a Cauchy sequence from $W^k(Q)$, i.e.,

$$\|f_s - f_m\|_{W^k(Q)}^2 = \sum_{|\alpha|\le k}\int_Q |D^\alpha f_s(x) - D^\alpha f_m(x)|^2 \, dx \to 0 \tag{9.3}$$

as $s, m \to \infty$. This implies that for every fixed α, $|\alpha| \le k$,

$$\int_Q |D^\alpha f_s(x) - D^\alpha f_m(x)|^2 \, dx \to 0$$

as $s, m \to \infty$ and so also specially for $\alpha = 0$

$$\int_Q |f_s(x) - f_m(x)|^2 \, dx \to 0$$

as $s, m \to \infty$. Since the space $L_2(Q)$ is complete we obtain by the last relation that there exists $f \in L_2(Q)$ such that $f_m \xrightarrow{L_2(Q)} f$ as $m \to \infty$. We obtain in an analogous way by (9.3) that for every α, $|\alpha| \le k$, there exists $f^{(\alpha)} \in L_2(Q)$ such that $D^\alpha f_m \xrightarrow{L_2(Q)} f^{(\alpha)}$ as $m \to \infty$. Since $D^\alpha f_m \in L_2(Q)$ ($|\alpha| \le k$, $m \in \mathbf{N}$), we have

$$(f_m | D^\alpha \varphi)_{L_2(Q)} = (-1)^{|\alpha|} (D^\alpha f_m | \varphi)_{L_2(Q)}$$

for every $\varphi \in C_0^k(Q)$. Letting $m \to \infty$ we obtain using the Lebesgue convergence theorem

$$(f | D^\alpha \varphi)_{L_2(Q)} = (-1)^{|\alpha|} (f^{(\alpha)} | \varphi)_{L_2(Q)}.$$

The last equality shows that $f^{(\alpha)}$ is the α–generalized derivative of f and so $f \in W^k(Q)$. It is easy to check that

$$\|f_m - f\|_{W^k(Q)} \to 0 \text{ as } m \to \infty.$$

Exercise 9.13 *Prove that the function $f(x) = |x|$ belongs to the space $W^1(-1,1)$ but not $W^2(-1,1)$.*

Example 9.14 *Suppose that for a bounded region Q and $\lambda > 0$ the region*

$$Q(\lambda) = \left\{ x \Big| \frac{x}{\lambda} \in Q \right\}$$

is starshaped. For $\lambda > 1$ and $f \in W^k(Q)$ we introduce a function f_λ by

$$f_\lambda(x) = f\left(\frac{x}{\lambda}\right) \qquad (x \in Q(\lambda)).$$

Prove that

a) $(D^\alpha f_\lambda)(x) = \frac{1}{\lambda^{|\alpha|}} (D^\alpha f)\left(\frac{x}{\lambda}\right)$, which implies $f_\lambda \in W^k(Q(\lambda))$;

b) $\lim\limits_{\lambda \to 1+0} \|f_\lambda - f\|_{W^k(Q)} = 0$.

Remark 9.14.1. A bounded region Q is *starshaped* if there exists $x_0 \in Q$ such that for every $\lambda > 1$ the set

$$\left\{ x \Big| x_0 + \frac{x - x_0}{\lambda} \in Q \right\}$$

is a subset of the set $\overline{Q}$.

Solution.

a) We have for $\varphi \in C_0^\infty(Q(\lambda))$

$$
\begin{aligned}
\int_{Q(\lambda)} D^\alpha f_\lambda(x)\overline{\varphi(x)}\, dx &= (-1)^{|\alpha|} \int_{Q(\lambda)} f(\frac{x}{\lambda})D^\alpha\overline{\varphi(x)}\, dx \\
&= (-1)^{|\alpha|} \int_Q f(z)\overline{(D^\alpha\varphi)(\lambda z)}\, dz \\
&= \frac{(-1)^{|\alpha|}}{\lambda^{|\alpha|}} \int_Q f(z)D_z^\alpha\overline{(\varphi(\lambda z))}\lambda^n\, dz \\
&= \frac{1}{\lambda^{|\alpha|}} \int_Q D^\alpha f(z)\overline{(\varphi(\lambda z))}\lambda^n\, dz \\
&= \frac{1}{\lambda^{|\alpha|}} \int_{Q(\lambda)} D^\alpha f(\frac{x}{\lambda})\overline{\varphi(x)}\, dx.
\end{aligned}
$$

b) First we shall show that $\lim_{\lambda \to 1+0} \|f_\lambda - f\|_{L_2(Q)} = 0$. $C_0(Q)$ is dense in $L_2(Q)$. Therefore for every $\varepsilon > 0$ there exists a function $g \in C_0(Q)$ such that $\|f - g\| < \varepsilon$. Using the substitution in the integral we have

$$
\begin{aligned}
\|f_\lambda - g_\lambda\|_{L_2(Q)} &\leq \|f_\lambda - g_\lambda\|_{L_2(Q(\lambda))} \\
&= (\int_{Q(\lambda)} |f(\frac{x}{\lambda}) - g(\frac{x}{\lambda})|^2\, dx)^{\frac{1}{2}} \\
&= (\lambda^n \int_Q |f(z) - g(z)|^2\, dz)^{\frac{1}{2}} \leq \lambda^{\frac{n}{2}}\varepsilon.
\end{aligned}
$$

Therefore

$$
\begin{aligned}
\|f_\lambda - f\|_{L_2(Q)} &\leq \|f_\lambda - g_\lambda\|_{L_2(Q(\lambda))} + \|g_\lambda - g\|_{L_2(Q)} + \|g - f\|_{L_2(Q(\lambda))} \\
&\leq \varepsilon(1 + \lambda^{\frac{n}{2}}) + \|g_\lambda - g\|_{L_2(Q)}.
\end{aligned}
$$

Taking $\lambda \to 1 + 0$ we obtain $\lim_{\lambda \to 1+0} \|f_\lambda - f\|_{L_2(Q)} = 0$, since

$$
\lim_{\lambda \to 1+0} \sup_{x \in \overline{Q}} |g_\lambda(x) - g(x)| = 0.
$$

Hence $\lim_{\lambda \to 1+0} \|g_\lambda - g\|_{L_2(Q)} = 0$.
In a quite analogous way we can prove that

$$
\lim_{\lambda \to 1+0} \|D^\alpha f_\lambda - D^\alpha f\|_{L_2(Q)} = 0.
$$

Finally by the definition of the norm in the space $W^k(Q)$ it follows

$$
\lim_{\lambda \to 1+0} \|f_\lambda - f\|_{W^k(Q)} = 0.
$$

Exercise 9.15 *Let Q be a starshaped locally quadratic bounded region. If $f \in C^1(Q)$ and $f|_{\partial Q} = 0$, then $f \in \overset{\circ}{W}{}^1(Q)$.*

Hints. Take the function f_λ from Exercises 9.14 for λ, $0 < \lambda < 1$. Then $\overline{Q(\lambda)} \subset Q$. Take $f_\lambda(x) = 0$ for $x \in Q \setminus Q(\lambda)$. Hence $f_\lambda \in C^1(\overline{Q(\lambda)})$ and $f_\lambda \in C^1(Q \setminus Q(\lambda))$. Use Example 9.14.

Example 9.16 *Let Q be a bounded region in $\mathbf{R}^n$ and the function $\Phi = (\Phi_1, \cdots, \Phi_n)$ is a C^k– dipheomorphism from $\overline{Q}$ onto $\overline{\Omega}$, where $\Omega = \Phi(Q)$. If $f \in W^k(Q)$, then $f \circ \Phi^{-1} \in W^k(\Omega)$ and there exists $C > 0$ such that*

$$\|f \circ \Phi^{-1}\|_{W^k(\Omega)} \leq C\|f\|_{W^k(\Omega)}.$$

Solution. $C^k(\overline{Q})$ is dense in the space $W^k(Q)$. Therefore for $f \in W^k(Q)$ there exists a sequence $\{f_i\}_{i \in \mathbf{N}}$ of functions from $C^k(Q)$ such that $\lim_{i \to \infty} \|f_i - f\|_{W^k(Q)} = 0$. Hence $\{f_i\}_{i \in \mathbf{N}}$ is a Cauchy sequence in $W^k(Q)$. Since $f_i \circ \Phi^{-1} \in C^k(\overline{\Omega})$ $(i \in \mathbf{N})$ we can apply the classical Leibniz rule on the product uv.

$$D^\alpha(uv) = \sum_{\beta \leq \alpha} a_\beta (D^\alpha \overline{u}^\beta)(D^\beta v).$$

Therefore by compactness of $\overline{Q}$ the sequence $\{f_i \circ \Phi^{-1}\}_{i \in \mathbf{N}}$ is a Cauchy sequence in the space $W^k(\Omega)$. By completeness of the space $W^k(Q)$ the sequence $\{f_i \circ \Phi^{-1}\}_{i \in \mathbf{N}}$ is convergent in $W^k(\Omega)$. Hence there exists $g \in W^k(\Omega)$ such that

$$\lim_{i \to \infty} \|f_i \circ \Phi^{-1} - g\|_{W^k(\Omega)} = 0. \tag{9.4}$$

Using the rule for changing the variable in a Lebesgue integral we have that if $f \in L_2(Q)$ then $f \circ \Phi^{-1} \in L_2(\Omega)$ and so

$$\lim_{i \to \infty} \|f_i \circ \Phi^{-1} - f \circ \Phi^{-1}\|_{L_2(\Omega)} = 0.$$

By (9.4) we have $g = f \circ \Phi^{-1}$ almost everywhere and so $f \circ \Phi^{-1} \in W^k(\Omega)$. Since $f_i \in C^k(\overline{Q}), f_i \circ \Phi^{-1} \in C^k(\overline{\Omega})$ and using the boundedness of the functions Φ_i and the chain rule we obtain that there exists $C > 0$ such that

$$\|f_i \circ \Phi^{-1}\|_{W^k(\Omega)} \leq C\|f_i\|_{W^k(Q)}.$$

Hence by (9.4) we obtain the desired inequality.

Example 9.17 *The system of functions*

$$\{(2\pi)^{-\frac{n}{2}} e^{\frac{\pi}{a} i \beta \cdot x}\}_{\beta \in \mathbf{Z}_+^n}$$

for $\beta \cdot x = \beta_1 x_1 + \cdots + \beta_n x_n$ is a complete orthonormal system of functions (see Section 10.1) in the space $L_2(P_a)$, where

$$P_a = \{x \mid |x_i| < a, i = 1, \cdots, n\} \text{ and } a > 0.$$

Solution. Since $\{(2\pi)^{-1}e^{i\frac{\pi}{a}x_ik}\}_{k\in\mathbf{Z}_+}$ is a complete orthonormal system of functions in the space $L_2(-a,a)$ (one-dimensional case), we obtain the desired conclusion by example from Chapter 8. on the product of orthonormal systems.

Example 9.18 *Let $f \in C_0^\infty(\mathbf{R}^n)$ be such a function that*

$$supp\, f \subset P = \{x|\ |x_i| < \pi, i = 1,\ldots,n\}.$$

Prove that the Fourier series of the function f with respect to the orthonormal system

$$\{\psi_\beta\}_{\beta\in\mathbf{Z}_+^n} = \{(2\pi)^{-\frac{n}{2}}e^{i\beta\cdot x}\}_{\beta\in\mathbf{Z}_+^n}$$

uniformly converges to f. Moreover, prove that for any arbitrary but fixed multi-index α applying the differential operator D^α on this Fourier series on every summand the new series converges uniformly on P to $D^\alpha f$.

Solution. By Example 9.17 the system $\{\psi_\beta\}_{\beta\in\mathbf{Z}_+^n}$ is complete and orthonormal. Therefore every function $f \in C_0^\infty(P) \subset L_2(P)$ has a representation with a Fourier series convergent in $L_2(P)$

$$f = \sum_\beta a_\beta \psi_\beta.$$

We shall show that the series $\sum_\beta a_\beta D^\alpha \psi_\beta$ converges absolutely and uniformly. Since

$$D^\alpha \psi_\beta(x) = (2\pi)^{-\frac{n}{2}} i^{|\alpha|} \beta^\alpha e^{i\beta\cdot x},$$

where $\beta^\alpha = \beta_1^{\alpha_1}\cdots\beta_n^{\alpha_n}$, we have

$$|D^\alpha \psi_\beta| = (2\pi)^{-\frac{n}{2}}|\beta^\alpha|.$$

Hence it is enough to prove that the series $\sum_\beta |a_\beta| \cdot |\beta^\alpha|$ is convergent. For that purpose we shall give an estimation for the Fourier coefficients a_β. Applying the partial integration with respect to x_k we obtain for $\beta_k \neq 0$

$$a_\beta = (2\pi)^{-\frac{n}{2}} \int_P f(x)e^{-i\beta\cdot x}\, dx = (2\pi)^{-\frac{n}{2}} \int_P D_k f(x)(i\beta_k)^{-1}e^{-i\beta\cdot x}\, dx.$$

Let $\gamma \in \mathbf{Z}_+^n$. Applying the preceding procedure γ_k–times with respect to the variable x_k, $k = 1,\cdots,n$, and the convention "$0^0 = 1$" we obtain

$$a_\beta = (2\pi)^{-\frac{n}{2}} i^{-|\gamma|} \frac{1}{\beta^\gamma} \int_P D^\gamma f(x)e^{-i\beta\cdot x}\, dx.$$

This equality implies for every $\gamma \in \mathbf{Z}_+^n$

$$|a_\beta| \leq (2\pi)^{-\frac{n}{2}} \frac{1}{\beta^\gamma} \int_P |D^\gamma f(x)|\, dx \quad (\beta \in \mathbf{Z}_+^n).$$

Taking a special multi-index $\gamma = (\gamma_1, \ldots, \gamma_n)$

$$\gamma_k = \left\{ \begin{array}{ll} 0 & \text{for } \beta_k = 0 \\ \alpha_k + 2 & \text{for } \beta_k \neq 0, \end{array} \right.$$

and using the last inequality we obtain that there exists a constant (with respect to β) $C > 0$ such that

$$|a_\beta| |\beta^\alpha| \leq \frac{C}{\displaystyle\prod_{k=1, \beta_k \neq 0}^{n} \beta_k^2}. \tag{9.5}$$

Since we have

$$\sum_\beta \frac{1}{\displaystyle\prod_{k=1, \beta_k \neq 0}^{n} \beta_k^2} \leq (1 + \sum_{\beta_1 \neq 0} \frac{1}{\beta_1^2}) \cdots (1 + \sum_{\beta_n \neq 0} \frac{1}{\beta_n^2})$$

and the series on the right side converge we obtain that the series

$$\sum_\beta \frac{1}{\displaystyle\prod_{k=1, \beta_k \neq 0}^{n} \beta_k^2}$$

also converges. Hence by (9.5) the series $\sum_\beta |a_\beta| |\beta^\alpha|$ converges, too.

Remark 9.18.1. We can prove the same statement in a quite analogous way for a more general case taking $P_a = \{x | \, |x_i| < a, i = 1, \cdots, n\}$ instead of P and the system

$$\{(2\pi)^{-\frac{n}{2}} e^{\frac{\pi}{a} i \beta \cdot x}\}_{\beta \in \mathbf{Z}_+^n}.$$

Exercise 9.19 *Prove that for every $f \in W^k(Q)$ and every subregion $Q' \subset\subset Q$ we have*

$$\lim_{h \to 0+0} \|f_h - f\|_{W^k(Q')} = 0.$$

Moreover if the function f is complementary finite in Q, then

$$\lim_{h \to 0+0} \|f_h - f\|_{W^k(Q)} = 0.$$

Example 9.20 *Prove that the space $W^k(P)$, where $P = \{x | \, |x_i| < \pi, \ i = 1, \cdots, n\}$ is a cube in $\mathbf{R}^n$, is separable, i.e., it has a countable dense subset.*

Solution. By Exercises 9.19 every function $f \in W^k(P)$ can be approximated by functions f_h from the set $C_0^\infty(P)$ with respect to the $W^k(P)$– norm. Every function f_h for enough small h can be represented by Exercises 9.18 by its uniformly convergent Fourier series (and so convergent also in $W^k(P)$) whose derivatives are also uniformly convergent. Therefore the function f_h can be approximated with respect to $W^k(P)$–norm by partial sums of the corresponding Fourier series, i.e.,

f_h can be approximated with respect to $W^k(P)$-norm by linear combinations of the system $\{(2\pi)^{-\frac{n}{2}} e^{i\beta \cdot x}\}$ and with coefficients whose real and imaginary parts are rational numbers. Therefore the countable set of all such linear combinations with rational coefficients of the system $\{(2\pi)^{-\frac{n}{2}} e^{i\beta \cdot x}\}_{\beta \in \mathbf{Z}_+^n}$ is dense in the space $W^k(P)$.

Example 9.21 *Let $a_i > 0$, $i = 1, \cdots, n$,*

$$P = \{x \mid x \in \mathbf{R}^n, a_i < x_i < 2a_i, i = 1, \cdots, n\},$$

$$P' = \{x \mid x \in \mathbf{R}^n, 0 < x_i < 3a_i, i = 1, \ldots, n\}.$$

Prove that for every function f from the space $W^k(P)$ there exists its extension $F \in W^1(P')$ such that

$$\|F\|_{W^1(P')} \leq C\|f\|_{W^1(P)},$$

where the constant C is independent of f. Specially, if $f \in C(\overline{P})$ then $F \in C(\overline{P'})$.

Solution. First we shall consider the special case $f \in C^1(\overline{P})$. We shall extend this function on P' and denote this extension by F such that it will be symmetric in all coordinates $i = 1, \cdots, n$ with respect to a_i and $2a_i$ $(i = 1, \cdots, n)$, i.e., $F(x') = f(x)$ where the mapping $t : P' \to P$ is defined by

$$x' = (x'_1, \ldots, x'_n) \mapsto x = (x_1, \ldots, x_n)$$

$$x_i = \begin{cases} 2a_i - x'_i & \text{for } 0 \leq x'_i \leq a_i \\ x'_i & \text{for } a_i < x'_i \leq 2a_i \\ 4a_i - x'_i & \text{for } 2a_i < x'_i \leq 3a_i. \end{cases}$$

Since the function t continuously maps $\overline{P'}$ onto $\overline{P}$ we have $F \circ t \in C(\overline{P'})$. The function F and the translation function for $(c_1, \ldots, c_n)$, where c_i is either 0 or a_i or $-a_i$ have continuous derivatives on $\overline{P}$. Since these derivatives of first order coincide with the generalized derivatives of first order we obtain $F \in W^1(P')$. Since

$$\int_P |F(x)|^2 \, dx = 3^n \int_P |f(x)|^2 \, dx$$

and

$$\int_{P'} |D_i F(x)|^2 \, dx = 3^n \int_P |D_i f(x)|^2 \, dx \quad (i = 1, \ldots, n)$$

we obtain

$$\|F\|_{W^1(P')} = 3^n \|f\|_{W^1(P)}^2,$$

i.e., the desired inequality reduces on the equality for this special case. We shall show now that the inequality is true for the general case, i.e., for $f \in W^1(P)$. We remark that the map $T : C^1(\underline{P}) \to W^1(P')$ given by $f \mapsto F$ is a linear and bounded operator on a dense subset $\overline{C^1}(P)$. Then we can extend the operator

T on the whole space $W^1(P)$ in a way that it remains to be bounded. We denote this extension also by T. Then for every $f \in W^1(P)$ and a sequence $\{f_j\}_{j \in \mathbb{N}}$ from $C^1(\overline{P})$ with the property $f_j \xrightarrow{W^1(P)} f$ we have $T(f_j) \xrightarrow{W^1(P')} T(f)$ as $j \to \infty$. Since the restriction of $T(f_j)$ on P is equal f_j it follows that the restriction of $T(f)$ on P is almost everywhere equal f and that the extension $T(f)$ is obtained in the same way as for the case $f \in C^1(\overline{P})$ (excluding a set of a measure zero).

Remark 9.21.1. Exercise 9.20 can be used in the proof of the following generalization of Exercise 9.20.

Theorem 9.8 *If Q and Q' are bounded regions such that $Q \subset\subset Q'$ and $\partial Q \in C^k$, then for every function $f \in W^k(Q)$ there exists a finite extension $F \in W^k(Q')$ such that*

$$\|F\|_{W^k(Q')} \leq C\|f\|_{W^k(Q)},$$

where the constant $C > 0$ depends only from Q and Q'.

Example 9.22 *Prove that the set $C_0^\infty(\overline{Q})$ is a dense subset of the space $W^k(Q)$, where Q is a bounded region whose boundary ∂Q belongs to C^k. This implies that $C^k(\overline{Q})$ is also a dense set in the space $W^k(Q)$, i.e., $\overline{C^k(\overline{Q})} = W^k(Q)$.*

Solution. Let O be a region with $Q \subset\subset O$. By Exercise 9.21 and corresponding Remark for every function $f \in W^k(Q)$ there exists its extension F on O which belongs to the space $W^k(O)$. Then by Exercise 9.19 we have

$$\lim_{h \to 0} \|F_h - f\|_{W^k(Q)} = \lim_{h \to 0} \|F_h - F\|_{W^k(Q)} = 0,$$

where

$$F_h(x) = \int_Q F(y)\delta_h(|x - y|)\, dy.$$

Since $F_h \in C_0^\infty(\overline{Q})$, we have proved the desired statement.

Example 9.23 *Let $f \in \overset{o}{W}{}^k(Q)$ and Q is a subregion of the region Q'. Prove that the extension by zero of the function f on Q' given by $f(x) = 0$ for $x \in Q' \setminus Q$ belongs to $\overset{o}{W}{}^k(Q')$.*

Solution. By the definition of the space $\overset{o}{W}{}^k(Q)$ there exists a sequence $\{\varphi_j\}_{j \in \mathbb{N}}$ from $C_0^\infty(Q)$ such that

$$\lim_{j \to \infty} \|\varphi_j - f\|_{\overset{o}{W}{}^k(Q)} = 0.$$

Since for every function φ_j its extension by zero on Q' belongs to $C_0^\infty(Q')$ and $\{\varphi_j\}_{j \in \mathbb{N}}$ is a Cauchy sequence in the space $\overset{o}{W}{}^k(Q')$ we obtain

$$\lim_{j \to \infty} \|\varphi_j - f\|_{\overset{o}{W}{}^k(Q')} = 0.$$

Hence $f \in \overset{o}{W}{}^k(Q')$.

Example 9.24 *Prove that if a function* $f \in \overset{\circ}{W}{}^{k}(Q)$ *is extended by zero on the whole* $\mathbf{R}^n$, *then for every multi-index* α, $|\alpha| \leq k$, *we have*

a) $D^\alpha \hat{f} \in L_2(\mathbf{R}^n)$;

b) $\widehat{D^\alpha f}(y) = (\imath y)^\alpha \hat{f}(y)$;

c) $\|f\|_{\overset{\circ}{W}{}^{k}(Q)} = \int\limits_{\mathbf{R}^n} \sum_{|\alpha| \leq k} |y^\alpha|^2 |\hat{f}(y)|^2 \, dy$.

Solution.

a) It is true for $f \in C_0^k(Q)$, but $C_0^k(Q)$ is dense in $\overset{\circ}{W}{}^{k}(Q)$.

b) By Example 9.23 the extension by zero of a function $f \in \overset{\circ}{W}{}^{k}(Q)$ on $\mathbf{R}^n$ belongs to $\overset{\circ}{W}{}^{k}(\mathbf{R}^n) \subset L_2(\mathbf{R}^n)$. Therefore we have for every $\varphi \in C_0^\infty(\mathbf{R}^n)$

$$(\widehat{D^\alpha f}|\hat{\varphi})_{L_2(\mathbf{R}^n)} = (-1)^{|\alpha|}(\hat{f}|\widehat{D^\alpha \varphi})_{L_2(\mathbf{R}^n)} = \int\limits_{P} (-i)^{|\alpha|} y^\alpha \hat{f}(y)\overline{\hat{\varphi}(y)} \, dy.$$

Since by Example 9.22 $C_0^\infty(\mathbf{R}^n)$ is dense in $L_2(\mathbf{R}^n)$ we obtain that $C_0^\infty(\hat{\mathbf{R}}^n)$ is also dense in $L_2(\mathbf{R}^n)$ ($f \mapsto \hat{f}$ is an isometrically isomorphism). Hence b).

c) Since $\overset{\circ}{W}{}^{k}(Q)$ is a Hilbert space we obtain c) by Parseval's identity.

Exercise 9.25 *For which sequences* $\{a_n\}_{n \in \mathbf{N}}$ *the function* f *given by*

$$f(x,y) = \sum_{n=1}^{\infty} a_n e^{-ny} \sin nx$$

for $0 \leq x \leq \pi$ *and* $y > 0$ *belongs to the space* $W^1(\{(x,y) : 0 < x < \pi, y > 0\})$?

Answer. The desired sequence $\{a_n\}_{n \in \mathbf{N}}$ has to satisfy the condition that the series $\sum\limits_{n=1}^{\infty} n|a_n|^2$ converges.

Example 9.26 *Prove that for every function* $f \in C^1[a,b]$

$$|f(a)|^2 \leq \frac{2}{b-a} \int\limits_{a}^{b} |f(x)|^2 \, dx + (b-a) \int\limits_{a}^{b} |f'(x)|^2 \, dx.$$

Solution. We have by Newton-Leibniz formula for $f \in C^1[a, b]$

$$f(x) - f(a) = \int\limits_a^x f'(z) \, dz.$$

Applying the Cauchy–Schwartz inequality we obtain

$$|f(a)|^2 \leq (|f(x)| + |\int\limits_a^x f'(z) \, dz|)^2$$

$$\leq 2|f(x)|^2 + 2|\int\limits_a^x f'(z) \, dz|^2$$

$$\leq 2|f(x)|^2 + 2(x - a) \int\limits_a^x |f'(z)|^2 \, dz$$

$$\leq 2|f(x)|^2 + 2(x - a) \int\limits_a^b |f'(z)|^2 \, dz.$$

Integrating over the interval $[a, b]$ we obtain

$$(b - a)|f(a)|^2 \leq 2 \int\limits_a^b |f(x)|^2 \, dx + (b - a)^2 \int\limits_a^b |f'(z)|^2 \, dz,$$

which implies the desired inequality.

Example 9.27 *Let P be the n-dimensional parallelepiped*

$$P = (a_1, b_1) \times \cdots \times (a_n, b_n)$$

with sides

$$P_1 = \{a_1\} \times (a_2, b_2) \times \cdots \times (a_n, b_n),$$

$$P_i = (a_1, b_1) \times \cdots \times (a_{i-1}, b_{i-1}) \times \{a_i\} \times (a_{i+1}, b_{i+1}) \times \cdots \times (a_n, b_n)$$

for $i = 1, \ldots, n$,

$$P_n = (a_1, b_1) \times \cdots \times (a_{n-1}, b_{n-1}) \times \{a_n\}.$$

Prove that for $f \in C^1(\overline{P})$

a)

$$\int\limits_{P_i} |f|^2 \leq \frac{2}{b_i - a_i} \int\limits_P |f(x)|^2 \, dx + (b_i - a_i) \int\limits_P |D_i f(x)|^2 \, dx;$$

b) *for a n-dimensional parallelepiped P_ε whose edges have the length $\varepsilon(b_i - a_i)$ $(i = 1, \ldots, n), 0 < \varepsilon < 1$, we have*

$$\int\limits_{\partial P} |f|^2 \leq \frac{C_1}{\varepsilon} \int\limits_{P} |f(x)|^2 \, dx + C_2 \varepsilon \sum_{i=1}^{n} \int\limits_{P} |D_i f(x)|^2 \, dx,$$

where $C_1 > 0$ and $C_2 > 0$ are constants.

Solution.

a) Using the same procedure as in Exercise 9.26 only starting now from the equality

$$f(x_1, \ldots, x_i, \ldots, x_n) - f(x_1, \ldots, a_i, x_{i+1}, \ldots, x_n)$$

$$= \int\limits_{a_i}^{x_i} D_i f(x_1, \ldots, x_{i-1}, z_i, x_{i+1}, \ldots, x_n) \, dz_i$$

we obtain the inequality

$$|f(x_1, \ldots, a_i, x_{i+1}, \ldots, x_n)|^2$$

$$\leq 2|f(x)|^2 + 2(x_i - a_i) \int\limits_{a_i}^{b_i} |D_i f(x_1, \ldots, x_{i-1}, z_i, x_{i+1}, \ldots, x_n)|^2 \, dz_i.$$

Integrating with respect to x on P we obtain the desired inequality.

b) Applying the inequality from a) on n-dimensional parallelepiped P_ε instead of P, holding the side P_i we obtain

$$\int\limits_{P_i} |f|^2 \leq \frac{2}{\varepsilon(b_i - a_i)} \int\limits_{P_\varepsilon} |f(x)|^2 \, dx + \varepsilon(b_i - a_i) \int\limits_{P_\varepsilon} |D_i f(x)|^2 \, dx$$

$$\leq \frac{2}{\varepsilon(b_i - a_i)} \int\limits_{P} |f(x)|^2 \, dx + \varepsilon(b_i - a_i) \int\limits_{P} |D_i f(x)|^2 \, dx$$

for $i = 1, \ldots, n$. Summing up all n inequalities we obtain

$$\int\limits_{\partial P} |f|^2 \leq \frac{C_1}{\varepsilon} \int\limits_{P} |f(x)|^2 \, dx + C_2 \varepsilon \sum_{i=1}^{n} \int\limits_{P} |D_i f(x)|^2 \, dx.$$

Example 9.28 (The trace of a function) *Let $Q \subseteq \mathbf{R}^n$ $(n \geq 2)$ be a bounded region and $S \subseteq \overline{Q}$ a locally quadratic $(n - 1)$–dimensional surface with the property that for every $x_0 \in S$ there exist regions $U(x_0)$ and V and C^1–dipheomorphism Φ from $\overline{U}$ onto $\overline{V}$ such that some subset of $Q \cap U$ is mapped on some n–dimensional parallelepiped $P \subset V$, and $S \cap \overline{U}$ is mapped on a side or union of sides of P. Then* *there exist constants $C_1 > 0$ and $C_2 > 0$ such that for every function $f \in C^1(\overline{Q})$ and $\varepsilon, 0 < \varepsilon < 1$ we have*

$$\int_S |f|^2 dS \leq \frac{C_1}{\varepsilon} \int_Q |f(x)|^2 \, dx + C_2 \varepsilon \sum_{i=1}^n \int_Q |D_i f(x)|^2 \, dx. \tag{9.6}$$

The preceding inequality (9.6) can be written also in the following form

$$\left(\int_S |f|^2 \, dS \right)^{\frac{1}{2}} \leq C_3 \|f\|_{W^1(Q)}$$

for some constant $C_3 > 0$.

Remark 9.28.1. If we define the operator $T : C^1(\overline{Q}) \to L_2(S)$ in the following way

$$T(f) = f|_{\partial Q}.$$

then by Exercise 9.28 there exists a unique linear and bounded extension $\overline{T} : W^1 \to L_2(S)$ of T. The function $\overline{T}(f) \in L_2(S)$ is the *trace* of the function $f \in W^1(Q)$ on the surface S, which we will denote by $f|_S$.

Solution. Since S is a compact set we can cover it with a finite number of regions U_i which C^1–dipheomorhisms Φ_i map on the regions V_i, respectively, and a subset of $Q \cap U_i$ is mapped on n–dimensional parallelepiped $P_i \subset V_i$, and $S \cap \overline{U}_i$ is mapped on a side or union of sides of the parallelepiped P_i. Let $\varphi_i \in C_0^\infty(U_i), \sum \varphi_i = 1$ be the partition of the unit in a neighborhood of S. By Exercise 9.27.b) we have for $f \in C^1(Q)$

$$
\begin{aligned}
\int_S |f|^2 \, dS &= \sum_{i=1}^s \int_{\Phi_i(S \cap U_i)} (\varphi_i |f|^2 \circ \Phi_i^{-1}) |J_i| \, dx \\
&\leq C_4 \sum_{i=1}^s \int_{\Phi_i(S \cap U_i)} |f \circ \Phi_i^{-1}|^2 \, dx \\
&\leq \sum_{i=1}^s \left(\frac{C_5}{\varepsilon} \int_{P_i} |f \circ \Phi_i^{-1}|^2 \, dx + C_6 \varepsilon \sum_{j=1}^n \int_{P_i} |D_j(f \circ \Phi_i^{-1})|^2 \, dx \right) \\
&= \sum_{i=1}^s \left(\frac{C_5}{\varepsilon} \int_{P_i} |f \circ \Phi_i^{-1}|^2 \, dx + C_6 \varepsilon \sum_{j=1}^n \int_{P_i} |\sum_{k=1}^n (D_k f \circ \Phi_i^{-1}) D_j(\Phi_i)_k^{-1}|^2 \, dx \right) \\
&\leq \sum_{i=1}^s \left(\frac{C_7}{\varepsilon} \int_{Q \cap U_i} |f|^2 \, dx + C_8 \varepsilon \sum_{j=1}^n \int_{Q \cap U_i} |D_j f|^2 \, dx \right) \\
&\leq \frac{C_7 s}{\varepsilon} \int_Q |f|^2 \, dx + C_8 s \varepsilon \sum_{j=1}^n \int_Q |D_j f|^2 \, dx.
\end{aligned}
$$

Remark 9.28.2. The trace of a function $f \in W^1(Q)$ we can interpret also in the following way. Since $C^1(\overline{Q})$ is dense in the space $W^1(Q)$ there exists a sequence $\{f_j\}_{j \in \mathbf{N}}$ from $C^1(\overline{Q})$ such that $\lim_{j \to \infty} \|f_j - f\|_{W^1(Q)} = 0$. Then by (9.6)

$$\|f_k - f_m\|_{L_2(S)} \leq C_1 \|f_k - f_m\|_{W^1(Q)} \quad (k, m \in \mathbf{N}).$$

Hence $\{f_j\}_{j \in \mathbf{N}}$ is a Cauchy sequence in the space $L_2(S)$. Since the space $L_2(S)$ is complete there exists $f_S \in L_2(S)$ such that

$$\lim_{j \to \infty} \|f_j - f_S\|_{L_2(S)} = 0.$$

The function f_S is the trace of the function $f \in W^1(Q)$ on the surface S.
For a continuous function f on $\overline{Q}$ its trace on a surface $S \subset \overline{Q}$ is a continuous function from $C(\overline{Q})$ on S which coincides almost everywhere with f.

Example 9.29 *Prove that $W^k(Q) \neq \overset{\circ}{W}{}^k(Q), k \geq 1$, for some region $Q \neq \mathbf{R}^n$.*

Solution. By Exercise 9.28 the trace $f|_{\partial Q}$ of a function $f \in \overset{\circ}{W}{}^k(Q)$ is zero (Remark 9.28.2). Taking the continuous function equal 1 on $\overline{Q}$ we have that it belongs $W^k(Q), k \geq 1$, but since it trace on ∂Q is 1 it does not belong to $\overset{\circ}{W}{}^k(Q)$.

Remark 9.29.1. A function from $W^k(Q)$ belongs to $\overset{\circ}{W}{}^k(Q)$ if and only if its trace on the border is zero.

Remark 9.29.2. If Q is a locally quadratic bounded region and $f \in W^1(Q \times (0, T))$ is a function whose trace on $Q \times [0, T]$ is zero, i.e., $f|_{\partial Q \times [0,T]} = 0$, then there exists a sequence $\{f_k\}_{k \in \mathbf{N}}$ from $C^1(\overline{Q \times (0, T)})$ such that

$$f_k|_{\partial Q \times [0,T]} = 0 \text{ and } f_k \overset{W^1(Q \times (0,T))}{\longrightarrow} f \text{ as } k \to \infty.$$

Example 9.30 *Prove that*

a) *for $f \in C_0^1(P), P = (a_1, b_1) \times \ldots \times (a_n, b_n)$, there exists a constant $C > 0$ such that*

$$\int_P |f(x)|^2 \, dx \leq C \int_P \sum_{j=1}^{n} |D_j f(x)|^2 \, dx;$$

b) *the preceding inequality holds also for every $f \in \overset{\circ}{W}{}^1(P)$;*

c) *for every bounded region $Q \subset \mathbf{R}^n$ and $f \in \overset{\circ}{W}{}^1(Q)$ there exists a constant $C > 0$ such that*

$$\int_Q |f(x)|^2 \, dx \leq C \int_Q \sum_{j=1}^{n} |D_j f(x)|^2 \, dx;$$

d) *the usual norm in the space* $\overset{\circ}{W}{}^{1}(Q)$ *and the norm*

$$\|f\|_1 = \left(\int_Q \sum_{j=1}^n |D_j f(x)|^2 \, dx \right)^{1/2}$$

are equivalent.

Solution.

a) By the Newton-Leibniz formula

$$f(x_1, x') = \int_{a_1}^{x_1} D_1 f(z_1, x') \, dz_1,$$

where $x' = (x_2, \ldots, x_n)$. Then by Cauchy-Schwartz inequality we obtain

$$|f(x_1, x')|^2 \le \int_{a_1}^{x_1} |D_1 f(z_1, x')| \, dz_1 \int_{a_1}^{x_1} dy_1 \le (x_1 - a_1) \int_{a_1}^{b_1} D_1 f(z_1, x') \, dz_1.$$

Therefore

$$\int_P |f(x_1, x')|^2 \, dx_1 dx' \le \int_{a_1}^{x_1} (x_1 - a_1) \, dx_1$$

$$\int_{a_2}^{b_2} \cdots \int_{a_n}^{b_n} \left(\int_{a_1}^{b_1} |D_1 f(z_1, x')|^2 \, dz_1 \right) dx_2 \ldots dx_n$$

$$= \frac{(b_1 - a_1)^2}{2} \int_P |D_1 f(x)|^2 \, dx.$$

b) Since $C_0^1(P)$ is dense in the space $\overset{\circ}{W}{}^{1}(P)$ there exists a sequence $\{f_j\}_{j \in \mathbb{N}}$ from $C_0^1(P)$ such that $f_k \xrightarrow{\overset{\circ}{W}{}^{1}(P)} f$ as $k \to \infty$. By a) we have

$$\int_P |f_k(x)|^2 \, dx \le C^2 \int \sum_{j=1}^n |D_j f_k(x)|^2 \, dx.$$

Since for each fixed $j, j = 1, \ldots, n$, we have

$$D_j f_k \xrightarrow{L_2(P)} D_j f \text{ as } k \to \infty,$$

taking in the preceding inequality $k \to \infty$ we obtain the desired inequality.

c) Let $P = (a_1, b_1) \times \cdots \times (a_n, b_n)$ be a n–dimensional parallelepiped with the property that $Q \subset P$. We extend the function $f \in \overset{\circ}{W}{}^{1}(Q)$ on P in such a way that it is zero almost everywhere on $P \setminus Q$. We denote this extension also by f. Since

$$\int_P |f(x)|^2 \, dx = \int_Q |f(x)|^2 \, dx \text{ and } \int_P \sum_{j=1}^n |D_j f(x)|^2 \, dx = \int_Q \sum_{j=1}^n |D_j f(x)|^2 \, dx,$$

we obtain by b) the desired inequality.

d) By c) we have for $f \in \overset{\circ}{W}{}^{1}(Q)$

$$\|f\|_{\overset{\circ}{W}{}^{1}(Q)} \le (1 + C^2) \int_Q \sum_{j=1}^n |D_j f(x)|^2 \, dx.$$

Since trivially

$$\left(\int_Q \sum_{j=1}^n |D_j f(x)|^2 \, dx \right)^{1/2} \le \|f\|_{\overset{\circ}{W}{}^{1}(Q)},$$

we have proved the equivalence of the norms $\|f\|_{\overset{\circ}{W}{}^{1}(Q)}$ and $\|f\|_1$.

Exercise 9.31 *Prove that for $f \in W^k(\mathbf{R})$*

$$\lim_{x \to +\infty} f(x) = \lim_{x \to -\infty} f(x).$$

Hints. Suppose the opposite of the statement. Take a sequence $\{f_k\}_{k \in \mathbf{N}}$ from $C^1(\mathbf{R})$ such that $f_k \overset{W^1(\mathbf{R})}{\longrightarrow} f$ as $k \to \infty$ and use

$$f_k(b) - f_k(a) = \int_a^b f_k'(x) \, dx.$$

Exercise 9.32 *Prove that for $f \in W^1([a, b])$*

$$f(b) - f(a) = \int_a^b f'(x) \, dx.$$

Hint. Use Exercise 9.31.

Exercise 9.33 *Does there exist any function $f \in C[a, b]$ such that $f \notin W^1([a, b])$?*

Exercise 9.34 *Prove that for every real function $f \in W^1([0, 2\pi])$*

$$\int\limits_0^{2\pi} f(x)^2 \, dx \leq \int\limits_0^{2\pi} f'(x)^2 \, dx + \left(\int\limits_0^{2\pi} f(x) \, dx \right)^2.$$

Exercise 9.35 *Prove that for every real function $f \in \overset{\circ}{W}{}^{1}([0, \pi])$*

$$\int\limits_0^{\pi} f(x)^2 \, dx \leq \int\limits_0^{\pi} f'(x)^2 \, dx.$$

Exercise 9.36 *Let W_0 be a subspace of $W^1([0, 2\pi])$ which contains functions f with the property*

$$\int\limits_0^{2\pi} f(x) \, dx = 0.$$

Prove that the scalar product in the space W_0 has the form

$$\int\limits_0^{2\pi} f'(x) g'(x) \, dx \qquad (f, g \in W_0).$$

Example 9.37 (Poincare inequality) *Prove that for an n–dimensional parallelepiped $P = (a_1, b_1) \times \ldots \times (a_n, b_n)$ and $f \in W^1(P)$ we have*

$$\int\limits_P |f(x)|^2 \, dx \leq \frac{1}{(b_1 - a_1) \cdots (b_n - a_n)} |\int\limits_P f(x) \, dx|^2 + \frac{n}{2} \int\limits_P \sum_{i=1}^n (b_i - a_i)^2 |D_i f(x)|^2 \, dx.$$

Solution. Each function $f \in W^1(P)$ can be approximated in the space $W^1(P)$ by a sequence $\{f_m\}_{m \in \mathbb{N}}$ of functions from $C^1(P)$, i.e., $f_m \overset{W^1(P)}{\longrightarrow} f$ as $m \to \infty$. By Exercise from Chapter 4. for f_m holds the Poincare inequality

$$\int\limits_P |f_m(x)|^2 \, dx \quad \leq \quad \frac{1}{(b_1 - a_1) \cdots (b_n - a_n)} |\int\limits_P f_m(x) \, dx|^2$$

$$+ \frac{n}{2} \int\limits_P \sum_{i=1}^n (b_i - a_i)^2 |D_i f_m(x)|^2 \, dx.$$

Therefore we have

$$\left| |\int\limits_P f_m(x) \, dx| - |\int\limits_P f(x) \, dx| \right| \quad \leq \quad |\int\limits_P f_m(x) \, dx - \int\limits_P f(x) \, dx|$$

$$\leq \int_P |f_m(x) - f(x)|\, dx$$

$$\leq \left(\int_P |f_m(x) - f(x)|^2\, dx \right)^{\frac{1}{2}} \cdot \int_P dx$$

$$= (b_1 - a_1) \cdots (b_n - a_n) \cdot \left(\int_P |f_m(x) - f(x)|^2\, dx \right)^{\frac{1}{2}}.$$

Exercise 9.38 (Sobolev lemma for $\overset{\circ}{W}{}^k(Q)$) *Let Q be a bounded region. Prove that for $k > s + \frac{n}{2}$, $s \in \mathbf{N} \cup \{0\}$, every function f from $\overset{\circ}{W}{}^k(Q)$ belongs also to $C^s(\overline{Q})$.*

Hints. Prove first that for every f from $C_0^\infty(Q)$ there exists $C > 0$ such that such that

$$\sup_{x \in Q} |f(x)| \leq C \|f\|_{\overset{\circ}{W}{}^k(Q)}. \tag{9.7}$$

This inequality (9.7) will imply

$$\sup_{x \in Q} |D^\alpha f(x)| \leq C \|D^\alpha f\|_{\overset{\circ}{W}{}^k(Q)}$$

for $|\alpha| \leq s$, and $s < k - \frac{n}{2}$. Therefore the embedding of $\overset{\circ}{W}{}^k(Q)$ in $C^s(\overline{Q})$ is continuous on a dense subspace $C_0^\infty(\overline{Q})$.

Exercise 9.39 *Prove that*

$$W^{k+1}((a,b)) \subset C^k[a,b] \qquad (k \in \mathbf{N} \cup \{0\})$$

and that the embedding is continuous.

Hint. Use Sobolev lemma 9.6.

Exercise 9.40 *Prove the formula for partial integration for $f, g \in W^1((a,b))$*

$$\int_a^b f'(x)g(x)\, dx \;=\; f|_b g|_b - f|_a g|_a - \int_a^b f(x)g'(x)\, dx$$

$$=\; f(b)g(b) - f(a)g(a) - \int_a^b f(x)g'(x)\, dx.$$

Hints. Use Example 9.28, since by Exercise 9.39 $f(a) = f|_a$.

Chapter 10

Some Elements from Functional Analysis

10.1   Hilbert Space

10.1.1   Preliminaries

Definition 10.1 *A Hilbert space is a vector space (real or complex) H endowed with a scalar product $(\cdot|\cdot)$, i.e., (real or complex) valued bilinear functional defined on $H \times H$ with the properties for all $x, y, z \in H$:*

(h1) $(ax|y) = a(x|y)$ for every scalar a;

(h2) $(x + y|z) = (x|z) + (y|z)$;

(h3) $(x|y) = \overline{(y|x)}$;

(h4) $(x|x) > 0$ for $x \neq 0$;

(h5) $(x|x) = 0$ for $x = 0$;

and H is a complete metric space with respect to the metric $\|x - y\|$ induced by the norm $\|x\| = \sqrt{(x|x)}$.

A subset B of H is bounded if and only if for every sequence $\{x_n\}_{n \in \mathbb{N}}$ from B and every sequence of numbers $\{\alpha_n\}_{n \in \mathbb{N}}$ which converges to zero, the sequence $\{\alpha_n x_n\}_{n \in \mathbb{N}}$ converges to zero.

Let H_1 and H_2 be Hilbert spaces.

Definition 10.2 *A linear operator $T : D(T) \to H_2$, $D(T) \subset H_1$, is bounded if there exists $M > 0$ such that*

$$\|T(x)\|_{H_2} \leq M\|x\|_{H_1} \qquad (x \in D(T)).$$

We denote by $L(H_1, H_2)$ the vector space of all bounded linear operators from H_1 into H_2 endowed with the norm $\|T\| = \sup_{\|x\|_{H_1} \leq 1} \|T(x)\|_{H_2}$.

Theorem 10.3 (Riesz representation) *For every continuous linear functional f on a Hilbert space H there exists a unique element $y_f \in H$ such that*

$$f(x) = (x|y_f) \qquad (x \in H),$$

and $\|f\| = \|y_f\|_H$.

Definition 10.4 *An orthonormal family $\{e_i\}_{i \in I}$ in a Hilbert space H is an* orthonormal base *in H if for every $x \in H$ the (Fourier) series*

$$\sum_{i \in I}(x|e_i)e_i$$

converges to x, where in the sum there are not more than countably many nonzero members.

We have for $x \in H$ the *Parseval identity* (for a countable base)

$$\sum_{i=1}^{\infty}|(x|e_i)|^2 = \|x\|^2.$$

If H is separable, i.e., has a dense countable subset, then a linear operator $T : H \to H$ has a matrix representation $[t_{ij}]_{i,j \in \mathbf{N}}$ in the following way

$$Tx = \sum_{j=1}^{\infty}(\sum_{i=1}^{\infty} t_{ij}x_i)e_j,$$

where $x = \sum_{k=1}^{\infty} x_k e_k$. Then the adjoint operator T^* has a matrix representation $[\bar{t}_{ji}]_{i,j \in \mathbf{N}}$.

A sequence $\{x_n\}_{n \in \mathbf{N}}$ from a Hilbert space H *weakly converges* to an element $x \in H$ if

$$\lim_{n \to \infty}(x_n|y) = (x|y) \qquad (y \in H).$$

Definition 10.5 *A linear operator $T : D(T) \to H_2, D(T) \subset H_1$, is* closed *if its graph*

$$G(T) = \{(x, T(x)) \mid x \in D(T)\}$$

is a closed set in $H_1 \times H_2$ with respect to the topology induced by the norm

$$\|(x,y)\| = \sqrt{\|x\|_{H_1}^2 + \|y\|_{H_2}^2},$$

which is induced by the scalar product

$$((x_1, y_1)|(x_2, y_2))_{H_1 \times H_2} = (x_1|x_2)_{H_1} + (y_1|y_2)_{H_2}.$$

We have the following characterization of the closed operators.

Theorem 10.6 *A linear operator* $T : D(T) \to H_2, D(T) \subset H_1$, *is closed if and only if for every sequence* $\{x_n\}_{n \in \mathbb{N}}$ *from* $D(T)$ *with the property that if it converges to* x *and* $T(x_n) \to y$ *as* $n \to \infty$, *then* $x \in D(T)$ *and* $T(x) = y$.

Definition 10.7 *Let* T *be a linear operator* $T : D(T) \to H_2$, *where* $D(T)$ *is a dense subspace of* H_1. *Then the adjoint operator* T^* *of the operator* T *has the domain*

$$D(T^*) = \{y \in H_2 \mid \text{ there exists } t \in H_1, \ (y|T(x))_{H_2} = (t|x)_{H_1} \ (x \in D(T))\}$$

and $T^* : D(T^*) \to H_1$ *is defined by* $T^*(y) = t$.

Theorem 10.8 (Closed Graph Theorem on Hilbert spaces) *Let* H_1 *and* H_2 *be Hilbert spaces. If* $T : H_1 \to H_2$ *is a linear closed operator, then* T *is bounded.*

Theorem 10.9 (Uniform Boundedness Theorem) *Let* $\mathcal{A}$ *be a family of additive and continuous operators from a Hilbert space* H_1 *into a Hilbert space* H_2. *If the family* $\mathcal{A}$ *is pointwise bounded on* H_1, *i.e., for* $x \in H_1$ *there exists* $M(x) > 0$ *such that*

$$\|A(x)\| \leq M(x) \qquad (A \in \mathcal{A}),$$

then it is also uniformly bounded on every bounded subset B *of* H_1, *i.e., there exists* $M > 0$ *such that*

$$\|A(x)\| \leq M \qquad (x \in B, A \in \mathcal{A}).$$

Definition 10.10 *A linear operator* $T : H_1 \to H_2$ *is compact if* $T(B)$, *for every bounded subset* B *of* X, *is a subset of some compact subset of* H_2.

10.1.2 Examples and Exercises

Example 10.1 *Prove that every bounded sequence in a Hilbert space has a weakly convergent subsequence.*

Solution. Let $\{x_n\}_{n \in \mathbb{N}}$ be a bounded sequence in a Hilbert space H, i.e., there exists $M > 0$ such that $\|x_n\| < M$ $(n \in \mathbb{N})$. Let $L(\{x_n\})$ be a closed subspace spanned by the sequence $\{x_n\}_{n \in \mathbb{N}}$. The sequence of numbers $\{(x_1|x_n)\}_{n \in \mathbb{N}}$ is bounded, since

$$|(x_1|x_n)| \leq \|x_1\|\|x_n\| < M\|x_1\|.$$

Therefore by Bolzano-Weierstrass theorem there exists a convergent subsequence $\{(x_1|x_n^1)\}_{n \in \mathbb{N}}$. The sequence $\{(x_2|x_n^1)\}_{n \in \mathbb{N}}$ has also a convergent subsequence $\{(x_2|x_n^2)\}_{n \in \mathbb{N}}$. Continuing this procedure, after k steps we get a convergent subsequence $\{(x_k|x_n^k)\}_{n \in \mathbb{N}}$. We choose the diagonal sequence $\{x_n^n\}_{n \in \mathbb{N}}$. Then the sequence

$\{(x_k|x_n^n)\}_{n\in N}$ converges for every k, since for $n > k$ this is a subsequence of the sequence $\{(x_k|x_n^k)\}_{n\in N}$. We shall show that $\{x_n^n\}_{n\in N}$ is the desired subsequence of $\{x_n\}_{n\in N}$, i.e., $\lim_{n\to\infty}(x|x_n^n) = f(x)$ for every $x \in H$ and f is a continuous linear functional.

It is obvious that the preceding limit exists for $l_n = \sum_{k=1}^{n}\alpha_k x_k$. Since the set of all linear combinations of the sequence $\{x_n\}_{n\in N}$ is dense in $L(\{x_n\})$ there exists a sequence of linear combinations $\{l_m\}_{m\in N}$ such that

$$\lim_{m\to\infty} l_m = x \quad \text{and there exists} \quad \lim_{m\to\infty}(l_m|x_n^n).$$

Therefore by the equality

$$(x|x_n^n) = (l_m|x_n^n) + (x - l_m|x_n^n)$$

and the inequality $|(x - l_m|x_n^n)| < M\|x - l_m\|$ we have that there exists $\lim_{n\to\infty}(x|x_n^n)$ for every $x \in L(\{x_n\})$. Since every element x from H can be represented in the form $x = y + h$ for $y \in L(\{x_n\})$ and h orthogonal on $L(\{x_n\})$ and $(h|x_n) = 0$ for every $n \in N$, we obtain that for every $x \in H$ there exists $\lim_{n\to\infty}(x|x_n^n)$ and it defines a linear functional f on H. This functional f is continuous, what easily follows from the inequality

$$|f(x)| = |\lim_{n\to\infty}(x|x_n^n)| < M\|x\|$$

for every $x \in H$. Therefore by Riesz representation theorem 10.3 there exists $y \in H$ such that

$$\lim_{n\to\infty}(x|x_n^n) = (x|y) \quad (x \in H).$$

Example 10.2 (Diagonal Theorem) *Let $[x_{ij}]_{i,j\in N}$ be an infinite matrix of non-negative real numbers such that*

$$\lim_{j\to\infty} x_{ij} = 0 \quad (i \in N), \quad \lim_{i\to\infty} x_{ij} = 0 \quad (j \in N), \quad \lim_{i\to\infty} x_{ii} = 0.$$

Prove that there exists an infinite subset I of N such that

$$\sum_{i\in I}\sum_{j\in I} x_{ij} < \infty.$$

Moreover, the elements of the set I can be ordered in a increasing sequence $\{p_i\}_{i\in N}$ such that

$$\lim_{i\to\infty}\sum_{j=1}^{\infty} x_{p_i p_j} = 0 \quad \text{and} \quad \lim_{j\to\infty}\sum_{i=1}^{\infty} x_{p_i p_j} = 0.$$

Solution. Let $i_0 = 0$. We shall choose a sequence $\{i_n\}_{n\in N}$ of natural numbers such that

(i) $i_{n-1} < i_n$ for every $n \in N$,

(ii) $x_{i_m i_k} < 2^{-k-m}$ for every $1 \leq m \leq n, 1 \leq k \leq n$ and $n \in \mathbf{N}$.

The proof goes by induction. Since $\lim_{i \to \infty} x_{ii} = 0$ there exists an index r such that $x_{ii} < 2^{-2}$ for $i \geq r$. Let $i_1 = r$. Then (i) and (ii) hold for $n = 1$. Suppose now that we have already find $i_1, \ldots, i_p$ such that (i) and (ii) hold for $1 \leq m \leq p, 1 \leq k \leq p$. Since $x_{i_m i} \to 0$ and $x_{i i_m} \to 0$ as $i \to \infty$ for $m = 1, \ldots, p$ and $x_{ii} \to 0$ as $i \to \infty$, there exists an index $i_{p+1} > i_p$ such that

$$x_{i_m i_{p+1}} < 2^{-p-m-1} \quad \text{and} \quad x_{i_{p+1} i_m} < 2^{-p-m-1}$$

for $m = 1, \ldots, p+1$. Therefore (i) and (ii) hold for $1 \leq m \leq p+1$ and $1 \leq k \leq p+1$. So we have constructed a sequence $\{i_n\}_{n \in \mathbf{N}}$ with the properties (i) and (ii). By (ii) we obtain

$$\sum_{m=1}^{\infty} \sum_{k=1}^{\infty} x_{i_m i_k} < 1.$$

Taking $I = \{i_1, i_2, \ldots\}$ we obtain the desired conclusion. Since the sequence of summands of a convergent series converges zero, we obtain

$$\lim_{m \to \infty} \sum_{k=1}^{\infty} x_{i_m i_k} = 0.$$

Take $p_k = i_k$ for $k \in \mathbf{N}$.

Exercise 10.3 Let $T : D(T) \to H_2$, for $D(T) \subset H_1$, be a linear operator with dense domain. Prove that

a) the adjoint operator T^* of T is a linear and closed operator;

b) if additionally T is closed operator, then $D(T^*)$ is dense in H_2.

Example 10.4 (Closed Graph Theorem on Hilbert spaces) Let H_1 and H_2 be Hilbert spaces and $T : H_1 \to H_2$ be a linear closed operator. Then T is a bounded operator.

Solution. Since T is a closed operator we have by Example 10.3 that $D(T^*)$ is dense in H_2. Hence $D(T^*) \neq \{0\}$ for non-trivial Hilbert spaces H_1 and H_2. Let $\{y_n\}_{n \in \mathbf{N}}$ be an arbitrary sequence from $D(T^*)$, $\{x_n\}_{n \in \mathbf{N}}$ an arbitrary but fixed sequence from H_1 such that $\|x_n\| \leq 1$ and $\{\alpha_n\}_{n \in \mathbf{N}}$ an arbitrary sequence of numbers such that $\lim_{n \to \infty} \alpha_n = 0$. We can represent the sequence $\{\alpha_n\}_{n \in \mathbf{N}}$ as a product $\alpha_n = t_n \cdot u_n$ where $t_n \geq 0$ and both sequences $\{t_n\}_{n \in \mathbf{N}}$ and $\{u_n\}_{n \in \mathbf{N}}$ converge to zero. We introduce an infinite matrix of nonnegative numbers $[x_{ij}]_{i,j \in \mathbf{N}}$ such that

$$x_{ij} = \begin{cases} t_i \dfrac{|(u_j x_j | T^*(y_i))|}{\|y_i\|} & \text{for } i \neq j, y_i \neq 0, \\ 0 & \text{for } i \neq j, y_i = 0, \\ 0 & \text{for } i = j. \end{cases}$$

We shall show that the matrix $[x_{ij}]_{i,j\in\mathbf{N}}$ satisfies the conditions from Example 10.2. Since $u_j x_j \to 0$ as $j \to \infty$ we obtain by the continuity of the scalar product that $x_{ij} \to 0$ as $j \to \infty$ for $i \in \mathbf{N}$. By the definition we have $x_{ii} = 0$. It remains to prove that $x_{ij} \to 0$ as $i \to \infty$ for $j \in \mathbf{N}$. Since we have

$$x_{ij} = t_i \frac{|(u_j x_j|T^*(y_i))|}{\|y_i\|} = t_i \frac{|(T(u_j x_j)|y_i)|}{\|y_i\|} \leq t_i \|T(u_j x_j)\|,$$

letting $i \to \infty$ we obtain $x_{ij} \to 0$ as $i \to \infty$ for arbitrary but fixed $j \in \mathbf{N}$.

Hence by Diagonal Theorem - Example 10.2 there exists an increasing sequence of integers $\{p_n\}_{n\in\mathbf{N}}$ such that

$$\lim_{i\to\infty} \sum_{j=1}^{\infty} x_{p_i p_j} = 0. \tag{10.1}$$

Since $u_j x_j \to 0$ as $j \to \infty$, we obtain by the completeness of H_1 that there exist a subsequence $\{s_j\}_{j\in\mathbf{N}}$ of $\{p_j\}_{j\in\mathbf{N}}$ and an element x from H_1 such that

$$\sum_{j=1}^{\infty} u_{s_j} x_{s_j} = x.$$

We have for every $p \in \mathbf{N}$ and every $y_{s_i} \neq 0$

$$t_{s_i} \frac{|(u_{s_i} x_{s_i}|T^*(y_{s_i}))|}{\|y_{s_i}\|} \leq \sum_{j=1, j\neq i}^{i+p} t_{s_i} \frac{|(u_{s_j} x_{s_j}|T^*(y_{s_i}))|}{\|y_{s_i}\|}$$
$$+ t_{s_i} \frac{|\sum_{j=1}^{i+p}(u_{s_j} x_{s_j}|T^*(y_{s_i}))|}{\|y_{s_i}\|}$$

for every $i \in \mathbf{N}$. Letting $p \to \infty$ in the preceding inequality we obtain by the continuity of the scalar product

$$t_{s_i} \frac{|(u_{s_i} x_{s_i}|T^*(y_{s_i}))|}{\|y_{s_i}\|} \leq \sum_{j=1}^{\infty} x_{s_i s_j} + t_{s_i} \frac{|(x|T^*(y_{s_i}))|}{\|y_{s_i}\|}$$
$$\leq \sum_{j=1}^{\infty} x_{s_i s_j} + t_{s_i} \|T(x)\|$$

for every $i \in \mathbf{N}$. Letting $i \to \infty$ we obtain by (10.1)

$$\alpha_{s_i} \frac{(x_{s_i}|T^*(y_{s_i}))|}{\|y_{s_i}\|} \to 0.$$

Therefore, by the Urysohn property of numbers: if for every subsequence $\{z_n\}_{n\in\mathbf{N}}$ of a given sequence of numbers $\{r_n\}_{n\in\mathbf{N}}$ there exists a subsequence $\{v_n\}_{n\in\mathbf{N}}$ such that $v_n \to 0$ as $n \to \infty$, then $r_n \to 0$, we obtain

$$\alpha_n \frac{(x_n|T^*(y_n))}{\|y_n\|} \to 0$$

as $n \to \infty$. Since the sequences $\{\alpha_n\}_{n\in\mathbf{N}}, \{x_n\}_{n\in\mathbf{N}}$ and $\{y_n\}_{n\in\mathbf{N}}$ were arbitrary sequences with the prescribed properties it follows (by the boundedness) the existence of $M > 0$ such that

$$\sup_{\|x\|\leq 1} \frac{|(x|T^*(y))|}{\|y\|} < M \qquad (y \in D(T^*), y \neq 0).$$

Therefore for an arbitrary element $x' \neq 0$ from H_1 and $y \neq 0$ we have

$$\frac{|(x'|T^*(y))|}{\|y\|} = \frac{|(\frac{x'}{\|x'\|}|T^*(y))|}{\|y\|}\|x'\| \leq M\|x'\|.$$

Taking $x' = T^*(y)$ we obtain

$$\frac{\|T^*(y)\|}{\|y\|} \leq M,$$

i.e,

$$\|T^*(y)\| \leq M\|y\| \qquad (y \in D(T^*)).$$

Since $D(T^*)$ is dense in H_2, the adjoint operator T^* has a bounded extension on the whole space H_2. Therefore T^* is a bounded linear operator from H_2 into H_1. Hence $T^{**} = (T^*)^*$ is a bounded operator on H_1 and since it is an extension of the operator T we have $T = T^{**}$. Hence T is a bounded operator on H_1.

Example 10.5 *Let H_1 and H_2 be Hilbert spaces. Prove that if $T : H_1 \to H_2$ is an injective linear operator then T^{-1} is a closed operator if and only if T is a closed operator. Moreover, if T maps H_1 onto H_2, then T^{-1} is a bounded linear operator.*

Solution. We have that $G(T)$ is closed in $H_1 \times H_2$ if and only if $G(T^{-1})$ is closed in $H_2 \times H_1$ with respect to the same norm $\sqrt{\|x_1\|_{H_1}^2 + \|x_2\|_{H_2}^2}$.
If T is a closed operator from H_1 onto H_2, then T^{-1} is a closed operator from H_2 onto H_1 and by the Closed Graph Theorem - Example 10.4 the operator T^{-1} is bounded.

Remark 10.5.1. The preceding result can be interpreted in the following way. If we know that the equation $T(u) = f$, where T is a closed linear operator, has for every $f \in H_2$ a unique solution, then this solution continuously depends from f.

Exercise 10.6 *Let H be a Hilbert space and $T : H \to H$ be a linear operator. Prove that*

a) *If there exists $m > 0$ such that $\|T(x)\| \geq m\|x\|$ $(x \in D(T))$, then T is closed operator if and only if the range $R(T)$ is a closed set;*

b) *If T is a closed operator, then T^{-1} is a bounded linear operator on H if and only if $R(T)$ is dense in H and there exists $m > 0$ such that $\|T(x)\| \geq m\|x\|$ $(x \in D(T))$.*

Hint. Consequences of Example 10.5.

Example 10.7 *A linear operator $T : H_1 \to H_2$ is compact if and only if T maps every weakly convergent sequence $\{x_n\}_{n\in\mathbf{N}}$ from H_1 on a strongly (norm) convergent sequence $\{Tx_n\}_{n\in\mathbf{N}}$ in H_2.*

Solution. Suppose that the sequence $\{x_n\}_{n\in\mathbf{N}}$ from H_1 weakly converges to $x \in H_1$. Then by theorem on Uniform Boundedness there exists $M > 0$ such that $\|x_n\| \leq M$ $(n \in \mathbf{N})$. By the definition of the compact operator the sequence $\{Tx_n\}_{n\in\mathbf{N}}$ belongs to a compact subset of H_2. Therefore for every subsequence of $\{Tx_n\}_{n\in\mathbf{N}}$ there exists a strongly convergent subsequence $\{Tx_{n_i}\}_{i\in\mathbf{N}}$ with the limit $v \in H_2$. We remark that v is independent of the choice of the subsequence of $\{Tx_n\}_{n\in\mathbf{N}}$. We shall show that $\lim_{n\to\infty} Tx_n = v$. Namely, we have for every $y \in H_2$

$$(v|y) = \lim_{i\to\infty}(Tx_{n_i}|y) = \lim_{i\to\infty}(x_{n_i}|T^*y) = (x|T^*y) = (Tx|y).$$

Therefore $v = Tx$. By the Urysohn property of the convergence in the Hilbert space we obtain that the sequence $\{Tx_n\}_{n\in\mathbf{N}}$ converges to v.
Suppose now that T maps every weakly convergent sequence $\{x_n\}_{n\in\mathbf{N}}$ from H_1 on a strongly (norm) convergent sequence $\{Tx_n\}_{n\in\mathbf{N}}$ in H_2. Since every bounded subset of a Hilbert space is weakly compact, see Example 10.1, we obtain that for every bounded subset B of H_1 the set $\overline{T(B)}$ is compact, i.e., T is a compact operator.

Exercise 10.8 *Prove that if a sequence $\{T_n\}_{n\in\mathbf{N}}$ of compact operators from $L(H_1, H_2)$ converges in the norm of operators to an operator T, then T is a compact operator.*

Exercise 10.9 *Let O be an open subset of $\mathbf{R}^n$ and $H = L_2(O)$. Prove that the operator T defined by*

$$(Tf)(x) = \int_O \frac{f(y)\,dy}{|x-y|^\alpha} \qquad (x \in O)$$

for $n - 1 < \alpha < n$ is a compact operator on $L_2(O)$.

Hint. Use Exercise 10.8.

Exercise 10.10 *Let K be a compact subset of $\mathbf{R}^n$ and $H = L_2(K)$. If the function k is continuous on the set $K \times K$ prove that the operator T defined by*

$$Tf(x) = \int_K k(x,y)f(y)\,dy$$

is a compact operator on $L_2(K)$.

Exercise 10.11 *Let O be an open subset of $\mathbf{R}^n$ and $H = L_2(O)$. If the function k is measurable and satisfies*

$$\int \int_{O \times O} |k(x,y)|^2 \, dx dy < \infty$$

prove that the operator T defined by

$$Tf(x) = \int_O k(x,y) f(y) \, dy$$

is a compact operator on $L_2(O)$.

Exercise 10.12 *A subset A of a n–dimensional Hilbert space is compact if and only if it is bounded.*

Hint. Show that an unbounded set A can not be compact, constructing a sequence in A which has no Cauchy subsequence.

Exercise 10.13 *Let H be a separable Hilbert space, $T : H \to H$ an linear bounded operator and $b_1, b_2, \ldots$ a base of the space H. Prove that the operator T is completely given by the matrix $[t_{ij}]_{i,j \in \mathbf{N}}$, where $t_{ij} = T(b_i | b_j)$.*

Hint. Show that $\sum_{i=1}^{\infty} |t_{ij}|^2 \le \|T\|$ $(j \in \mathbf{N})$, and then

$$T(x) = \sum_{j=1}^{\infty} (\sum_{i=1}^{\infty} t_{ij} x_i) b_j,$$

where $x = \sum_{s=1}^{\infty} x_s b_s$.

Exercise 10.14 *A subset A of a separable Hilbert space H is compact if and only if it is bounded and for every $\varepsilon > 0$ there exists $n_0 \in \mathbf{N}$ such that $\|P''_{n_0}(x)\| \le \varepsilon$ for every $x \in A$, where P''_{n_0} is the projection operator on a n_0–dimensional subspace of the space H, and $P''_{n_0} = I - P'_{n_0}$.*

Exercise 10.15 *A linear bounded operator $T : H \to H$, where H is a separable Hilbert space, is compact if and only if for every $\varepsilon > 0$ there exist $n \in \mathbf{N}$ and linear operators T_1 and T_2 such that T_1 is n-dimensional and $\|T_2\| \le \varepsilon$ and $T = T_1 + T_2$.*

Hints. The necessity follows by the decomposition

$$T(x) = P'_n T(x) + P''_n T(x)$$

and Exercise 10.14. Take $T_1 = P'_n T$ and $T_2 = P''_n T$.

In order to prove the sufficiency of the condition, we have to prove that for the sequence $\{T(x_k)\}_{k \in \mathbf{N}}$, for $\|x_k\| < M$ for some $M > 0$, there exists a Cauchy subsequence (use Exercise 10.12).

Exercise 10.16 *Let a linear operator* $T : H \to H$, *where* H *is a separable Hilbert space, be a contraction, i.e.,* $\|T\| < 1$. *Prove that there exists* $(I - T)^{-1} : H \to H$ *and*

$$\|(I - T)^{-1}\| \le \frac{1}{1 - \|T\|}.$$

Hint. Prove that the unique solution of the equation $(I - T)x = y$ for arbitrary $y \in H$ is of the following form

$$x = \sum_{k=1}^{\infty} T^k(y).$$

Exercise 10.17 *Prove that if* $T : H \to H$ *is a compact operator on a separable Hilbert space* H, *then its adjoint operator* T^* *is also a compact operator.*

Hints. Using Exercise 10.15, it follows

$$T^* = T_1^* + T_2^* \quad \text{and} \quad \|T_2^*\| = \|T_2\| \le \varepsilon.$$

Prove that T_1^* is a finite dimensional operator.

Example 10.18 *A linear operator* T *on a complex Hilbert space* H *is symmetric if its domain* $D(T)$ *is dense in* H *and*

$$(T(x)|y) = (x|T(y)) \qquad (x, y \in D(T)). \tag{10.2}$$

A linear operator T *on* H *is positive if* $(T(x)|x) \ge 0 \quad (x \in D(T))$.
Prove that

a) *T is symmetric if and only if* $(T(x)|x) \in \mathbf{R} \quad (x \in D(T))$.

b) *if* T *is symmetric (positive) operator, then all its eigenvalues, i.e.,* $\lambda \in \mathbf{C}$ *for which* $T(x) = \lambda x$ *for some* $x \in H, x \ne 0$, *are real (positive) numbers and the eigenvectors which corresponds to different eigenvalues are orthogonal.*

Solution.

a) If $(T(x)|x) \in \mathbf{R} \quad (x \in D(T))$, then we have for $x, y \in D(T)$

$$\Re((T(y)|x) - (T(x)|y)) = \Re\frac{1}{\imath}((T(x + \imath y|x + \imath y) - (T(x)|x) - (T(y)|y)) = 0 \tag{10.3}$$

and

$$\Im((T(y)|x) - (T(x)|y)) = \Im((T(x+y)|x+y) - (T(x)|x) - (T(y)|y)) = 0 \tag{10.4}$$

We obtain by (10.3) and (10.4)

$$(T(x)|y) = \Re(T(x)|y) + i\Im(T(x)|y) = \Re(T(y)|x) - i\Im(T(y)|x)$$
$$= \overline{(T(y)|x)} = (x|T(y)),$$

i.e., T is a symmetric operator.

If we suppose that T is a symmetric operator, then we have by (10.2)

$$(T(x)|x) = (x|T(x)) = \overline{(T(x)|x)} \qquad (x \in D(T)),$$

i.e., $(T(x)|x) \in \mathbf{R}$ for $x \in D(T)$.

b) Let $\lambda \in \mathbf{C}$ be an eigenvalue of a symmetric operator T and x its corresponding eigenvector. Then we have

$$(T(x)|x) = (\lambda x|x) = \lambda\|x\|^2 = \lambda. \qquad (10.5)$$

By a) $(T(x)|x)$ is always real (positive) and therefore (10.5) implies that the eigenvalue is also real (positive).

If λ_1 and λ_2 are two different eigenvalues for operator T and x_1, x_2 the corresponding eigenvectors, respectively, then by (10.2) we have

$$\lambda_1(x_1|x_2) = (\lambda_1 x_1|x_2) = (T(x_1)|x_2) = (x_1|T(x_2)) = (x_1|\lambda_2 x_2) = \lambda_2(x_1|x_2).$$

Since $\lambda_1 \neq \lambda_2$ the preceding equality implies $(x_1|x_2) = 0$, i.e., x_1 and x_2 are orthogonal vectors.

10.2 The Fredholm Alternatives

10.2.1 Preliminaries

Theorem 10.11 (Fredholm alternative) *Let* $T : H \to H$ *be a linear compact operator, where* H *is a Hilbert space. Then for* $\lambda \neq 0$, *the equation*

$$(\lambda I - T)x = 0 \qquad (10.6)$$

either has the only trivial solution $x = 0$, *and then the equation*

$$(\lambda I - T)x = y \qquad (10.7)$$

has exactly one solution $x = (\lambda I - T)^{-1}y$,

or the homogeneous equation (10.6) has a solution $x \neq 0$, *and then the nonhomogeneous equation (10.7) has solution if and only if* $(y|x^*) = 0$ *for every solution* x^* *of the equation*

$$(\overline{\lambda} I - T^*)x^* = 0.$$

10.2.2   Examples and Exercises

Exercise 10.19 *Let T be a mapping $H \to H$, where H is a separable Hilbert space and let $I : H \to H$ be identical mapping. Then the equation*

$$(I - T)x = y$$

can be written in the form:

$$h - T_1(I - T_2)^{-1}h = y,$$

where $T = T_1 + T_2$, $h = (I - T_2)x$, $x = (I - T_2)^{-1}h$ and T_1 and T_2 are defined as in Example 10.15 for adjoint operator T^* of the operator T. The adjoint equation

$$(I - T^*)x^* = y^*$$

can be written in the form

$$x^* - [(I - T_2)^{-1}]^* T_1^* x^* = z^*,$$

where $(I - T_2^*)^{-1}y^* = z^*, y^* = (I - T_2^*)z^*$ and $T^* = T_1 + T_2^*$. (Note: according to Example 10.16, there exists $(I - T_2^*)^{-1} = [(I - T_2)^{-1}]^*$.)

Hints. Use Examples 10.15 and 10.16. See the solution of Example 10.20.

Example 10.20 (First Fredholm alternative) *Let H be a separable Hilbert space, $T : H \to H$ a compact operator on H, $I : H \to H$ the identical operator and*

$$(I - T)x = y \quad (x, y \in H), \tag{10.8}$$

$$(I - T^*)x^* = y^* \quad (x^*, y^* \in H^*), \tag{10.9}$$

where T^* is the adjoint operator of the operator T. If one of the above equations has a solution for an arbitrary right–hand side, then the other equation has a solution for an arbitrary right–hand side of the equation, and those solutions are unique, i.e., the homogeneous equations

$$(I - T)x = 0 \quad (x \in H)$$

$$(I - T^*)x^* = 0 \quad (x^* \in H),$$

have only zero solutions.

*   If one of the above homogeneous equations has only zero solution, then the other one has only zero solution, and equations (10.8) and (10.9) are uniquely solvable for arbitrary $y, y^* \in H$, i.e., the inverse operators of the operators $(I - T)$ and $(I - T^*)$ are defined on H and they are bounded.*

Solution. Example 10.15 implies that for each $\epsilon > 0$ there exist $n \in \mathbf{N}$ and the linear operators T_1 and T_2, such that T_1 is $n-$ dimensional, $\|T_2\| < \epsilon$ and $T = T_1 + T_2$. Therefore we obtain that the following equations are equivalent with (10.8)

$$(I - T_2)x - T_1 x = y, \tag{10.10}$$

$$h - T_1 (I - T_2)^{-1} h = y, \tag{10.11}$$

where $h = (I - T_2)x$ (from Example 10.16 it follows that $x = (I - T_2)^{-1}h$). The operator $T_1(I - T_2)^{-1}$ is $n-$ dimensional operator (since it is T_1). Let $[t_{ij}]$ be its matrix representation in an orthonormal basis $e_k, k = 1, 2, \dots$. Under the assumption that the space generated by $e_1, e_2, \dots, e_n$ is equal with the range of the operator $T_1(I - T_2)^{-1}$ it holds $t_{ij} = 0$, $j \geq n + 1$, $i \geq 1$ and for each j we have $\sum_{i=1}^{\infty} |t_{ij}|^2 \leq \|T_1(I - T_2)^{-1}\|^2$. It follows that (10.11) is equivalent to:

$$\sum_j h_j e_j - \sum_j (\sum_i h_i t_{ij}) e_j = \sum_j y_j e_j,$$

i.e., to

$$h_j - \sum_{i=1}^{\infty} t_{ij} h_i = y_j, j \leq n$$

$$h_j = y_j, \quad j > n.$$

Since

$$h_j = y_j, \quad j > n, \tag{10.12}$$

the last system of equations reduces on the system for $j \leq n$

$$h_j - \sum_{i=1}^{n} t_{ij} h_i = y_j - \sum_{i=n+1}^{\infty} t_{ij} y_j, \quad j = 1, 2, \dots, n. \tag{10.13}$$

The equality $T = T_1 + T_2$ implies $T^* = T_1^* + T_2^*$. Therefore the following equations are equivalent with (10.9)

$$x^* - [(I - T_2)^{-1}]^* T_1^* x^* = (I - T_2^*)^{-1} y^* \tag{10.14}$$

$$x^* - [(I - T_2)^{-1}]^* T_1^* x^* = z^* \tag{10.15}$$

(from Example 10.16 it follows $(I - T_2^*)^{-1} = [(I - T_2)^{-1}]^*$), where $z^* = (I - T_2^*)^{-1} y^*$, and $y^* = (I - T_2^*) z^*$). Since $[(I - T_2)^{-1}]^* T_1^*$ is adjoint operator for the operator $T_1(I-T_2)^{-1}$ its matrix representation is $[\bar{t}_{ji}]$. Therefore equation (10.15) is equivalent to

$$x_j^* - \sum_{i=1}^{n} \bar{t}_{ji} x_i^* = z_j^*, \quad j = 1, 2, \dots, n. \tag{10.16}$$

$$x_j^* = z_j^* + \sum_{i=1}^{n} \bar{t}_{ji} x_i^*, \quad j > n, \tag{10.17}$$

The matrices of the system of equations (10.13) and (10.16) are Hermit-conjugate, implying that the absolute values of theirs determinants are equal. Hence for them the analogue of the finite Fredholm theorem holds.

Let equation (10.8), i.e., (10.10), (or (10.9)) be solvable for each $y \in H$ (or $y^* \in H$). This assumption is equivalent to the assumption that equation (10.11) (or (10.15)) is solvable for each $y \in H$ (or $y^* \in H$). Specially it is solvable for each y from the space induced by $e_1, e_2, ..., e_n$, and therefore system (10.13) (or (10.16)) is solvable for the arbitrary right hand side. Thus, the determinant of the system is different of zero and the same is true for the determinant of system (10.16). It follows that (10.13) and (10.16), with an arbitrary right hand side, has one and only one solution and therefore (10.17) has one and only one solution. As the system (10.17) and (10.16), is equivalent to (10.15), i.e., (10.9), we conclude that (10.9) with an arbitrary right hand side, has one and only one solution. Therefore the homogeneous equations (10.13) and (10.16) have only zero solution. Then by (10.12) and (10.17) the homogeneous equations (10.8) and (10.9) have only zero solutions.

The opposite statement follows analogously.

Let us prove that operators $(I - T)^{-1}$ and $(I - T^*)^{-1}$ are bounded. Let the system (10.13) has one and only one solution (the determinant is nonzero) and let $(h_1, h_2, ..., h_n)$ be a solution. Then (on the base of the Cramer rule) it follows that there is a constant $c > 0$ such that

$$\sum_{j=1}^{n} |h_j|^2 \le c^2 \sum_{j=1}^{n} |y_j + \sum_{i=n+1}^{\infty} t_{ij} y_i|^2. \tag{10.18}$$

Since

$$\sum_{j=1}^{n} |y_j + \sum_{i=n+1}^{\infty} t_{ij} y_i|^2 \le 2 \sum_{j=1}^{n} \left(|y_j|^2 + \sum_{i=n+1}^{\infty} |t_{ij}|^2 \sum_{i=n+1}^{\infty} |y_i|^2 \right)$$

$$\le 2 \sum_{j=1}^{n} |y_j|^2 + 2 \sum_{j=1}^{n} (\sum_{i=1}^{\infty} |t_{ij}|^2) \sum_{i=n+1}^{\infty} |y_i|^2$$

$$\le 2\|y\|^2 + 2n\|T_1.(I - T_2)^{-1}\|^2 \|y\|^2$$

$$\le \|y\|^2 (2 + 2n\|T_1 \cdot (I - T_2)^{-1}\|^2) = c_1^2 \|y\|^2,$$

it holds

$$\sum_{j=1}^{n} |h_1|^2 \le (cc_1)^2 \|y\|^2,$$

and therefore

$$\|h\|^2 \sum_{j=1}^{n} |h_j|^2 + \sum_{j=n+1}^{\infty} |y_j|^2 \le (1 + c^2 c_1^2) \|y\|^2 = c_2^2 \|y\|^2,$$

(since $h_j = y_j$ for $j > n$). Since $x = (I - T_2)^{-1}h$, it holds

$$\|x\| = \|(I - T_2)^{-1}h\| \leq \|(I - T_2)^{-1}\|\|h\| \leq c_3\|y\|,$$

where $c_3 > 0$ is a constant which does not depend on y.

Example 10.21 (Second Fredholm alternative)
Let H be a separable space, let $T : h \to H$ a compact operator, and

$$(I - T)x = 0, \tag{10.19}$$

$$(I - T)^*x^* = 0, \tag{10.20}$$

where $I : H \to H$ is the identical operator and $x, x^*, y, y^* \in H$. If equation (10.19) has nonzero solutions, only finite many of them are linearly independent, and equation (10.20) has the same number of linearly independent solutions.

Solution. The matrices $B = [b_{ij}]$ and $B^* = [\bar{b}_{ji}]$, where

$$b_{ij} = \begin{cases} -t_{ij}, & i \neq j, \\ 1 - t_{ii}, & i = j, \end{cases} \qquad i, j = 1, 2, ..., n,$$

have the same rank. Therefore, the homogeneous systems (10.13) and (10.16) have the same number k, $k \leq n$ of linearly independent solutions. From Example 10.20, it follows that there are k linearly independent solutions of homogeneous equations (10.19) and (10.20).

Example 10.22 (Third Fredholm alternative) *Let H be a separable Hilbert space, $T : H \to H$ compact operator and*

$$(I - T)x = y \tag{10.21}$$

$$(I - T^*)x^* = y^*. \tag{10.22}$$

Equation (10.21) has a solution if and only if y is orthogonal on all solutions of the homogeneous equation (10.22) ($y^* = 0$). Among the solutions of equation (10.21), there exists a unique solution x, which is orthogonal on all solution of homogeneous equation (10.21). Every solution of equation (10.21) is the sum of x and a solution of a homogeneous equation (10.21) (i.e., $y = 0$). There exists a constant c, which does not depend on y, such that $\|x\| \leq c\|y\|$.

Solution. Let $x^1, x^2, ..., x^k$ be linearly independent solutions of homogeneous equation (10.21), and $x^{1*}, x^{2*}, ..., x^{k*}$ the linearly independent solutions of homogeneous equation (10.22). The system $x^1, x^2, ..., x^k$, as well as $x^{1*}, x^{2*}, ..., x^{k*}$, can be considered as orthogonal. Suppose that the solution of equation (10.8) exists. This implies the existence of the solution of the system

$$h_j - \sum_{i=1}^{n} t_{ij}h_i = y_j - \sum_{i=n+1}^{\infty} t_{ij}y_j, \quad j = 1, 2, ..., n. \tag{10.23}$$

Let $B = [b_{ij}]_{i,j \in \mathbf{N}}$, where

$$b_{ij} = \begin{cases} 1 - t_{ij}, & i = j, \\ -t_{ij}, & i \neq j, \end{cases} \qquad i, j = 1, 2, ..., n.$$

The rank of the matrix B is equal to $n - k$. Let us denote by R_{n-k} the subspace of the n–dimensional vector space, which is generated by the columns of the matrix B, i.e., by the vectors $B_i = (b_{i1}, ..., b_{in})$. The system (10.23) has a solution if and only if the vector on the right–hand side belongs to R_{n-k}, or equivalently, if it is orthogonal on the space $R_{n-k}^{\perp}$.

Let us characterize $R_{n-k}^{\perp}$. The vector $\tilde{x}^* = (x_1^*, ..., x_n^*)$ belongs to $R_{n-k}^{\perp}$ if and only if $(\tilde{x}^* | B_i) = 0$, for $i = 1, 2, ..., n$, i.e., if and only if $\sum_{i=1}^{\infty} \bar{b}_{ji} x_i^* = 0$ for $j = 1, 2, ..., n$. The solutions of homogeneous equations given by (10.16) belong to $R_{n-k}^{\perp}$. If x^* is an arbitrary solution of homogeneous equation (10.22)

$$x^* = x_1^* e_1 + \cdots + x_n^* e_n + x_{n+1}^* e_{n+1} + \cdots,$$

where $\tilde{x}^* = (x_1^*, \ldots, x_n^*)$ is a solution of the homogeneous system

$$x_j^* - \sum_{i=1}^{n} \bar{t}_{ji} x_i^* = z_j^*, \quad j = 1, 2, ..., n, \tag{10.24}$$

(then $\tilde{x}^* \in R_{n-k}^{\perp}$), and $x_j^* = \sum_{i=1}^{n} \bar{t}_{ji} x_i^*$, for $j > n$. Since $\bar{x}^* \in R_{n-k}^{\perp}$ it holds

$$\sum_{j=1}^{n} (y_j + \sum_{i=n+1}^{\infty} t_{ij} y_i) \bar{x}_j^* = 0.$$

It follows

$$0 = \sum_{j=1}^{n} (y_j + \sum_{i=n+1}^{\infty} t_{ij} y_i) \bar{x}_j^* = \sum_{j=1}^{n} y_j \bar{x}_j^* + \sum_{j=1}^{n} (\sum_{i=n+1}^{\infty} t_{ij} y_i) \bar{x}_j^*$$

$$= \sum_{j=1}^{n} y_j \bar{x}_j^* + \sum_{i=n+1}^{\infty} y_i \bar{x}_i^* = (y | x^*),$$

i.e., y is orthogonal on all solutions of the homogeneous equation (10.22).

Let us prove the opposite statement. If y is orthogonal on all solutions x^* of the homogeneous equation (10.22), then vector $y_j + \sum_{i=n+1}^{\infty} t_{ij} y_i, j = 1, 2, ..., n$, is orthogonal on all solutions $\tilde{x}^*$ of the homogeneous system (10.24), which implies that systems (10.23) and (10.21) have solutions.

Let x_0 be a solution of homogeneous equation (10.21), and $x^1, x^2, ..., x^k$ be an orthonormal system of solutions of homogenous equations (10.21). It follows that $x = x_0 - (x_0 | x^1) x^1 - \cdots - (x_0 | x^k) x^k$ is a solution of equation (10.21). It is unique (prove that). Let x' be an arbitrary solution of equation (10.21); then $x' - x = x''$ is a solution of the homogeneous equation, i.e., $x' = x + x''$.

Let us prove the inequality $||x|| \leq c||y||$. Let h be an element of H such that $(I - T_2)x = h$; then h is a solution of the equation

$$h - T_1(I - T_2)^{-1}h = y,$$

where T_1 and T_2 are linear operators, such that T_1 is $n(\epsilon)$–dimensional, $||T_2|| \leq \epsilon$ and $T = T_1 + T_2$. Then h satisfies the following k conditions:

$$0 = (x|x^i) = ((I - T_2)^{-1}h|x^i) = (h|x^i) = (h|(I - T_2^*)^{-1}x^i), \qquad (10.25)$$

where $i = 1, 2, ..., k$.

Since the rank $n - k$ of the extended matrix of system (10.23) is the same as the rank of the matrix B, if follows that in system (10.23) there are k equations, which are linear combinations of the rest of $n - k$ equations, they can be excluded from the system.

The vector $(h_1, h_2, ..., h_n)$ is a solution of the system of n linear equations ($n - k$ of them are linearly independent equations from the system (10.23) and k of them from the system (10.25)), whose coefficients are independent of the right hand side in system (10.23). The uniqueness of x implies that $(h_1, h_2, ..., h_n)$ is a unique solution of the system, i.e., the determinant of the system is not equal to zero. The vector $(h_1, h_2, ..., h_n)$ can be evaluated by the Cramer rule, and therefore it holds

$$\sum_{j=1}^{n} |h_j|^2 \leq c^2 \sum_{j=1}^{n} |y_j + \sum_{i=n+1}^{\infty} t_{ij}y_i|^2. \qquad (10.26)$$

which implies $||x|| \leq c||y||$.

Example 10.23 (Fourth Fredholm alternative) *Let H be a separable Hilbert space and $T : H \to H$ compact operator. For an arbitrary constant $M > 0$ in the circle $\{\mu \in C| |\mu| < M\}$ of the complex plane, there exist only finitely many characteristic values (i.e., they are of the form $\frac{1}{\lambda}$, where λ is an eigenvalue) of the operator T, i.e. outside the circle $\{\mu \in C| |\lambda| < 1/M\}$ there can be only finitely many eigenvalues.*

Solution. Let us suppose that there exist infinitely many numbers $\mu_1, \mu_2, ..., \mu_n, ...$, which are characteristic values of the operator T, such that $\mu_i \neq \mu_j$ for $i \neq j$, which are elements of the set $\{\mu \in C| |\mu| < M\}$. By e_i we denote the eigenvector determined by the characteristic value μ_i, $i = 1, 2, ...$.

For arbitrary $n \geq 1$, the system $e_1, ..., e_n$ is linearly independent. We will prove this assertion by induction. For $n = 1$ the assertion is obvious. Let us suppose that the assertion is true for $n = m - 1$, and that $e_1, e_2, ..., e_m$ are linearly dependent vectors. It follows that there are nonzero constants $c_1, c_2, ..., c_{m-1}$, such that $e_m = c_1e_1 + \cdots + c_{m-1}e_{m-1}$, and therefore:

$$Te_m = \frac{e_m}{\mu_m} = c_1\frac{e_1}{\mu_1} + \cdots + c_{m-1}\frac{e_{m-1}}{\mu_{m-1}},$$

and

$$c_1(1 - \frac{\mu_m}{\mu_1})e_1 + \cdots + c_{m-1}(1 - \frac{\mu_m}{\mu_{m-1}})e_{m-1} = 0.$$

Therefore, $1 - \mu_m/\mu_k = 0$, $k = 1, 2, ..., m - 1$, which is a contradiction. So the assertion holds also in the case $n = m$. Denote by R_n the space generated by $\{e_1, e_2, \ldots, e_n\}$. Then we have $R_1 \subset R_2 \subset \ldots \subset R_n \subset \ldots$ and $R_n \neq R_{n-1}$, for each $n \in \mathbf{N}$. For each $n \in \mathbf{N}$ there exists $x_n \in R_n$, such that x_n is orthogonal on R_{n-1} and $\|x_n\| = 1$. Since $\{x_1, x_2, ..., x_n, ...\}$ is a bounded set and T compact operator, the sequence $Tx_1, Tx_2, ...Tx_n, ...$ has a Cauchy subsequence. We will prove that this contradicts to the our assumption that there exist infinitely many numbers $\mu_1, \mu_2, \ldots$.

If $m < n$, then

$$Tx_n - Tx_m = \frac{1}{\mu_n}x_n + \frac{1}{\mu_n}(\mu_n Tx_n - x_n) - Tx_m = \frac{1}{\mu_n}x_n + \sigma_n,$$

where $\sigma_n \in R_{n-1}$. This holds since $Tx_m \in R_m \subset R_{n-1}$ and

$$\mu_n Tx_n - x_n = \mu_n T(c_1 e_1 + \cdots + c_n e_n) - (c_1, e_1 + \cdots + c_n e_n)$$

$$= c_1(\frac{\mu_n}{\mu_1} - 1)e_1 + \cdots + c_{n-1}(\frac{\mu_n}{\mu_{n-1}} - 1)e_{n-1} \in R_{n-1}.$$

Therefore

$$\|Tx_n - Tx_m\|^2 = \|\frac{1}{\mu_n}x_n + \sigma_n\|^2 = (\frac{1}{\mu_n}x_n + \sigma_n|\frac{1}{\mu_n}x_n + \sigma_n)$$

$$= \frac{1}{|\mu_n|^2}(x_n|x_n) + \frac{1}{\mu_n}(x_n|\sigma_n) + \frac{1}{\mu_n}(\sigma_n|x_n) + (\sigma_n|\sigma_n)$$

$$= \frac{1}{|\mu_n|^2}\|x_n\|^2 + \|\sigma_n\|^2 \geq \frac{\|x_n\|^2}{|\mu_n|^2} \geq \frac{1}{M^2}.$$

The above inequality implies that the sequence $Tx_1, ..., Tx_n, ...$ has not a Cauchy subsequence. A contradiction.

Exercise 10.24 *The equation*

$$x - \mu Tx = y, \tag{10.27}$$

has for each $y \in H$ a solution if and only if μ is not a characteristic value of the operator T.

If μ is a characteristic value of the operator T, its multiplicity is finite and $\overline{\mu}$ is characteristic value of the operator T^*, with the same multiplicity. Equation (10.27) is in that case solvable if and only if y is orthogonal on all eigenvectors of the operator T^*, which correspond to eigenvalue $\frac{1}{\overline{\mu}}$. If equation (10.27) is solvable then *there exists unique solution of the equation, which is orthogonal on all eigenvectors of operator T, which correspond to the eigenvalue $\dfrac{1}{\mu}$.*

A compact operator has not more than countably many characteristic values. Let

$$\mu_1, \mu_2, ..., \qquad |\mu_i| < |\mu_{i+1}|, \ i = 1, 2, ..., \tag{10.28}$$

be a sequence of characteristic values of a compact operator (if they exist), where each number μ_i appears in the sequence $k(i)$ - times, where $k(i)$ is its multiplicity. If the sequence (10.28) is infinite, then $|\mu_n| \to \infty$, as $n \to \infty$. Then the sequence (10.28) determines a sequence of eigenvectors $e_1, e_2, ...$, which are linearly independent.

Hint. Follows by the first, second and third Fredholm alternatives.

10.3 Normed Vector Spaces

10.3.1 Preliminaries

Let X and Y be normed vector spaces.

Definition 10.12 *A linear operator $T : D(T) \to Y, D(T) \subset X$, is bounded if there exists $M > 0$ such that*

$$\|T(x)\|_Y \le M \|x\|_X \qquad (x \in D(T)).$$

We denote by $L(X, Y)$ the vector space of all bounded linear operators from X into Y endowed with the norm $\|T\| = \sup_{\|x\|_X \le 1} \|T(x)\|_Y$.

Definition 10.13 *A linear operator $T : D(T) \to Y, D(T) \subset X$, is closed if its graph*

$$G(T) = \{(x, T(x)) \mid x \in D(T)\}$$

is a closed set in $X \times Y$ with respect to the topology induced by the norm

$$\|(x, y)\| = \sqrt{\|x\|_X^2 + \|y\|_Y^2}.$$

We have the following characterization of the closed operators.

Theorem 10.14 *A linear operator $T : D(T) \to Y, D(T) \subset X$, is closed if and only if for every sequence $\{x_n\}_{n \in \mathbb{N}}$ from $D(T)$ with the property that it converges to x and $T(x_n) \to y$ as $n \to \infty$, then $x \in D(T)$ and $T(x) = y$.*

Definition 10.15 *Let T be a linear operator $T : D(T) \to Y$, where $D(T)$ is a dense subspace of X. Then the adjoint operator T^* of the operator T has the domain*

$$D(T^*) = \{y' \mid y' \in Y', y'T \text{ is continuous on } D(T)\}$$

and $T^* : D(T^*) \to X'$ is defined by $T^*(y') = y'T$, where X' and Y' are the dual spaces (spaces of continuous linear functionals) of X and Y, respectively.

We are using for $x' \in X'$ and $x \in X$ also the notation $< x', x > = x'(x)$.

Definition 10.16 *A sequence $\{x_n\}_{n\in\mathbb{N}}$ from a normed vector space X converges weakly to $x \in X$ if*

$$\lim_{n\to\infty} < x', x_n > = < x', x >$$

for every $x' \in X'$.

Definition 10.17 *A sequence $\{x'_n\}_{n\in\mathbb{N}}$ from the dual space X'*

 (i) *converges weakly to $x' \in X'$ if*

$$\lim_{n\to\infty} < x'', x'_n > = < x'', x' >$$

  for every $x'' \in X''$;

 (ii) *converges $*$-weakly to $x' \in X'$ if*

$$\lim_{n\to\infty} < x, x'_n > = < x, x' >$$

  for every $x \in X \subset X''$.

A Banach space X is reflexive if $X = X''$ (in the sense of the canonical map $x \mapsto x''$).

Theorem 10.18 (Closed Graph Theorem) *Let X and Y be Banach spaces. If $T : X \to Y$ is a linear closed operator, then T is bounded.*

Theorem 10.19 (Uniform Boundedness Theorem) *Let $\mathcal{A}$ be a family of additive and continuous operators from a Banach space X into a normed vector space Y. If the family $\mathcal{A}$ is pointwise bounded on X, i.e., for $x \in X$ there exists $M(x) > 0$ such that*

$$\|A(x)\| \le M(x) \qquad (A \in \mathcal{A}),$$

then it is also uniformly bounded on every bounded subset B of X, i.e., there exists $M > 0$ such that

$$\|A(x)\| \le M \qquad (A \in \mathcal{A}, x \in B).$$

Theorem 10.20 (Banach-Steinhaus) *Let X and Y be Banach spaces and $\{A_n\}_{n\in\mathbb{N}}$ a sequence from $L(X,Y)$. The sequence $\{A_n\}_{n\in\mathbb{N}}$ converges strongly to an operator $A \in L(X,Y)$, i.e., $\lim_{n\to\infty} A_n(x) = A(x)$ for every $x \in X$, if and only if*

 a) $M = \sup\{\|A_n\| \mid n \in \mathbb{N}\} < \infty$,

 b) *$\{A_n\}_{n\in\mathbb{N}}$ is a Cauchy sequence for every x from a set $E \subset X$ such that $\overline{L(E)} = X$, where $L(E)$ is a vector space spanned by E.*

Definition 10.21 *A linear operator $T : X \rightarrow Y$ is compact if $T(B)$ for every bounded subset B of X is a subset of some compact subset of Y.*

Theorem 10.22 (Banach fixed point theorem) *If X is a Banach space and $T : X \rightarrow X$ is a contraction, i.e., there exists $d, 0 \leq d < 1$, such that*

$$\|T(x) - T(y)\| \leq d\|x - y\| \qquad (x, y \in X),$$

then T has a unique fixed point x_0, i.e., $T(x_0) = x_0$.

10.3.2 Examples and Exercises

Example 10.25 *Let T be a linear operator with the domain $D(T)$ in a normed vector space X and with values in a Banach space Y. Prove that if T is bounded and closed operator, then $D(T)$ is a closed set with respect to the norm.*

Solution. Let x be an arbitrary but fixed element of $D(T)$ and $\{x_n\}_{n \in \mathbb{N}}$ a sequence from $D(T)$ which converges to x. The inequality

$$\|T(x_k) - T(x_j)\| \leq \|T\|\|x_k - x_j\|$$

implies that $\{T(x_k)\}_{k \in \mathbb{N}}$ is a Cauchy sequence. Since Y is complete the sequence $\{T(x_k)\}_{k \in \mathbb{N}}$ converges to some element y from Y. By the closedness of the operator T it follows $x \in D(T)$ and $y = T(x)$. Hence $D(T)$ is a closed set.

Example 10.26 *Let X and Y be normed vector spaces. If $T : X \rightarrow Y$ is a linear operator, then prove that*

    a) its adjoint operator $T^* : D(T^*) \rightarrow X'$ is closed,

    b) $D(T^*)$ is weakly $*$-dense in Y'.

Solution.

    a) Let $\{y'_n\}_{n \in \mathbb{N}}$ be a convergent sequence from $D(T^*)$, i.e., $\lim_{n \to \infty} y'_n = y'$, and $T^*(y_n) = z'_n \rightarrow z'$. We have by the definition of the adjoint operator

$$< T(x), y'_n >=< x, z'_n > \qquad (x \in X, n \in \mathbb{N}).$$

    Hence by the continuity of the functionals $< T(x), \cdot >$ and $< x, \cdot >$

$$< T(x), y' >=< x, z' > \qquad (x \in X).$$

    Therefore $y' \in D(T^*)$ and $T^*(y') = z'$.

    b) Follows by the definitions of $D(T^*)$ and weak $*$-convergence.

Example 10.27 (Adjoint Theorem) *Let X and Y be normed vector spaces. If X is a Banach space and $T : X \to Y$ a linear operator, then prove that its adjoint operator $T^* : D(T^*) \to X'$ is a bounded operator.*

Solution. Since T^* is a closed operator the domain $D(T^*)$ of the adjoint operator T^* is dense in Y' and therefore $D(T^*) \neq \{0\}$ for non-trivial spaces X and Y.
Let $\{y'_n\}_{n \in \mathbf{N}}$ be an arbitrary sequence from $D(T^*)$ with the property $\|y'_n\| \leq 1$. We shall prove that the sequence $\{T^*(y'_n)\}_{n \in \mathbf{N}}$ is bounded what will imply the desired conclusion.
We choose a sequence $\{x_n\}_{n \in \mathbf{N}}$ from X such that $\|x_n\| = 1$ and

$$\|T^*(y'_n)\| \leq |T^*(y'_n)(x_n)| + 1 \qquad (n \in \mathbf{N}). \tag{10.29}$$

Let $\{\alpha_n\}_{n \in \mathbf{N}}$ be an arbitrary sequence of numbers such that $\lim_{n \to \infty} \alpha_n = 0$. We can represent the sequence $\{\alpha_n\}_{n \in \mathbf{N}}$ as a product $\alpha_n = t_n \cdot u_n$ where $t_n \geq 0$ and both sequences $\{t_n\}_{n \in \mathbf{N}}$ and $\{u_n\}_{n \in \mathbf{N}}$ converge to zero.
We introduce an infinite matrix of nonnegative numbers $[x_{ij}]_{i,j \in \mathbf{N}}$ such that

$$x_{ij} = \begin{cases} t_i |T^*(y'_i)(u_j x_j)| & \text{for } i \neq j, \\ 0 & \text{for } i = j. \end{cases}$$

We shall show that the matrix $[x_{ij}]_{i,j \in \mathbf{N}}$ satisfies the conditions from Example 10.2. Since $u_j x_j \to 0$ as $j \to \infty$ we obtain by the continuity of the functional that $x_{ij} \to 0$ as $j \to \infty$ for $i \in \mathbf{N}$. By the definition we have $x_{ii} = 0$. It remains to prove that $x_{ij} \to 0$ as $i \to \infty$ for $j \in \mathbf{N}$. Since we have

$$x_{ij} = t_i |T^*(y'_i)(u_j x_j)| = t_i |y'_i(T(u_j x_j))| \leq t_i \|T(u_j x_j)\|,$$

letting $i \to \infty$ we obtain $x_{ij} \to 0$ as $i \to \infty$ for arbitrary but fixed $j \in \mathbf{N}$.
Hence by Diagonal Theorem - Example 10.2 there exists an increasing sequence of integers $\{p_n\}_{n \in \mathbf{N}}$ such that

$$\lim_{i \to \infty} \sum_{j=1}^{\infty} x_{p_i p_j} = 0. \tag{10.30}$$

Since $u_j x_j \to 0$ as $j \to \infty$, we obtain by the completeness of X that there exist a subsequence $\{s_j\}_{j \in \mathbf{N}}$ of $\{p_j\}_{j \in \mathbf{N}}$ and an element x from X such that

$$\sum_{j=1}^{\infty} u_{s_j} x_{s_j} = x.$$

On the other side, we have for every $p \in \mathbf{N}$ and every $y_{s_i} \neq 0$

$$t_{s_i} |T^*(y'_{s_i})(u_{s_i} x_{s_i})| \leq \sum_{j=1, j \neq i}^{i+p} t_{s_i} |T^*(y'_{s_i})(u_{s_j} x_{s_j})|$$

$$+ t_{s_i} |T^*(y'_{s_i})(\sum_{j=1}^{i+p} (u_{s_j} x_{s_j})|$$

for every $i \in \mathbf{N}$. Letting $p \to \infty$ in the preceding inequality we obtain by the continuity of the functionals

$$
\begin{aligned}
t_{s_i}|T^*(u_{s_i}x_{s_i})| &\leq \sum_{j=1}^{\infty} x_{s_is_j} + t_{s_i}|T^*(y'_{s_i})(x)| \\
&\leq \sum_{j=1}^{\infty} x_{s_is_j} + t_{s_i}\|y_{s_i}\|\|T(x)\|
\end{aligned}
$$

for every $i \in \mathbf{N}$. Letting $i \to \infty$ we obtain by (10.30)

$$
\alpha_{s_i}|T^*(y'_{s_i})(x_{s_i})| \to 0.
$$

Therefore, by the Urysohn property of numbers: if for every subsequence $\{z_n\}_{n \in \mathbf{N}}$ of a given sequence of numbers $\{r_n\}_{n \in \mathbf{N}}$ there exists a subsequence $\{v_n\}_{n \in \mathbf{N}}$ such that $v_n \to 0$ as $n \to \infty$, then $r_n \to 0$, we obtain

$$
\alpha_n|T^*(y'_n)(x_n)| \to 0
$$

as $n \to \infty$. Therefore by (10.29) we obtain

$$
\alpha_n\|T^*(y'_n)\| \to 0
$$

as $n \to \infty$. Since the sequences $\{\alpha_n\}_{n \in \mathbf{N}}, \{x_n\}_{n \in \mathbf{N}}$ and $\{y'_n\}_{n \in \mathbf{N}}$ were arbitrary sequences with the prescribed properties it follows that T^* is a bounded operator on its domain $D(T^*)$.

Example 10.28 (Hellinger-Toeplitz) *Let H be a Hilbert space. Prove that if a linear operator $T : H \to H$ is selfadjoint, i.e., $(T(x)|y) = (x|T(y))$ $(x,y \in H)$, then T is a bounded operator.*

Solution. Since $T = T^*$ we obtain the desired conclusion by Example 10.27.

Example 10.29 (Closed Graph Theorem for normed spaces) *Let X be a Banach space and Y a reflexive Banach space. If a linear operator $T : X \to Y$ is closed, then prove*

a) *that $D(T^*) = Y'$,*

b) *that T is a continuous operator.*

Solution.

a) By Example 10.26 b) $D(T^*)$ is weakly dense in Y', since Y is a reflexive space. By Example 10.26 a) and Example 10.27 the adjoint operator T^* is closed and continuous. Therefore by Example 10.25 $D(T^*)$ is a closed subspace with respect to the norm. Since $D(T^*)$ is a subspace the closures for weak topology and norm topology coincides and therefore $D(T^*) = Y'$.

b) By a) and the inequality

$$\|T(x)\| = \sup_{\|y'\|\leq 1} | < y', T(x) > | = \sup_{\|y'\|\leq 1} | < T^*y', x > |$$

$$\leq \|x\| \sup_{\|y'\|\leq 1} \|T^*(y')\| = \|T^*\|\|x\|$$

we obtain the continuity of the operator T.

Example 10.30 (Banach-Steinhaus Theorem) *Let X and Y be Banach spaces and $\{A_n\}_{n\in\mathbb{N}}$ a sequence of operators from $L(X,Y)$. Prove that the sequence $\{A_n\}_{n\in\mathbb{N}}$ strongly converges to an operator $A \in L(X,Y)$, i.e., $\lim_{n\to\infty} A_n x = A x$ $(x \in X)$ if and only if the following conditions are satisfied*

(i) $M = \sup\{\|A_n\| \,|n \in \mathbb{N}\} < \infty$;

(ii) the sequence $\{A_n x\}_{n\in\mathbb{N}}$ is a Cauchy sequence for every x from a subset E of X such that $\overline{L(E)} = X$, where $L(E)$ is the vector space generated by E.

Solution. Suppose that $\lim_{n\to\infty} A_n x = A x$ $(x \in X)$. Therefore

$$\sup\{\|A_n x\| \,|n \in \mathbb{N}\} < \infty   (x \in X).$$

Therefore by the Uniform Boundedness Theorem 10.19 follows (i). Since $\{A_n x\}_{n\in\mathbb{N}}$ is a convergent sequence it follows (ii).

Suppose now that (i) and (ii) hold. Let $x \in X$ and $\varepsilon > 0$. We choose $x' \in L(E)$ such that $\|x - x'\| < \varepsilon$. Since $\{A_n x'\}_{n\in\mathbb{N}}$ is a Cauchy sequence in Y, there exists $n_0 \in \mathbb{N}$ such that for every $n, m \geq n_0$

$$\|A_n x' - A_m x'\| < \varepsilon.$$

Therefore we have for every $n, m \geq n_0$

$$\begin{aligned}\|A_n x - A_m x\| &\leq \|A_n x - A_n x'\| + \|A_n x' - A_m x'\| + \|A_m x' - A_m x\| \\ &< \varepsilon\|A_n\| + \varepsilon + \varepsilon\|A_m\| < (2M + 1)\varepsilon.\end{aligned}$$

Hence $\{A_n x\}_{n\in\mathbb{N}}$ is a Cauchy sequence in Y. Since Y is a Banach space the sequence $\{A_n x\}_{n\in\mathbb{N}}$ converges to $v \in Y$. Denote by A the correspondence $x \mapsto v$. The operator is obviously linear and by Uniform Boundedness Theorem 10.19 there exists $M > 0$ such that $\|A_n\| \leq M$ $(n \in \mathbb{N})$. Then for x such that $\|x\| \leq 1$ we have

$$\|A x\| = \lim_{n\to\infty} \|A_n x\| \leq \|x\|\|A_n\| \leq M.$$

Exercise 10.31 (Riemann-Lebesgue lemma) *The Fourier coefficients*

$$a_n = \frac{1}{\pi} \int_{-\pi}^{\pi} f(x) \cos nx \, dx \text{ and } b_n = \frac{1}{\pi} \int_{-\pi}^{\pi} f(x) \sin nx \, dx \quad (n \in \mathbb{N})$$

for a function $f \in L_1[-\pi, \pi]$ converges to zero as $n \to \infty$.

Hints. Consider the Fourier coefficients as sequences of bounded linear functionals on $L_1[-\pi, \pi]$, e.g.,

$$a_n(f) = \frac{1}{\pi} \int_{-\pi}^{\pi} f(x) \cos nx \, dx,$$

and apply Banach-Steinhaus theorem - Exercise 10.30 , where .

$$E = \{1, \cos x, \sin x, \cos 2x, \sin 2x, \ldots\}.$$

Example 10.32 *Prove that in the Banach theorem on fixed point*

a) *the inequality can not be changed to strict inequality, i.e., to $\|T(x) - T(y)\| < \|x - y\|$ $(x, y \in X, x \neq y)$,*

b) *if T^k is a contraction, then T have not to be continuous.*

Solution.

a) Counterexample: $X = Y = \mathbb{R}$ and T is given by $T(x) = \frac{\pi}{2} + x - \arctan x$.

b) Counterexample: $X = Y = [0, 2]$ and

$$T(x) = \begin{cases} 0 & \text{for } x \in [0, 1], \\ 1 & \text{for } x \in (1, 2]. \end{cases}$$

Exercise 10.33 *The integral equation*

$$u(x) = 1 + \lambda \int_x^1 u(t - x) u(t) \, dt \tag{10.31}$$

for $x \in [0, 1]$ and $\lambda \in (0, 3/8)$ has a solution in the space $C[0, 1]$.

Hint. Show that the solution u of equation (10.31) satisfies

$$\lambda I(u)^2 - 2I(u) + 2 = 0, \tag{10.32}$$

where $I(u) = \int_0^1 u(x) \, dx$. Examine (10.31) with respect to the parameter λ. Then apply the Banach fixed point theorem for

$$X = \{u \mid u \in C[0, 1], |u(x)| \geq 1 \ (x \in [0, 1]), I(u) = \frac{1}{\lambda}(1 - \sqrt{1 - 2\lambda})\}$$

and operator T given by

$$T(u)(x) = 1 + \lambda \int_x^1 u(t - x) u(t) \, dt.$$

Example 10.34 (Uniform Boundedness Theorem) *Let $\mathcal{A}$ be a family of additive and continuous operators from a Banach space X into a normed vector space Y. Prove that if the family $\mathcal{A}$ is pointwise bounded on X, i.e., for $x \in X$ there exists $M(x) > 0$ such that $\|A(x)\| \le M(x)$ $(A \in \mathcal{A})$, then it is also uniformly bounded on every bounded subset B of X, i.e., there exists $M > 0$ such that $\|A(x)\| \le M$ $(A \in \mathcal{A}, x \in B)$.*

Solution. Let $\{A_n\}_{n \in \mathbb{N}}$ be a sequence of operators from $\mathcal{A}$, $\{x_n\}_{n \in \mathbb{N}}$ a sequence of elements from a bounded subset B of X and $\{\alpha_n\}_{n \in \mathbb{N}}$ a sequence of numbers which converges to zero. We have to prove that $\alpha_n A_n(x_n) \to 0$ as $n \to \infty$.

There exists a sequence $\{r_n\}$ of natural numbers such that $r_n \to \infty$ and $\alpha_n r_n \to 0$. We have by the additivity of A_n

$$\alpha_n A_n(x_n) = \alpha_n r_n A_n(r_n^{-1} x_n) \qquad (n \in \mathbb{N}). \tag{10.33}$$

Suppose the theorem were not true. Then there exist $\varepsilon > 0$ and two increasing sequence of natural numbers $\{m_i\}_{i \in \mathbb{N}}$ and $\{n_i\}_{i \in \mathbb{N}}$ such that

$$\|\alpha_{m_i} r_{m_i} A_{m_i}(r_{n_i}^{-1} x_{n_i})\| > \varepsilon \qquad (i \in \mathbb{N}), \tag{10.34}$$

where we have used (10.33).

We introduce an infinite matrix $[x_{ij}]_{i,j \in \mathbb{N}}$ in the following way $x_{ij} = \|\alpha_{m_i} r_{m_i} A_{m_i}(r_{n_j}^{-1} x_{n_j})\|$ for $i \ne j$ and $x_{ii} = 0$. By the suppositions we obtain

$$\lim_{i \to \infty} x_{ij} = 0 \quad (j \in \mathbb{N}) \quad \text{and} \quad \lim_{j \to \infty} x_{ij} = 0 \quad (i \in \mathbb{N}).$$

Therefore by the Diagonal Theorem - Example 10.2 there exists an increasing sequence of natural numbers $\{p_i\}_{i \in \mathbb{N}}$ such that

$$\lim_{i \to \infty} \sum_{j=1}^{\infty} x_{p_i p_j} = 0. \tag{10.35}$$

Since X is a Banach space, there exists a subsequence $\{s_i\}_{i \in \mathbb{N}}$ of $\{p_i\}_{i \in \mathbb{N}}$ such that $\lim_{n \to \infty} \sum_{j=1}^{n} y_{s_j} = y$ for some $y \in X$, where $y_j = r_{n_j}^{-1} x_{n_j}$. We have

$$\|\alpha_{s_i} A_{s_i}(y_{s_i})\| \le \sum_{j=1, j \ne i}^{i+p} \|\alpha_{s_i} A_{s_i}(y_{s_j})\| + \|\sum_{j=1}^{i+p} \alpha_{s_i} A_{s_i}(y_{s_j})\|$$

for $p \in \mathbb{N}$. Letting $p \to \infty$ we obtain

$$\|\alpha_{s_i} A_{s_i}(y_{s_i})\| \le \sum_{j=1, j \ne i}^{\infty} x_{s_i s_j} + \|\alpha_{s_i} A_{s_i}(y)\| \qquad (i \in \mathbb{N}).$$

Letting $i \to \infty$ we obtain by (10.35) $\lim_{i \to \infty} \|\alpha_{s_i} A_{s_i}(y_{s_i})\| = 0$. A contradiction with (10.34).

Chapter 11

Functional Analysis Methods in PDEs

11.1 Generalized Dirichlet Problem

11.1.1 Preliminaries

The equation

$$\sum_{i=1}^{n}\sum_{j=1}^{n} a_{ij}(x)\frac{\partial^2 u}{\partial x_i \partial x_j} + \sum_{i=1}^{n} b_i(x)\frac{\partial u}{\partial x_i} + c(x)u = F(x)$$

is *uniformly elliptic* in the region $Q \subset \mathbf{R}^n$ if there exist constants $C_1 > 0$ and $C_2 > 0$ such that

$$C_1|z|^2 \leq \langle A(x)z, z \rangle \leq C_2|z|^2 \quad (x \in \overline{Q}, z \in \mathbf{R}^n)$$

where $A(x) = [a_{ij}]_{n \times n}$.

A *differential operator of order 2k* is given by

$$L(g) = \sum_{|\alpha|,|\beta| \leq k} (-1)^{|\alpha|} D^\alpha(a_{\alpha\beta} D^\beta g) \quad (g \in C^{2k}(Q)), \tag{11.1}$$

where $a_{\alpha\beta} \in C^\infty(Q), a_{\alpha\beta} \neq 0$ for some α and β such that $|\alpha| = |\beta| = k$. The corresponding adjoint operator L^* is given by

$$L^*(g) = \sum_{|\alpha|,|\beta| \leq k} (-1)^{|\alpha|} D^\alpha(\overline{a_{\beta\alpha}} D^\beta g) \quad (g \in C^{2k}(Q)). \tag{11.2}$$

The corresponding bilinear form is given by

$$B(f,g) = \sum_{|\alpha|,|\beta| \leq k} (a_{\alpha\beta} D^\alpha f | D^\beta g)_{L_2(Q)}.$$

329

The generalized Dirichlet problem for the equation $L(u) = f \in L_2(Q)$ means that for a given $h_W \in \overset{\circ}{W}{}^k(Q)$ (corresponding to F) we have to find $u \in \overset{\circ}{W}{}^k(Q)$ such that

$$B(u,g) = (h_W|g)_{\overset{\circ}{W}{}^k(Q)} \qquad (g \in \overset{\circ}{W}{}^k(Q)).$$

The bilinear form B is *coercitive* if there exists a constant $c > 0$ such that

$$\Re B(f,f) \geq c\|f\|^2_{\overset{\circ}{W}{}^k(Q)} \qquad (g \in \overset{\circ}{W}{}^k(Q)).$$

Theorem 11.1 (Gårding inequality) *Let the operator L given in Preliminaries be strongly (uniformly) elliptic, i.e., there exists $c' > 0$ such that for every $z \in \mathbf{R}^n$*

$$\Re \sum_{|\alpha|=|\beta|=k} z^\alpha a_{\alpha\beta}(x)z^\beta \geq c'|z|^{2k} \quad (x \in \overline{Q}).$$

Then there exist constants $c > 0$ and $a \in \mathbf{R}$ such that

$$\Re B(g,g) \geq c\|g\|^2_{\overset{\circ}{W}{}^k(Q)} - a\|g\|^2_{L_2(Q)}.$$

11.1.2   Examples and Exercises

Example 11.1 *Let L be a linear differential operator defined by*

$$L(u) = \sum_{|\alpha|\leq k} a_\alpha(x)D^\alpha u, \quad k \geq 1,$$

defined on bounded region Q of $\mathbf{R}^n$.
   Prove

a) *that if $a_\alpha(x)$ are continuous on $\overline{Q}$, then the operator $L : C(Q) \to C(Q)$ is not bounded, but as an operator $L : C^k(\overline{Q}) \to C(\overline{Q})$ is bounded.*

b) *that if a_α are bounded measurable functions, then the operator $L : L_2(Q) \to L_2(Q)$ is not bounded, but as an operator $L : W^k(Q) \to L_2(Q)$ is bounded.*

Solution.

a) We prove first that the operator D^α is not a bounded operator from $C(Q)$ to $C(Q)$. Namely, if we take the sequence of functions

$$f_s(x) = \exp(is(x_1 + \cdots + x_n)) \quad (s \in \mathbf{N})$$

which belongs to $C^k(\overline{Q})$ and which is bounded in $C(\overline{Q})$, then the operator D^α maps it on the sequence $(is)^{|\alpha|}\exp(is(x_1 + \cdots + x_n))$ $(s \in \mathbf{N})$, which is not bounded in the space $C(\overline{Q})$. This follows by

$$\|(is)^{|\alpha|}\exp(is(x_1+\cdots+x_n))\|_{C(\overline{Q})} = \max_{x\in\overline{Q}}|(is)^{|\alpha|}\exp(is(x_1+\cdots+x_n))| = s^{|\alpha|} \to \infty$$

as $s \to \infty$.

On the other hand, since the functions a_α are continuous on $\overline{Q}$, they are bounded functions in the space $C(\overline{Q})$. Hence there exists a constant $C > 0$ such that

$$\|L(u)\|_{C(\overline{Q})} \le C\|u\|_{C^k(\overline{Q})} \quad (u \in C^k(Q)),$$

where

$$\|u\|_{C^k(\overline{Q})} = \sum_{|\alpha|\le k} \max_{x\in\overline{Q}}|D^\alpha u(x)|.$$

b) Taking the same sequence of functions $\{f_s\}_{s\in\mathbf{N}}$ as in a), we obtain that the operator D^α is not a bounded operator from $L_2(Q)$ into $L_2(Q)$.

On the other hand, the inequality

$$\|L(u)\|_{L_2(Q)} \le C\|u\|_{W^k(Q)} \quad (u \in W^k(Q))$$

implies that L is a bounded operator from $W^k(Q)$ into $L_2(Q)$.

Example 11.2 *Prove that*

$$(L(f)|g)_{L_2(Q)} = (f|L^*(g))_{L_2(Q)} \quad (f, g \in C_0^\infty(Q)),$$

where L and L^* are given by (11.1) and (11.2), respectively.

Solution. Applying few times the partial integration we obtain

$$
\begin{aligned}
(L(f)|g)_{L_2(Q)} &= \sum_{|\alpha|,|\beta|\le k}(-1)^{|\alpha|}\int_Q D^\alpha(a_{\alpha\beta}D^\beta g)\overline{f}\,dx \\[2mm]
&= \sum_{|\alpha|,|\beta|\le k}\int_Q (a_{\alpha\beta}D^\beta f)D^\alpha\overline{g}\,dx \\[2mm]
&= \sum_{|\alpha|,|\beta|\le k}(-1)^{|\beta|}\int_Q f\,\overline{D^\beta(\overline{a}_{\alpha\beta}D^\alpha g)}\,dx \\[2mm]
&= \sum_{|\alpha|,|\beta|\le k}(-1)^{|\alpha|}\int_Q f\,\overline{D^\alpha(\overline{a}_{\beta\alpha}D^\beta g)}\,dx \\[2mm]
&= (f|L^*(g))_{L_2(Q)}.
\end{aligned}
$$

Example 11.3 *Let L be a linear differential operator defined by*

$$L(u) = \sum_{|\alpha| \le k} a_\alpha(x) D^\alpha u, \quad k \ge 1,$$

defined on a bounded region Q of $\mathbf{R}^n$ with coefficients $a_\alpha \in C^{s-k}(\overline{Q})$, where $s \ge k$. Prove that L is a continuous linear operator from $C^s(\overline{Q})$ into $C^{s-k}(\overline{Q})$.

Solution. Since L is a linear operator it is sufficient to prove that it is bounded, i.e., that there exists $M > 0$ such that

$$\|L(u)\|_{C^{s-k}(\overline{Q})} \le M \|u\|_{C^s(\overline{Q})}, \text{ where } \|u\|_{C^s(\overline{Q})} = \sum_{|\alpha| \le s} \max_{x \in \overline{Q}} |D^\alpha u|.$$

We have by the Leibniz formula

$$D^\alpha L(u) = \sum_{|\beta| \le k} \sum_{\alpha_1 \le \alpha} \binom{\alpha}{\alpha_1} D^{\alpha_1} a_\beta D^{\alpha - \alpha_1} D^\beta u.$$

Therefore, taking

$$M' = \max_{\substack{|\beta| \le k \\ |\alpha_1| \le s-k}} \max_{x \in \overline{Q}} |D^{\alpha_1} a_\beta|,$$

since $|\alpha_1| \le |\alpha| \le s - k$, we have

$$
\begin{aligned}
\sum_{|\alpha| \le s-k} \max_{x \in \overline{Q}} |D^\alpha L(u)| &\le \sum_{|\alpha| \le s-k} \sum_{|\beta| \le k} \sum_{\alpha_1 \le \alpha} \binom{\alpha}{\alpha_1} \max_{x \in \overline{Q}} |D^{\alpha_1} a_\beta| \max_{x \in \overline{Q}} |D^{\alpha + \beta - \alpha_1} u| \\
&\le M' \sum_{|\alpha| \le s-k} \sum_{|\beta| \le k} \sum_{\alpha_1 \le \alpha} \max_{x \in \overline{Q}} |D^{\alpha + \beta - \alpha_1} u|.
\end{aligned}
$$

Therefore by the inequality

$$|\alpha + \beta - \alpha_1| = |\alpha| + |\beta| - |\alpha_1| \le k + s - k = s$$

we obtain

$$\sum_{|\alpha| \le s-k} \max_{x \in \overline{Q}} |D^\alpha L(u)| \le M \sum_{|\gamma| \le s} \max_{x \in \overline{Q}} |D^\gamma L(u)|.$$

Exercise 11.4 *Prove that the differential operator from Example 11.3 is a continuous operator from $W^s(Q)$ into $W^{s-k}(Q)$.*

  Hint. Prove that there exists $M > 0$ such that

$$\|L(u)\|_{W^{s-k}(Q)} \le M \|u\|_{W^s(Q)}.$$

11.1. GENERALIZED DIRICHLET PROBLEM
333

Example 11.5 *Let Q be a bounded region of $\mathbf{R}^n$ with an enough regular boundary ∂Q. Prove that the bilinear form*

$$B(f,g) = \sum_{|\alpha|,|\beta|\leq k} (a_{\alpha\beta}D^\alpha f | D^\beta g)_{L_2(Q)} \qquad (f,g \in \overset{\circ}{W}{}^k(Q))$$

is bounded on $\overset{\circ}{W}{}^k(Q) \times \overset{\circ}{W}{}^k(Q)$, i.e., there exists $M > 0$ such that

$$|B(f,g)| \leq M\|f\|_{\overset{\circ}{W}{}^k(Q)}\|g\|_{\overset{\circ}{W}{}^k(Q)} \qquad (f,g \in \overset{\circ}{W}{}^k(Q)).$$

Solution. Since $a_{\alpha\beta} \in C(\overline{Q})$ $(\alpha,\beta \in \mathbf{Z}_+^n)$, we have $\|a_{\alpha\beta}\|_{L_2(Q)} < C$ for some $C > 0$. Therefore we have for $f,g \in \overset{\circ}{W}{}^k(Q)$

$$\begin{aligned}
|B(f,g)| &\leq \sum_{|\alpha|,|\beta|\leq k} \|a_{\alpha\beta}D^\alpha f\|_{L_2(Q)}\|D^\beta g\|_{L_2(Q)} \\
&\leq C \sum_{|\alpha|,|\beta|\leq k} \|D^\alpha f\|_{L_2(Q)}\|D^\beta g\|_{L_2(Q)} \\
&\leq M\|f\|_{\overset{\circ}{W}{}^k(Q)}\|g\|_{\overset{\circ}{W}{}^k(Q)}.
\end{aligned}$$

Example 11.6 *Prove that for every $F \in L_2(Q)$ there exists a unique $h_W \in \overset{\circ}{W}{}^k(Q)$ such that*

$$(F|g)_{L_2(Q)} = (h_W|g)_{\overset{\circ}{W}{}^k(Q)} \qquad (g \in \overset{\circ}{W}{}^k(Q)).$$

Solution. For a fixed but arbitrary $F \in L_2(Q)$ we have that the functional h defined by

$$h(g) = (F|g)_{L_2(Q)} \qquad (g \in \overset{\circ}{W}{}^k(Q))$$

is continuous on the Hilbert space $\overset{\circ}{W}{}^k(Q)$, since we have by Cauchy-Schwartz inequality

$$|h(g)| = |(F|g)_{L_2(Q)}| \leq \|f\|_{L_2(Q)}\|g\|_{L_2(Q)} \leq \|F\|_{L_2(Q)}\|g\|_{\overset{\circ}{W}{}^k(Q)}.$$

Therefore by Riesz representation theorem there exists a unique $h_W \in \overset{\circ}{W}{}^k(Q)$ such that

$$(h_W|g)_{\overset{\circ}{W}{}^k(Q)} = h(g) = (F|g)_{L_2(Q)}.$$

Example 11.7 *Prove the equivalence of the following two problems*

(i) For a given $F \in L_2(Q)$ find $u \in \overset{\circ}{W}{}^k(Q)$ such that

$$B(u,g) = (F|g)_{L_2(Q)} \qquad (g \in \overset{\circ}{W}{}^k(Q));$$

(ii) For a given $h_W \in \overset{\circ}{W}{}^k(Q)$ find $u \in \overset{\circ}{W}{}^k(Q)$ such that

$$B(u,g) = (h_W|g)_{\overset{\circ}{W}{}^k(Q)} \qquad (g \in \overset{\circ}{W}{}^k(Q)),$$

where B is from Example 11.5.

Solution. Follows by Example 11.5.

Example 11.8 (Lax-Milgram) *Prove that for every bounded bilinear form B on* $\overset{\circ}{W}{}^k(Q) \times \overset{\circ}{W}{}^k(Q)$

a) *there exists a unique linear continuous operator* $T : \overset{\circ}{W}{}^k(Q) \to \overset{\circ}{W}{}^k(Q)$ *such that*

$$B(f,g) = (Tf|g)_{\overset{\circ}{W}{}^k(Q)} \qquad (f, g \in \overset{\circ}{W}{}^k(Q));$$

b) *if additionally B is coercitive, then there exists T^{-1} and it is a continuous operator on* $\overset{\circ}{W}{}^k(Q)$.

Solution.

a) Since for an arbitrary but fixed $f \in \overset{\circ}{W}{}^k(Q)$ the functional $h(g) = B(f,g)$ is antilinear (for complex case) and continuous by Example 11.5, there exists by Riesz representation theorem a unique function $w \in \overset{\circ}{W}{}^k(Q)$ such that

$$(w|g)_{\overset{\circ}{W}{}^k(Q)} = h(g) = B(f,g) \qquad (g \in \overset{\circ}{W}{}^k(Q))$$

and $\|w\|_{\overset{\circ}{W}{}^k(Q)} = \|h\|$. Then the desired operator T is defined by

$T(f) = w$ $(f \in \overset{\circ}{W}{}^k(Q))$. The operator T is linear and $B(f,g) = (Tf|g)_{\overset{\circ}{W}{}^k(Q)}$. The boundedness of the operator T follows by

$$\|Tf\|_{\overset{\circ}{W}{}^k(Q)} = \|w\|_{\overset{\circ}{W}{}^k(Q)} = \|h\| \le M\|f\|_{\overset{\circ}{W}{}^k(Q)}$$

for some $M > 0$.

b) Since B is coercitive there exists $c > 0$ such that

$$\|Tf\|_{\overset{\circ}{W}{}^k(Q)}\|f\|_{\overset{\circ}{W}{}^k(Q)} \ge |B(f,f)| \ge \Re B(f,f) \ge c\|f\|^2_{\overset{\circ}{W}{}^k(Q)}.$$

Therefore

$$\|Tf\|_{\overset{\circ}{W}{}^k(Q)} \ge c\|f\|_{\overset{\circ}{W}{}^k(Q)}.$$

If g is orthogonal on the range $R(T)$ then

$$B(g,g) = (Tg|g)_{\overset{o}{W}{}^{k}(Q)} = 0.$$

Hence by the coercitivity of B we have $g = 0$. Therefore $R(T) = \overset{o}{W}{}^{k}(Q)$. This implies by Closed Graph Theorem that T^{-1} is a continuous operator on $\overset{o}{W}{}^{k}(Q)$.

Example 11.9 *Let the corresponding bilinear form B to the differential operator L of the order $2k$ in the generalized Dirichlet problem from Exercise 11.7 be coercitive. Prove that for every $F \in L_2(Q)$ there exists a unique solution.*

Solution. By Example 11.7 and Example 11.8 we obtain

$$(Tu|g)_{\overset{o}{W}{}^{k}(Q)} = (h_W|g)_{\overset{o}{W}{}^{k}(Q)} \qquad (g \in \overset{o}{W}{}^{k}(Q)).$$

Therefore $T(u) = h_W$. Since by Example 11.8 b) the operator T^{-1} is continuous and linear on the space $\overset{o}{W}{}^{k}(Q)$. Therefore $u = T^{-1}h_W$ gives the unique solution of the considered generalized solution of the generalized Dirichlet problem.

Remark 11.9.1 It is important the question of the regularity of the solution of the generalized Dirichlet problem, i.e., when the obtained generalized solution from the space $\overset{o}{W}{}^{k}(Q)$ is also the classical solution. We give here only a general theorem in this direction.

Theorem 11.2 *Let Q be an open set of $\mathbf{R}^n$ and the operator L of the order $2k$ is given by*

$$L(u) = \sum_{|\alpha| \leq 2k} a_\alpha D^\alpha u,$$

where a_α are constants for $|\alpha| = 2k$ and $a_\alpha \in C^\infty(Q)$ for others α. If L is strongly elliptic, $F \in W^s(Q)$ and u is the generalized solution in $L_2(Q)$, then $u \in W^{2k+s}(Q')$ for every bounded open subset Q' of Q.

If additionally $m < 2k + s - \frac{n}{2}$ then $u \in C^m(Q')$. If additionally $F \in C^\infty(Q)$, then $u \in C^\infty(Q')$.

Example 11.10 *Prove under same suppositions as in Theorem 11.1 that the bilinear form B_a for $a \in \mathbf{R}$ given by*

$$B_a(f,g) = B(f,g) + a(f|g)_{L_2(Q)}$$

is coercitive.

Solution. By Theorem 11.1 we obtain

$$\Re b_a(g,g) = \Re B(g,g) + a\|g\|^2_{L_2(Q)}$$
$$\geq c\|g\|^2_{\overset{o}{W}{}^k(Q)} - a\|g\|^2_{L_2(Q)} + a\|g\|^2_{\overset{o}{W}{}^k(Q)} = c\|g\|^2_{\overset{o}{W}{}^k(Q)}$$

Example 11.11 *The solution u of the generalized Dirichlet problem from Example 11.7 continuously depends on $F \in L_2(Q)$.*

Solution. We denote by U the embedding operator $U : \overset{o}{W}{}^k(Q) \to L_2(Q)$ and by U^* its adjoint operator. The operators U and U^* are continuous. We have

$$(F|g)_{L_2(Q)} = (F|U(g))_{L_2(Q)} = (U^*(F)|g)_{\overset{o}{W}{}^k(Q)} \quad (F \in L_2(Q), g \in \overset{o}{W}{}^k(Q)). \quad (11.3)$$

By Example 11.7 and Example 11.8 the solution is $T^{-1}(h_W)$, where

$$(h_W|g)_{\overset{o}{W}{}^k(Q)} = (F|g)_{L_2(Q)} \quad (g \in \overset{o}{W}{}^k(Q)).$$

Therefore by (11.3) we have $u = T^{-1}U^*(F)$. Since $T^{-1}U^*$ is a continuous operator we obtain the desired conclusion.

Remark 11.11.1 We shall call the operators

$$G = T^{-1}U^* : L_2(Q) \to \overset{o}{W}{}^k(Q),$$
$$\tilde{G} = UT^{-1}U^* : L_2(Q) \to L_2(Q)$$

Green operators. If the corresponding bilinear form is coercitive then the Green operators are continuous.

Example 11.12 *Prove that for the bounded region Q the Green operator $\tilde{G}_a : L_2(Q) \to L_2(Q)$ which corresponds to the bilinear form B_a from Example 11.10 is a compact operator.*

Solution. By Theorem 9.7 the embedding $U : \overset{o}{W}{}^k(Q) \to \overset{o}{W}{}^0(Q) = L_2(Q)$ is a compact operator. Since T_a^{-1} and U^* are continuous operators, where T_a is the corresponding operator to B_a from Example 11.8, we have that $\tilde{G}_a = UT_a^{-1}U^*$ is a compact operator.

Example 11.13 *Prove the following analogy to Fredholm alternative.*
If L is a strongly elliptic operator from Preliminaries on a bounded region Q, then either the generalized Dirichlet problem from Example 11.7 has exactly one solution for every $F \in L_2(Q)$,
or the zero is the characteristic value, i.e., λ is the characteristic value if there exists a function $u \neq 0$ from $\overset{o}{W}{}^k(Q)$ such that

$$B(u,g) = \lambda(u|g)_{L_2(Q)} \quad (g \in \overset{o}{W}{}^k(Q)).$$

Solution. By Theorem 11.1 and Example 11.10 there exists a real number a such that the Green operator $\tilde{G}_a$ is a compact operator. Since the generalized Dirichlet problem is equivalent with the equation

$$u = aG_a\tilde{u} + \tilde{F},$$

where $\tilde{F}\tilde{G}_a(F)$ and $\tilde{u} = U(u)$ (to prove that it is enough to add $a(\tilde{u}|g)$ to both sides in the equality $B(u,g) = (h_W|g)_{\overset{\circ}{W}{}^k(Q)}$ and apply Remark 11.11.1), we obtain the desired conclusion by theorem on Fredholm alternative.

Example 11.14 *Prove that the following bilinear form*

$$B(f,g) = \int_Q (\nabla f(x)\nabla \overline{g(x)} + (k + p(x))f(x)\overline{g(x)})\, dx$$

for $k \in \mathbf{R}$, $p \in C^\infty(\overline{Q})$ and $k \geq 1 - p_0$, where $p_0 = \inf_{x\in Q} \Re p(x)$, is coercitive.

Solution. The coercitivity of B follows by

$$
\begin{aligned}
\Re B(f,f) &= \Re \int_Q (\nabla f(x)\nabla \overline{f(x)} + (k + p(x))f(x)\overline{f(x)})\, dx \\
&= \|f\|^2_{W^1(Q)} - \|f\|_{L_2(Q)} + \Re((k+p)f|f)_{L_2(Q)} \\
&\geq \|f\|^2_{W^1(Q)} + (k + p_0 - 1)\|f\|^2_{L_2(Q)} \geq \|f\|^2_{W^1(Q)},
\end{aligned}
$$

since $k \geq 1 - p_0$.

Example 11.15 *With the same notations as in Example 11.14 prove that for the generalized Dirichlet problem for $L = -\Delta + k + p$ there exists a unique solution for $k \geq 1 - p_0$, where $p_0 = \inf_{x\in Q} \Re p(x)$.*

Solution. By Example 11.14 we have that for the operator L the corresponding bilinear form

$$B(f,g) = \int_Q (\nabla f(x)\nabla \overline{g(x)} + (k + p(x))f(x)\overline{g(x)})\, dx$$

is coercitive. Then by Example 11.9 the given generalized Dirichlet problem has a unique solution.

Exercise 11.16 *Let Q be a bounded region with smooth boundary ∂Q. If a function $u \in C(\overline{Q}) \cap C^2(Q)$ is the solution of the Dirichlet problem*

$$L(u) = \sum_{i=1}^{n}\sum_{k=1}^{n} a_{ik}(x)\frac{\partial^2 u}{\partial x_i \partial x_k} + \sum_{i=1}^{n} a_i(x)\frac{\partial u}{\partial x_i} + au = f \text{ on } Q$$

$$u|_{\partial Q} = \varphi,$$

where

$$a_{ik}, a_i, a \in C(\overline{Q}), \; a_{ik} = a_{ki} \quad (x \in \overline{Q}),$$

$$\varphi \in C(\partial Q), \; a(x) \le 0 \quad (x \in Q)$$

and L is strongly (uniform) elliptic, then there exists a constant $C > 0$ such that we have the following a-priori inequality

$$\|u\|_{\overline{Q}} \le \|\varphi\|_{\partial Q} + C\|f\|_{\overline{Q}},$$

where

$$\|f\|_Q = \sup_{x \in Q} |f(x)| \; and \; \|\varphi\|_{\partial Q} = \max_{x \in \partial Q} |\varphi(x)|.$$

Hints. Take $x_1 \ge 0$ in $\overline{Q}$. Choose a number $z > x_1 \quad (x \in \overline{Q})$ and a number $\alpha > 0$ enough big that the following inequalities hold

$$c\alpha^2 - k(\alpha + 19) \ge 1 \; and \; e^{\alpha z} > 2\max_{x \in \overline{Q}} e^{\alpha x_1}.$$

Then introduce the function

$$h(x) = \|\varphi\|_{\partial Q} + (e^{\alpha z} - e^{\alpha x_1})\|f\|_{\overline{Q}}$$

and prove that $-L(h) \ge \|f\|_Q$. Then show by the Maximum Principle that $|u(x)| \le h(x) \quad (x \in \overline{Q})$. Taking

$$C = \max_{x \in \overline{Q}}(e^{\alpha z} - e^{\alpha x_1})$$

prove by the last inequality the desired inequality.

Example 11.17 *Let*

$$L = -\sum_{i=1}^{n}\sum_{j=1}^{n} \frac{\partial}{\partial x_j}\left(p\frac{\partial}{\partial x_i}\right) + q,$$

be a differential operator on a bounded region Q, where $p \in C^1(\overline{Q}), q \in C(\overline{Q})$; $p(x) > 0, q(x) \ge 0 \quad (x \in \overline{Q})$. The domain $D(L)$ of the operator L consists of the functions $u \in C^2(Q) \cap C^1(\overline{Q})$ and $L(u) \in L_2(Q)$ and on the boundary ∂Q they satisfy the condition

$$\left(\alpha u + \beta\frac{\partial u}{\partial \mathbf{n}}\right)\Big|_{\partial Q} = 0, \tag{11.4}$$

where $\alpha, \beta \in C(\partial Q)$ and

$$\alpha(x) \ge 0, \; \beta(x) \ge 0, \; \alpha(x) + \beta(x) > 0 \qquad (x \in \partial Q).$$

Prove that

   a) the operator L is symmetric and positive;

   b) eigenvalues of the operator L are nonnegative and the corresponding eigenfunctions to different eigenvalues are orthogonal.

Solution.

a) We obtain by Green identity

$$\int_Q (vL(u) - uL(v))\, dx = \int_{\partial Q} P\left(u\frac{\partial v}{\partial \mathbf{n}} - v\frac{\partial u}{\partial \mathbf{n}}\right) dS.$$

Therefore taking $\overline{v} \in D(L)$ instead of v and $u \in D(L)$ we obtain

$$\int_Q (\overline{v}L(u) - u\overline{L(v)})\, dx = (L(u)|v)_{L_2(Q)} - (u|L(v))_{L_2(Q)}$$

$$= \int_{\partial Q} P\left(u\frac{\partial \overline{v}}{\partial \mathbf{n}} - \overline{v}\frac{\partial u}{\partial \mathbf{n}}\right) dS. \tag{11.5}$$

The functions u and $\overline{v}$ satisfy the boundary condition (11.4), i.e.,

$$\left(\alpha u + \beta \frac{\partial u}{\partial \mathbf{n}}\right)\big|_{\partial Q} = 0, \ \left(\alpha \overline{v} + \beta \frac{\partial \overline{v}}{\partial \mathbf{n}}\right)\big|_{\partial Q} = 0.$$

The condition $\alpha(x) + \beta(x) > 0$ $(x \in \partial Q)$ enshures that the preceding homogeneous system of linear equations has nontrivial solution (α, β). Therefore the determinant of this system is equal zero, i.e.,

$$\begin{vmatrix} u & \frac{\partial u}{\partial \mathbf{n}} \\ \overline{v} & \frac{\partial \overline{v}}{\partial \mathbf{n}} \end{vmatrix}_{\partial Q} = \left(u\frac{\partial \overline{v}}{\partial \mathbf{n}} - \overline{v}\frac{\partial u}{\partial \mathbf{n}}\right)\big|_{\partial Q} = 0.$$

Putting this in (11.5) we obtain $(L(u)|v)_{L_2(Q)} = (u|L(v))_{L_2(Q)}$, i.e., the operator L is symmetric.

We shall prove now that the operator L is positive. We obtain by the Green formula

$$(L(u)|u)_{L_2(Q)} = \int_Q P\sum_{j=1}^n \left|\frac{\partial u}{\partial x_j}\right|^2 dx - \int_{\partial Q} pu\frac{\partial u}{\partial \mathbf{n}}\, dS + \int_Q q|u|^2\, dx. \tag{11.6}$$

Since (11.4) implies

$$\frac{\partial u}{\partial \mathbf{n}} = -\frac{\alpha}{\beta}u \quad \text{for} \quad \beta(x) > 0 \quad (x \in \partial Q)$$

and

$$u = 0 \quad \text{for} \quad \beta(x) = 0 \quad (x \in \partial Q),$$

we obtain by (11.6)

$$(L(u)|u)_{L_2(Q)} = \int_Q (p \sum_{j=1}^n |\frac{\partial u}{\partial x_j}|^2 + q|u|^2)\, dx + \int_{\substack{\partial Q \\ \alpha>0,\beta>0}} p\frac{\alpha}{\beta}|u|^2\, dS \quad (u \in D(L)).$$

$$(11.7)$$

Since all summands in the preceding equality are nonnegative we obtain $(L(u)|u)_{L_2(Q)} \geq 0$, i.e., the operator L is positive.

b) By a) the operator $L : D(L) \to L_2(Q)$ is symmetric and positive and therefore the desired conclusion follows by Example 10.18.

Remark 11.17.1 By (11.7) we obtain

$$(L(u)|u)_{L_2(Q)} \geq \int_Q p \sum_{j=1}^n |\frac{\partial u}{\partial x_j}|^2\, dx \geq \min_{x \in \overline{Q}} p(x) \int_Q \sum_{j=1}^n |\frac{\partial u}{\partial x_j}|^2\, dx,$$

since the function p is continuous on the compact set $\overline{Q}$.

Example 11.18 (Hörmander) *If a linear differential operator*

$$P(D) = P(-i\frac{\partial}{\partial x_1}, \cdots, -i\frac{\partial}{\partial x_n})$$

with constant coefficients is hypoelliptic , i.e., for every generalized solution $u \in L_{2,loc}(Q)$ of the equation $P(D)u = F$ for $F \in C^\infty$ is almost everywhere equal to a function from C^∞, then for every constant $C_1 > 0$ there exists a constant $C_2 > 0$ such that every solution $z = \xi + i\eta = (z_1,\ldots,z_n)$ of the algebraic equation $P(z) = 0$ satisfies the condition:

$$if\ |\eta| = (\sum_{j=1}^n |\eta_j|^2)^{1/2} < C_2,\ then\ |z| = (\sum_{j=1}^n |z_j|^2)^{1/2} < C_1. \quad (11.8)$$

Solution. Let U be the set of all generalized solutions $u \in L_2(Q')$, where Q is an open subset of the region Q, for the equation $P(D)u = 0$, i.e.,

$$(u|P^*(D)\varphi)_{L_2(Q')} = \int_{Q'} u \cdot P^*(D)\varphi\, dx = 0 \text{ for every } \varphi \in C_0^\infty(Q'),$$

where P^* is the adjoint operator for P, $P^*(z) = P(-z_1,\ldots,-z_n)$. Since $P(D)$ is a linear operator U is a vector subspace of the space $L_2(Q')$. We shall show that U is a closed subspace. Let $\{u_n\}_{n \in \mathbb{N}}$ be a sequence from U which converges in the space $L_2(Q')$ to $u \in L_2(Q')$. Since we have

$$\int_{Q'} u_n \cdot P^*(D)\varphi\, dx = 0 \qquad (n \in \mathbf{N}),$$

we obtain by the continuity of the scalar product $(\cdot|\cdot)_{L_2(Q')}$ that

$$\lim_{n \to \infty} (u_n | P^*(D)\varphi)_{L_2(Q')} = (u | P^*(D)\varphi)_{L_2(Q')}.$$

Hence $(u|P^*(D)\varphi)_{L_2(Q')} = 0$, i.e., $u \in U$. Therefore U as a closed subspace of the Banach space $L_2(Q')$ itself is a Banach space.

By the hypoellipticity of the operator $P(D)$ it follows that every function u from the space U belongs to $C^\infty(Q')$. If $Q'_1 \subset\subset Q'$, then $\dfrac{\partial u}{\partial x_j} \in C^\infty(Q'_1)$ $(j = 1, \ldots, n)$ for every function $u \in U$. The operators $T_j : U \to L_2(Q'_1)$, $j = 1, \ldots, n$, given by $T_j(u) = \dfrac{\partial u}{\partial x_j}$ are closed operators, which by the Closed Graph theorem (U is a Banach space and $L_2(Q'_1)$ is a Hilbert space) are bounded operators. Therefore there exists a constant $C > 0$ such that

$$\int_{Q'_1} \sum_{j=1}^{n} |\frac{\partial u}{\partial x_j}|^2 \, dx \le C \int_{Q'} |u|^2 \, dx \qquad (u \in U). \tag{11.9}$$

If $z = \xi + i\eta$ is a solution of the algebraic equation $P(z) = 0$, then applying the inequality (11.9) on the special function $u(x) = \exp(i \sum_{i=1}^{n} x_i z_i)$ we obtain

$$\sum_{j=1}^{n} |z_j|^2 \int_{Q'_1} \exp(-2 \sum_{i=1}^{n} x_i \eta_i) \, dx \le C \int_{Q'} \exp(-2 \sum_{i=1}^{n} x_i \eta_i) \, dx.$$

This implies that if for some $C_2 > 0$ we have $|\eta| < C_2$, i.e., it is bounded, then there exists $C_1 > 0$ such that $|z| < C_1$.

Remark 11.18.1 It is true also the statement in the opposite way, which gives a characterization of the hypoelliptic operators. As a special case of this opposite statement we obtain

Theorem 11.3 (Weyl lemma) *Every generalized solution $u \in L_2$ of the Laplace equation*

$$\Delta u = F \in L_2$$

is a function from C^∞ except on the set of measure zero from the region where $F \in C^\infty$.

Follows by the fact that the zeroes of the equation $- \sum_{j=1}^{n} z_j^2 = 0$ satisfy the condition (11.8).

Exercise 11.19 *Let A be a closed subspace of a Hilbert space H. Prove that for every $x \in H$ there exists a unique $x_A \in A$ such that*

$$\|x - x_A\| = \inf_{z \in A} \|x - z\| = d.$$

Hints. Consider the set $x - A$ and a sequence $\{x_n\}_{n \in N}$ from $x - A$ such that

$$\lim_{n \to \infty} \|x_n\| = d,$$

and prove using the parallelogram law

$$\|x - y\|^2 + \|x + y\|^2 = 2(\|x\|^2 + \|y\|^2) \qquad (x, y \in H),$$

that $\{x_n\}_{n \in N}$ is a Cauchy sequence with the limit $x - x_A$, where x_A is the desired unique element.

Remark 11.19.1 This exercise gives as a consequence

Theorem 11.4 (Projection theorem) *Every element $x \in H$ can be represented in the following form*

$$x = x_A + y \ \text{for some } y \in A^{\perp},$$

where $A^{\perp}$ is the subspace of all orthogonal elements on A.

Example 11.20 *Let Q be a bounded region with $\partial Q \in C^{\infty}$. Let $p \in C^{\infty}(\overline{Q})$ and $F \in L_2(Q)$. Prove*

a) *that*

$$(u|g)_E = \int_Q (\nabla u \nabla \overline{g} + pu\overline{g}) \, dx \qquad (u, g \in \overset{\circ}{W}{}^1(Q))$$

is a scalar product on $\overset{\circ}{W}{}^1(Q)$ which induces an equivalent norm with the norm $\| \cdot \|_{\overset{\circ}{W}{}^1(Q)}$;

b) *if $u \in \overset{\circ}{W}{}^1(Q)$ is a solution of the equation*

$$\int_Q (\nabla u \nabla \overline{g} + pu\overline{g}) \, dx = (F|g)_{L_2(Q)} \qquad (g \in \overset{\circ}{W}{}^1(Q)), \qquad (11.10)$$

then for an arbitrary closed subspace A of $\overset{\circ}{W}{}^1(Q)$ the function $u_A \in A$ (see Exercise 11.19) is the unique solution of the equation

$$(u_A|g)_E = (F|g)_{L_2(Q)} \qquad (g \in A). \qquad (11.11)$$

If A is finite dimensional we call U_A the Ritz approximate solution.

Solution.

a) See for more general case Example 11.17.

b) By Exercise 11.19 we have for $u \in \overset{\circ}{W}{}^1(Q)$

$$u = u_A + (u - u_A) \text{ for } u_A \in A \text{ and } u - u_A \in A^\perp. \qquad (11.12)$$

Putting this function u from (11.12) into (11.10) we obtain that for every $g \in A$

$$(F|g)_{L_2(Q)} = (u_A + u - u_A|g)_E = (u_A|g)_E + (u + u_A|g)_E = (u_A|g)_E,$$

i.e., (11.11). If we suppose that u'_A is another solution of the equation (11.11) in A, then we have $(u_A - u'_A|g)_E = 0 \quad (g \in A)$. Hence $u_A - u'_A \in A^\perp$. Therefore $u_A - u'_A = 0$, i.e., $u_A = u'_A$.

Remark 11.20.1 For $n-$dimensional A we have

Theorem 11.5 (Ritz) *If $b_1, \ldots, b_n$ is a base of the space A, then the system of equations*

$$\sum_{j=1}^{n} c_j (b_j|b_i)_E = (F|b_i)_E \qquad (i = 1, \ldots, n)$$

has a unique solution and the approximate Ritz solution u_A in A for the corresponding Dirichlet problem has the following form

$$u_A = \sum_{i=1}^{n} c_i b_i$$

and the following estimation holds

$$\|u - u_A\| \leq \|u - g\| \qquad (g \in A).$$

Exercise 11.21 *Prove that the following systems of functions $\{\varphi_n\}_{n \in \mathbf{N}}$ are orthonormal bases in the space $L_2(I)$, where I is the corresponding interval in $\mathbf{R}$*

a) $\quad \varphi_n(x) = \dfrac{e^{inx}}{\sqrt{2\pi}}, \; n = 0, \pm 1, \pm 2, \ldots, \text{ for } x \in I = (-\pi, \pi) \text{ (the Fourier system)};$

b) $\quad \varphi_n(x) = (n + \dfrac{1}{2})^{1/2} \cdot 2^{-n} \sum_{i=0}^{[n/2]} (-1)^i \binom{n}{i} \binom{2n - 2i}{n} x^{n-2i},$

$\quad n = 0, \pm 1, \pm 2, \ldots, \text{ for } x \in I = (-1, 1), \text{ (the Legendre system)};$

c) $\quad \varphi_n(x) = (2^n n! \sqrt{\pi})^{-1/2} e^{x^2/2} H_n(x),$

$\quad n = 0, \pm 1, \pm 2, \ldots, \text{ for } x \in I = (-\infty, \infty), \text{ where } H_n \text{ is the Hermite polynomial given by}$

$$H_n(x) = n! \sum_{i=0}^{[n/2]} \frac{(-1)^i (2x)^{n-2i}}{i!(n - 2i)!},$$

(the Hermite system);

d) $\varphi_n(x) = (\frac{2}{\pi})^{1/2} \sin nx$, $n = 1, 2, \ldots$, for $x \in I = (0, \pi)$.

Exercise 11.22 *Prove that the systems of functions $\{\varphi_n\}_{n \in \mathbf{N}}$ from Exercise 11.21 a)- d) are the corresponding eigenfunctions for the following differential operators with the corresponding eigenvalues $\{\lambda_n\}_{n \in \mathbf{N}}$:*

a) $A = -iD$, $\lambda_n = n$, where $D = \dfrac{d}{dx}$;

b) $A = D(x^2 - 1)D$, $\lambda_n = n(n+1)$;

c) $A = e^{\frac{x^2}{2}} D e^{-x^2} D e^{\frac{x^2}{2}} = D^2 - x^2 + 1$, $\lambda_n = -2n$;

d) $A = D^2$, $\lambda_n = -n^2$.

Exercise 11.23 *Let $Q \subset \mathbf{R}^n$ be a bounded region with smooth boundary ∂Q. Prove that the equation*

$$\sum_{i=1}^{n} \sum_{j=1}^{n} a_{ij}(x) \frac{\partial^2 u}{\partial x_i \partial x_j} + \sum_{i=1}^{n} b_i(x) \frac{\partial u}{\partial x_i} + cu = F \text{ on } Q, \tag{11.13}$$

where $a_{ij} \in C^1(\overline{Q}), b_j, c \in C(\overline{Q})$ and $F \in C(Q)$, can be written also in the following form

$$\sum_{i=1}^{n} \sum_{j=1}^{n} \frac{\partial}{\partial x_i} (a_{ij}(x)) \frac{\partial u}{\partial x_j} + \sum_{i=1}^{n} \frac{\partial (a_i u)}{\partial x_i} + \sum_{i=1}^{n} B_i \frac{\partial u}{\partial x_i} + Cu = F, \text{ on } Q, \tag{11.14}$$

where $a_i \in C^1(\overline{Q}), B_i, C \in C(\overline{Q})$. Prove that the equation (11.14) can be written in the form (11.13).

Exercise 11.24 *Let Q be a locally quadratic bounded region and*

$$a_{ij}, a_i \in C^1(\overline{Q}), \quad b_i, c \in C(\overline{Q}), \quad F \in C(\overline{Q}) \cap L_2(Q)$$

and $g \in C(\partial Q)$. Prove that the (classical) solution $u \in C^2(\overline{Q}) \cap C(Q)$ of the Dirichlet problem for the uniform elliptic PDE

$$\sum_{i=1}^{n} \sum_{j=1}^{n} \frac{\partial}{\partial x_i} \left(a_{ij}(x) \frac{\partial u}{\partial x_j} \right) + \sum_{i=1}^{n} \frac{\partial (a_i u)}{\partial x_i} + \sum_{i=1}^{n} b_i \frac{\partial u}{\partial x_i} cu = F, \ u|_{\partial Q} = g, \tag{11.15}$$

which belongs also to $C^1(\overline{Q})$ satisfies:

a) *for every $\varphi \in C_0^1(Q)$ the following equality*

$$-\sum_{i=1}^{n} \sum_{j=1}^{n} \int_Q a_{ij}(x) \frac{\partial u}{\partial x_j} \frac{\overline{\partial \varphi}}{\partial x_i} \, dx - \sum_{i=1}^{n} \int_Q a_i u \frac{\overline{\partial \varphi}}{\partial x_i} \, dx$$

$$+ \sum_{i=1}^{n} \int_Q b_i(x) \frac{\partial u}{\partial x_i} \overline{\varphi} \, dx + \int_Q cu\overline{\varphi} \, dx = \int_Q F\overline{\varphi} \, dx. \tag{11.16}$$

b) for every $\varphi \in \overset{o}{W}{}^1 (Q)$ the equality (11.16);

c) the trace of the function u on ∂Q (see Remark 11.17.1) is equal to the function g.

Remark 11.24.1 (The generalized solution) Let $a_{ij}, a_i, b_i, c \in L_\infty(Q), a_{ij} = a_{ji}, F \in L_2(Q), g \in L_2(\partial Q)$ in (11.15) (the generalized Dirichlet problem). A function $u \in W^1(Q)$ for which the equality (11.16) holds for every $\varphi \in \overset{o}{W}{}^1 (Q)$ and its trace on the boundary ∂Q coincides with g is called the generalized solution of the problem (11.15).

Exercise 11.25 *Prove:*

a) that the classical solution $u \in C^1(\overline{Q})$ of the problem (11.15) is also a generalized solution;

b) that the generalized solution u of the problem (11.15) which satisfies $u \in C^1(\overline{Q}) \cap C^2(Q)$ is also the classical solution.

Hints. a) follows by Exercise 11.24 a), c).

Example 11.26 *Formulate the generalized Dirichlet problem for $u|_{\partial Q} = 0$.*

Solution. Since the conditions $u \in W^1(Q)$ and $u|_{\partial Q} = 0$ are equivalent with $u \in \overset{o}{W}{}^1 (Q)$, we obtain that the Dirichlet problem (11.15) for $g = 0$ reduces on the problem of finding a function $u \in \overset{o}{W}{}^1 (Q)$ which satisfies (11.16) for every $\varphi \in \overset{o}{W}{}^1 (Q)$.

Example 11.27 *Which of the following differential operators are uniform elliptic*

a) $-\Delta = -\sum_{i=1}^n D_i^2$; c) $D_1 - D_2^2$;

b) $-(x_1 D_1^2 + D_2^2)$; d) $(-1)^k \Delta^k$?

Solution.

a) The equality

$$-\sum_{i=1}^n z_i(-1)z_i = \sum_{i=1}^n z_i^2,$$

the differential operator $-\Delta$ is uniform elliptic.

b) The operator $-(x_1 D_1^2 + D_2^2)$ is uniform elliptic on $\{(x_1, x_2)| x_1 \geq a\}$ for $a > 0$, but not for $a = 0$.

c) Since $z_2 \cdot 1 \cdot z_2 \leq |z|^2 = z_1^2 + z_2^2$ the heat transfer operator is not uniform elliptic.

d) This operator is uniform elliptic.

Example 11.28 *Let us consider the following PDE on a bounded region Q*

$$\sum_{i=1}^{n}\sum_{j=1}^{n} D_i(a_{ij} D_j u) + cu = F, \qquad (11.17)$$

where $a_{ij}, c \in C^{\infty}(\overline{Q})$, $a_{ij} = a_{ji}$, $F \in L_2(Q)$ and $c(x) \le 0$ $(x \in Q)$ and the differential operator in the equation (11.17) is uniform elliptic. Prove that

a) the bilinear functional

$$(f|g)_E = \sum_{i=1}^{n}\sum_{j=1}^{n}\int_Q a_{ij}(D_i f)(D_j \overline{g})\,dx - \int_Q f\overline{g}\,dx \qquad (f,g \in \overset{o}{W}{}^{1}(Q))$$

is a scalar product on the space $\overset{o}{W}{}^{1}(Q)$ which induce a norm equivalent to the norm induced by the usual scalar product $(f|g)_{\overset{o}{W}{}^{1}(Q)}$;

b) the functional

$$h(\varphi) = -\int_Q \overline{F}\varphi\,dx \qquad (\varphi \in \overset{o}{W}{}^{1}(Q))$$

is linear and continuous on $\overset{o}{W}{}^{1}(Q)$ and there exists a function h_W from $\overset{o}{W}{}^{1}(Q)$ such that $h(\varphi) = (\varphi|h_W)_E$ and for some $M > 0$ we have

$$(h_W|h_W)_E = \|h\|^2 \le M^2 \|F\|^2_{L_2(Q)}.$$

c) there exists a unique solution $u \in \overset{o}{W}{}^{1}(Q)$ of the generalized Dirichlet problem for the equation (11.17), i.e., we have for every $\varphi \in \overset{o}{W}{}^{1}(Q)$

$$-\sum_{i=1}^{n}\sum_{j=1}^{n}\int_Q a_{ij}(D_i u)(D_j \overline{\varphi})\,dx + \int_Q f\overline{\varphi}\,dx = \int_Q F\overline{\varphi}\,dx.$$

d) there exists $C > 0$ such that

$$\|u\|_{\overset{o}{W}{}^{1}(Q)} \le C\|F\|_{L_2(Q)}.$$

Solution.

a) It is easy to check that

$$(f|g)_E = \sum_{i=1}^{n}\sum_{j=1}^{n}\int_Q a_{ij}(D_i f)(D_j \overline{g})\,dx - \int_Q cf\overline{g}\,dx \qquad (f,g \in \overset{o}{W}{}^{1}(Q))$$

is a scalar product on the space $\overset{\circ}{W}{}^1(Q)$ (the property $(f|f) \geq 0$ will follow from (11.18)).

We shall prove that the induced norm $\|f\|_E = \sqrt{(f|f)_E}$ is equivalent to the usual norm

$$\|f\|_{\overset{\circ}{W}{}^1(Q)} = \left(\int_Q |f|^2\, dx + \sum_{i=1}^n \int_Q |D_i f|^2\, dx\right)^{\frac{1}{2}}.$$

Since the differential operator in the equation (11.17) is uniform elliptic and $c \leq 0$ we obtain

$$\|f\|_E^2 = (f|f)_E = \sum_{i=1}^n \sum_{j=1}^n \int_Q a_{ij}(D_j f)(\overline{D_i f})\, dx - \int_Q c|f|^2\, dx$$

$$\geq C_1 \int_Q \sum_{j=1}^n |D_j f|^2\, dx - \int_Q c|f|^2\, dx \geq C_1 \int_Q \sum_{j=1}^n |D_j f|^2\, dx. \tag{11.18}$$

Since by Example 9.14 d) the norm

$$\|f\|_1 = \sqrt{\sum_{j=1}^n \int_Q |D_j f|^2\, dx}$$

is equivalent to the usual norm $\|f\|_{\overset{\circ}{W}{}^1(Q)}$ there exists $C_2 > 0$ such that

$$\|f\|^2 \geq C_2 \|f\|_{\overset{\circ}{W}{}^1(Q)}^2.$$

Therefore by (11.18) we have

$$\|f\|_E^2 \geq C_1 C_2 \|f\|_{\overset{\circ}{W}{}^1(Q)}^2. \tag{11.19}$$

We shall prove that there exists a constant $K > 0$ such that

$$\|f\|_E^2 \leq K \|f\|_{\overset{\circ}{W}{}^1(Q)}^2.$$

Since $a_{ij}, c \in C^\infty(\overline{Q})$ thre exists $C_1' > 0$ such that $|a_{ij}(x)| \leq C_1'$ and $|c(x)| \leq C_1'$ for $x \in \overline{Q}$. Therefore we have by the Cauchy–Schwartz inequality

$$\|f\|_E = \sum_{i=1}^n \sum_{j=1}^n \int_Q a_{ij}(D_j f)(\overline{D_i f})\, dx - \int_Q c|f|^2\, dx$$

$$\leq C_1' \sum_{i=1}^n \sum_{j=1}^n \int_Q |D_j f||D_i f|\, dx + C_1' \int_Q |f|^2\, dx$$

$$\leq C_1' \sum_{i=1}^n \sum_{j=1}^n \left(\int_Q |D_j f|^2\, dx\right)^{\frac{1}{2}} \cdot \left(\int_Q |D_i f|^2\, dx\right)^{\frac{1}{2}} + C_1'\int_Q |f|^2\, dx$$

$$= C_1' \left(\sum_{i=1}^n \left(\int_Q |D_i f|^2\, dx\right)^{\frac{1}{2}}\right)^2 + C_1' \int_Q |f|^2\, dx.$$

Therefore, using the inequality $(\sum_{i=1}^{n} s_i)^2 \leq n \sum_{i=1}^{n} s_i^2$ for real numbers $s_i \geq 0$, $(i = 1, \ldots, n)$, we obtain

$$\|f\|_E \leq C_1'n \sum_{i=1}^{n} \int_Q |D_if|^2 \, dx + C_1' \int_Q |f|^2 \, dx \leq K\|f\|_{\overset{\circ}{W}^1(Q)}$$

for $K \geq C_1'n$.

b) It is obvious that h is a linear functional on the space $\overset{\circ}{W}^1(Q)$. We shall show that it is a bounded functional. Using the Cauchy-Schwartz inequality and (11.19) we can find a constant $M > 0$ such that

$$|h(\varphi)| = |\int_Q \overline{F}\varphi \, dx| \leq \|F\|_{L_2(Q)}\|\varphi\|_{L_2(Q)} \leq \|F\|_{L_2(Q)}\|\varphi\|_{\overset{\circ}{W}^1(Q)} \leq \|F\|_{L_2(Q)}\|\varphi\|_E.$$
$$(11.20)$$

Therefore applying Riesz representation theorem on the scalar product $(\varphi|\varphi)_E$ we obtain that there exists a unique $h_W \in \overset{\circ}{W}^1(Q)$ such that $h(\varphi) = (\varphi|h_W)$. Then by (11.20) we obtain

$$(h_W|h_W)_E^2 = \|h\|^2 \leq M^2\|F\|_{L_2(Q)}^2. \qquad (11.21)$$

c) Using the scalar product introduced in a) we can rewrite the Dirichlet problem for the equation (11.17) in the following form

$$(u|\varphi)_E = -\int_Q F\overline{\varphi} \, dx \qquad (\varphi \in \overset{\circ}{W}^1(Q)).$$

By b) there exists $h_W \in \overset{\circ}{W}^1(Q)$ such that the previous equality can be rewritten in the form

$$(u|\varphi)_E = \overline{(\varphi|h_W)}, \text{ i.e., } (u - h_W|\varphi) = 0$$

for every $\varphi \in \overset{\circ}{W}^1(Q)$. Hence $u - h_W = 0$, i.e., $u = h_W$, what means that the considered generalized Dirichlet problem has a unique solution $u = h_W$.

d) The inequality (11.21) implies

$$\|u\|_{\overset{\circ}{W}^1(Q)} = \|h_W\|_{\overset{\circ}{W}^1(Q)} \leq M\|h_W\|_E \leq M^2\|F\|_{L_2(Q)}.$$

Remark 11.28.1 By Example 11.28 and Exercise 11.24 a) we obtain also the uniqueness of the classical solution for $F \in L_2(Q)$ for a locally quadratic region , in $C^1(\overline{Q})$.

Exercise 11.29 *Prove that for the generalized Dirichlet problem*

$$-\sum_{i=1}^{n}\sum_{j=1}^{n}\int_{Q}a_{ij}(D_{j}u)(D_{i}\overline{\varphi})\,dx + \int_{Q}cu\overline{\varphi}\,dx = \int_{Q}F\overline{\varphi}\,dx \quad (\varphi \in \overset{o}{W}{}^{1}(Q)), \quad (11.22)$$

$$u|_{\partial Q} = g, \quad (11.23)$$

for $a_{ij}, c \in C^{\infty}(\overline{Q}), a_{ij} = a_{ji}, F \in L_{2}(Q), c(x) \leq 0 \quad (x \in Q)$ and $g \in L_{2}(\partial Q)$ such a function that there exists $g_{W} \in W^{1}(Q)$ such that the trace $g_{W}|_{\partial Q} = g$, always there exists a unique generalized solution $u \in W^{1}(Q)$ and a constant $M > 0$ such that

$$\|u\|_{W^{1}(Q)} \leq M(\|F\|_{L_{2}(Q)} + \inf_{g_{W}|_{\partial Q}=g}\|g_{W}\|_{W^{1}(Q)}).$$

Hints. Use the preceding Example 11.28, since $u \in W^{1}(Q)$ is a solution of the generalized Dirichlet problem (11.22) and (11.23) if and only if the function $u_{1} = u - g_{W}$ is the generalized solution from the space $\overset{o}{W}{}^{1}(Q)$ of the equation

$$-\sum_{i=1}^{n}\sum_{j=1}^{n}\int_{Q}a_{ij}(D_{j}u_{1})(D_{i}\overline{\varphi})\,dx + \int_{Q}cu_{1}\overline{\varphi}\,dx$$

$$= \int_{Q}F\overline{\varphi}\,dx + \sum_{i=1}^{n}\sum_{j=1}^{n}\int_{Q}a_{ij}(D_{j}g_{W})(D_{i}\overline{\varphi})\,dx - \int_{Q}cg_{W}\overline{\varphi}\,dx \quad (\varphi \in \overset{o}{W}{}^{1}(Q)).$$

Taking the right side as a linear continuous functional $-h$ we can obtain the solution in an analogous way as in Example 11.28.

Example 11.30 *Prove that*

a) *for every bounded sequence $\{f_{m}\}_{m\in\mathbb{N}}$ of functions from the space $W^{1}(P)$, where $P = (a_{1}, b_{1}) \times \cdots \times (a_{n}, b_{n})$, there exists its subsequence which is convergent in the space $L_{2}(Q)$;*

b) *for a bounded region $Q \subset \mathbf{R}^{n}$ every bounded sequence $\{f_{m}\}_{m\in\mathbb{N}}$ from the space $\overset{o}{W}{}^{1}(Q)$ there exists its subsequence which is convergent in the space $L_{2}(Q)$, i.e., the embedding map of the space $\overset{o}{W}{}^{1}(Q)$ into the space $L_{2}(Q)$ is a compact operator;*

c) *for a locally quadratic bounded region $Q \subset \mathbf{R}^{n}$ the embedding map of the space $W^{1}(Q)$ into the space $L_{2}(Q)$ is a compact operator.*

Solution.

a) We divide the parallelepiped $P = (a_1, b_1) \times \cdots \times (a_n, b_n)$ on smaller parallelepipeds with sides

$$\frac{b_1 - a_1}{s}, \ldots, \frac{b_n - a_n}{s},$$

for some $s \in \mathbf{N}$. The total number of these parallelepipeds is s^n, and therefore we denote them by $P_1, \ldots, P_{s^n}$.

We shall show that the sequence $\{f_m\}_{m \in \mathbf{N}}$ has a Cauchy subsequence in the space $L_2(P)$. For that purpose we apply the Poincaire inequality from Chapter 9. on the function $f_m - f_k$ on one of the parallelepipeds, say P_j

$$\int_{P_j} |f_m - f_k|^2 \, dx \;\leq\; \frac{s^n}{(b_1^j - a_1^j) \cdots (b_n^j - a_n^j)} |\int_{P_j} (f_m - f_k) \, dx|^2$$

$$+ \frac{n}{2} \int_{P_j} \sum_{i=1}^{n} (\frac{b_i - a_i}{s})^2 |D_i f_m - D_i f_k|^2 \, dx \quad (j = 1, \ldots, s^n).$$

Adding all these inequalities we obtain

$$\int_{P} |f_m - f_k|^2 \, dx \leq \sum_{j=1}^{s^n} \frac{s^n}{(b_1^j - a_1^j) \cdots (b_n^j - a_n^j)} |\int_{P_j} (f_m - f_k) \, dx|^2$$

$$+ \frac{n}{2s^2} \int_{P} \sum_{i=1}^{n} (b_i - a_i)^2 |D_i f_m - D_i f_k|^2 \, dx. \qquad (11.24)$$

By the boundedness of the sequence $\{f_m\}_{m \in \mathbf{N}}$ there exists a constant $C > 0$ such that

$$\int_{P} \sum_{i=1}^{n} (b_i - a_i)^2 |D_i f_m - D_i f_k|^2 \, dx < C.$$

Therefore for every $\varepsilon > 0$ there exists s_0 (the number from the dividing the parallelepiped) such that

$$\frac{n}{2s_0} \int_{P} \sum_{i=1}^{n} (b_i - a_i)^2 |D_i f_m - D_i f_k|^2 \, dx < \frac{\varepsilon}{2}. \qquad (11.25)$$

On the other side, since the sequence $\{f_m\}_{m \in \mathbf{N}}$ is bounded in the space $W^1(P)$ it is bounded also in the space $L_2(Q)$. By Example 10.1.2 on weak compactness, there exists a subsequence $\{f_{r_m}\}_{m \in \mathbf{N}}$ of $\{f_m\}_{n \in \mathbf{N}}$ which weakly converges in the space $L_2(P)$, i.e., the sequence of numbers $\{\int_{P} f_{r_m} \chi_j \, dx\}_{m \in \mathbf{N}}$ $(j = 1, \ldots, s^n)$ converges, where χ_j is the characteristic function of the parallelepiped P_j. Therefore there exists $m_0 \in \mathbf{N}$ such that for every $m \geq m_0$

$$|\int_{P_j} (f_{r_m} - f_{r_k}) \, dx| = |\int_{P} (f_{r_m} - f_{r_k}) \chi_j \, dx| < \frac{\varepsilon}{2M}, \qquad (11.26)$$

where

$$M = \sum_{j=1}^{s^n} \frac{s^n}{(b_1^j - a_1^j) \cdots (b_n^j - a_n^j)}.$$

Applying the inequalities (11.24), (11.25) and (11.26) on the subsequence $\{f_{r_m}\}_{n \in N}$ we obtain

$$\int_P |f_{r_m} - f_{r_k}|^2 \, dx < \varepsilon \quad \text{for} \quad m, k \geq m_0.$$

Hence $\{f_{r_m}\}_{m \in N}$ is a Cauchy sequence in the space $L_2(Q)$ and therefore also a convergent sequence.

b), c) It is enough to take a parallelepiped $P \supset Q$, since the bounded sequence can be extended with $\{F_m\}_{m \in N}$ on P such that it remains bounded on P (taking $F_m(x) = 0$ for $x \in P \setminus Q$ and $F_m(x) = f_m(x)$ for $x \in Q$).

Example 11.31 *Prove that the problem of finding the generalized eigenvalues for homogeneous boundary problem for uniform elliptic equation*

$$\sum_{i=1}^{n} \sum_{j=1}^{n} D_i(a_{ij} D_j u) + cu = F,$$

where $a_{ij}, c \in C^\infty(Q), a_{ij} = a_{ji}, c(x) \leq 0 \ (x \in Q), F \in L_2(Q)$, which consists in finding nontrivial solutions $u \in \overset{\circ}{W}^1(Q)$ for the equation

$$\sum_{i=1}^{n} \sum_{j=1}^{n} D_i(a_{ij} D_j u) + cu + \lambda u = F \tag{11.27}$$

($\lambda \in C$ is the eigenvalue) reduces on the equation

$$u - \lambda T(u) = h \tag{11.28}$$

for $h \in \overset{\circ}{W}^1(Q)$, where $T : \overset{\circ}{W}^1(Q) \to \overset{\circ}{W}^1(Q)$ is a compact selfadjoint positive operator.

Solution. The generalized problem (11.27) reduces on finding a function $u \in \overset{\circ}{W}^1(Q)$ such that

$$-\sum_{i=1}^{n} \sum_{j=1}^{n} \int_Q a_{ij}(D_j u)(D_i \bar{\varphi}) \, dx + \int_Q cu\bar{\varphi} \, dx + \lambda \int_Q u\bar{\varphi} \, dx = \int_Q F\bar{\varphi} \, dx.$$

Using the solution of Example 11.28 introducing the scalar product

$$(f|g)_E = \sum_{i=1}^{n} \sum_{j=1}^{n} \int_Q a_{ij}(D_i f)(D_j \bar{g}) \, dx - \int_Q f\bar{g} \, dx \qquad (f, g \in \overset{\circ}{W}^1(Q)) \tag{11.29}$$

and then we can rewrite the preceding equation ($h = h_W \in \overset{\circ}{W}^1(Q)$ from Example 11.28)

$$(h|\varphi)_E = (u|\varphi)_E - \lambda \int_Q u\overline{\varphi}\,dx \qquad (\varphi \in \overset{\circ}{W}^1(Q)). \qquad (11.30)$$

We shall represent $\int_Q u\overline{\varphi}\,dx$ also by the scalar product (11.29). Namely, take the linear functional h_1 for an arbitrary but fixed $u \in L_2(Q)$

$$h_1(\varphi) = \int_Q u\overline{\varphi}\,dx \qquad (\varphi \in \overset{\circ}{W}^1(Q)).$$

It is continuous, since we have

$$|h_1(\varphi)| \leq \|u\|_{L_2(Q)}\|\varphi\|_{\overset{\circ}{W}^1(Q)} \leq M\|u\|_{L_2(Q)}\|\varphi\|_E.$$

By Riesz representation theorem there exists a unique element $H_1 \in \overset{\circ}{W}^1(Q)$ such that

$$h_1(\varphi) = (\varphi|H_1)_E \qquad (11.31)$$

and

$$\|h_1\|^2 = (H_1|H_1)_E \leq M^2\|u\|_{L_2(Q)}. \qquad (11.32)$$

By the definition of the functional h_1 and (11.31) we have

$$(H_1|\varphi)_E = \int_Q u\overline{\varphi}\,dx. \qquad (11.33)$$

Let $T_1 : L_2(Q) \to \overset{\circ}{W}^1(Q)$ be an operator which maps the element u on the element H_1. By (11.33) T_1 is linear, and by (11.32) it is bounded, since we have

$$\|T_1(u)\|_E^2 = (H_1|H_1)_E \leq M^2\|u\|_{L_2(Q)}.$$

By (11.33) we obtain

$$(T_1(u)|\varphi)_E = \int_Q u\overline{\varphi}\,dx \qquad (\varphi \in \overset{\circ}{W}^1(Q)). \qquad (11.34)$$

Therefore the equation (11.30) reduces on the form

$$(u|\varphi)_E - \lambda(T_1(u)|\varphi)_E = (h|\varphi)_E \qquad (\varphi \in \overset{\circ}{W}^1(Q)).$$

Hence $u - \lambda T(u) = h$, where we denote by T the restriction of the operator T_1 to $\overset{\circ}{W}^1(Q)$.

The operator T is selfadjoint, since by (11.34) we have

$$(T(u)|\varphi)_E = \int_Q u\overline{\varphi}\,dx = \overline{\int_Q \varphi\overline{u}\,dx} = \overline{(T(\varphi)|u)_E} = (u|T(\varphi))_E.$$

On the other side, since

$$(T(u)|u)_E = \int_Q |u|^2 \, dx > 0$$

if the function u is not almost everywhere equal to zero, we obtain that the operator T is strictly positive.

We shall prove that the operator T is compact. Namely, since $T = T_1 T_0$, where $T_0 : \overset{\circ}{W}^1(Q) \to L_2(Q)$ is the embedding operator, which is by Example 11.30 compact, and $T_1 : L_2(Q) \to \overset{\circ}{W}^1(Q)$ is a bounded operator, we obtain that their composition T is a compact operator.

Example 11.32 *Let Q be a bounded region in $\mathbf{R}^n$. Prove:*

a) *that the equation (11.27) from Example 11.31 with the same conditions has countable many positive eigenvalues $\lambda_1, \lambda_2, \ldots$, which as a sequence $\{\lambda_n\}_{n \in \mathbf{N}}$ tends to $+\infty$;*

b) *that the corresponding eigenfunctions $u_1, u_2, \ldots$, form a base of the space $W^1(Q)$;*

c) *that the eigenfunctions $u_1, u_2, \ldots$, form a complete orthogonal system in the space $L_2(Q)$;*

d) *that holds*

$$(u_i | u_i) = \lambda_i \qquad (11.35)$$

for normed $\{u_i\}_{i \in \mathbf{N}}$ in $L_2(Q)$.

Solution.

a), b) By Example 11.31 the problem of eigenvalues of the equation (11.27) is reduced on the equation (11.28) with a selfadjoint compact positive operator T. Then there exist a sequence of eigenvalues $\{\alpha_i\}_{i \in \mathbf{N}}$ of the operator T which converges to zero and a sequence of eigenfunctions. Taking $\lambda_i = \frac{1}{\alpha_i}$ we obtain the desired conclusions.

c) By (11.34) we have

$$\int_Q u_i \overline{u_j} \, dx = (T(u_i)|u_j)_E = (\alpha_i u_i|u_j)_E = \alpha_i(u_i|u_j)_E = \frac{1}{\lambda_i}(u_i|u_j)_E. \quad (11.36)$$

Since for $i \neq j$ we have $(u_i|u_j)_E = 0$ we obtain by (11.36) for $i \neq j$

$$\int_Q u_i \overline{u_j} \, dx = 0,$$

i.e., $\{u_i\}_{i \in \mathbb{N}}$ is a orthogonal system of functions in the space $L_2(Q)$. The completeness of the system $\{u_i\}$ in the space $L_2(Q)$ follows by the facts that the space $\overset{\circ}{W}^1(Q)$ is dense in the space $L_2(Q)$ and that the set of all linear combinations of the sequence $\{u_i\}_{i \in \mathbb{N}}$ is dense in the space $L_2(Q)$.

d) Follows by (11.36).

Example 11.33 *Let $\{\lambda_i\}_{i \in \mathbb{N}}$ be the sequence of generalized eigenvalues and $\{u_i\}_{i \in \mathbb{N}}$ the sequence of generalized eigenfunctions as a base in $L_2(Q)$ corresponding to Example 11.32 and u is the solution of the generalized boundary problem . Prove:*

a) that for $\lambda \neq \lambda_i$ $(i \in \mathbb{N})$

$$u = \sum_{i=1}^{n} \frac{c_i}{\lambda - \lambda_i} u_i, \; where \; c_i = \int_Q F \overline{u_i} \, dx,$$

and the series converges in the space $\overset{\circ}{W}^1(Q)$;

b) that for $\lambda = \lambda_i$ $(i \in \mathbb{N})$

$$u = \sum_{j=1}^{n} \frac{c_j}{\lambda - \lambda_j} u_j + c_{i_1} u_{i_1} + \cdots + c_{i_s} u_{i_s},$$

where $u_{i_1}, \ldots, u_{i_s}$ are the eigenfunctions for $\lambda = \lambda_i$. The series converges in the space $\overset{\circ}{W}^1(Q)$.

Solution. We expand the function $u \in \overset{\circ}{W}^1(Q)$ in the following convergent series in the space $\overset{\circ}{W}^1(Q)$

$$u = \sum_{i=1}^{\infty} a_i u_i, \; where \; a_i = \frac{(u|u_i)}{(u_i|u_i)},$$

and we omit the index in the scalar product. By (11.30) and (11.34) we obtain

$$(h|\varphi) = (u|\varphi) - \lambda(T(u)|\varphi) \qquad (\varphi \in \overset{\circ}{W}^1(Q)).$$

Taking $\varphi = u_1$ and using that $(h|\varphi) = -\int_Q F\overline{\varphi} \, dx$ we obtain

$$-\int_Q F\overline{\varphi} \, dx = (u|u_i) - \lambda(T(u)|u_i). \tag{11.37}$$

Since the operator T is selfadjoint we have

$$(T(u)|u_i) = (u|T(u_i)) = (u|\frac{1}{\lambda_i} u_i) = \frac{1}{\lambda_i}(u|u_i).$$

Putting this in (11.37) we obtain

$$-\int_Q F\overline{\varphi}\,dx = (u|u_i) - \frac{\lambda}{\lambda_i}(u|u_i).$$

Therefore for $\lambda \neq \lambda_i$

$$(u|u_i) = \frac{\lambda_i}{\lambda - \lambda_i}\int_Q F\overline{u_i}\,dx.$$

This implies by $(u_i|u_i) = \lambda_i$ (see (11.35)) and the expansion $\sum_{i=1}^{\infty} a_i u_i$ the conclusions in a) and b).

11.2 The Generalized Mixed Problems

11.2.1 Examples and Exercises

Example 11.34 *Let $Q \subset \mathbf{R}^n$ ($n \geq 2$) be a bounded region and $S \subset \overline{Q}$ a locally quadratic $(n-1)$-dimensional surface with the property that for every point $x_0 \in S$ there exist regions $U(x_0)$ and V and a C^1-dipheomorphism Φ from U onto $\overline{V}$ such that a subset of $Q \cap U$ is mapped on a n- dimensional parallelepiped $P \subset V$, and $S \cap \overline{U}$ is mapped on a side or union of sides of P.*
Prove that:

    a) there exist constants $C_1 > 0$ and $C_2 > 0$ such that for every $f \in W^1(Q)$ and every $\varepsilon, 0 < \varepsilon < 1$,

$$\int_S |f|_S|^2\,dS \leq \frac{C_1}{\varepsilon}\int_Q |f|^2\,dx + C_2\varepsilon\sum_{j=1}^{n}\int_Q |D_j f|^2\,dx, \tag{11.38}$$

    i.e.,

$$\|f|_S\|_{L_2(S)} \leq C_3\|f\|_{W^1(Q)}.$$

    b) If additionally Q is a locally quadratic region, then the operator $T : W^1(Q) \to L_2(Q)$ given by $T(f) = f|_S$ is compact.

Solution.

    a) There exists a sequence $\{f_k\}_{k\in\mathbf{N}}$ from $C^1(\overline{Q})$ such that

$$f_k \xrightarrow{W^1(Q)} f. \tag{11.39}$$

    By the definition of the trace of the function f on S (see Chapter 9.)

$$f_k \xrightarrow{L_2(S)} f|_S. \tag{11.40}$$

We have (see Chapter 9.)

$$\int_S |f_k|^2 \, dS \leq \frac{C_1}{\varepsilon} \int_Q |f_k|^2 \, dx + C_2 \varepsilon \sum_{j=1}^{n} |D_j f_k|^2 \, dx.$$

Letting $k \to \infty$ we obtain by (11.39) and (11.40) the inequality (11.38).

b) We have to prove that the operator T maps every bounded sequence $\{f_k\}_{k \in \mathbb{N}}$ from $W^1(Q)$ on a sequence from the space $L_2(Q)$ which has a convergent subsequence. Since the embedding operator of the space $W^1(Q)$ into the space $L_2(Q)$ is compact there exists a subsequence $\{f_{n_k}\}_{k \in \mathbb{N}}$ of the sequence $\{f_k\}_{k \in \mathbb{N}}$ which is convergent in the space $L_2(Q)$. We shall prove that $\{f_{n_k}|_S\}_{k \in \mathbb{N}}$ is the desired subsequence, which converges in $L_2(Q)$.
By (11.38) we have

$$\int_S |f_{n_k} - f_{n_m}|^2 \, dS \leq \frac{C_1}{\varepsilon} \int_Q |f_{n_k} - f_{n_m}|^2 \, dx + C_2 \varepsilon \sum_{j=1}^{n} |D_j (f_{n_k} - f_{n_m})|^2 \, dx. \quad (11.41)$$

Since $\{f_{n_k}\}_{k \in \mathbb{N}}$ is a bounded sequence in the space $W^1(Q)$ we have that for every $\delta > 0$ there exists $\varepsilon' > 0$ such that

$$C_2 \varepsilon' \sum_{j=1}^{n} |D_j (f_{n_k} - f_{n_m})|^2 \, dx < \frac{\delta}{2} \qquad (k, m \in \mathbb{N}). \qquad (11.42)$$

On the other side, since $\{f_{n_k}\}_{k \in \mathbb{N}}$ is a convergent sequence in the space $L_2(Q)$ there exists $k_0 \in \mathbb{N}$ such that

$$\frac{C_1}{\varepsilon} \int_Q |f_{n_k} - f_{n_m}|^2 \, dx < \frac{\delta}{2} \text{ for } k, m \geq k_0. \qquad (11.43)$$

We obtain by (11.41), (11.42) and (11.43)

$$\int_S |f_{n_k} - f_{n_m}|^2 \, dS < \varepsilon.$$

Since $L_2(S)$ is a complete space, we have that $\{f_{n_k}|_S\}_{n \in \mathbb{N}}$ converges in the space $L_2(S)$.

Example 11.35 *Let Q be a locally quadratic bounded region in $\mathbb{R}^n$ and for $0 < T < +\infty$ we introduce*

$$\Gamma_T = \partial Q \times [0, T), \quad Q_0 = Q \times \{0\}, \quad Q_T = Q \times \{T\}.$$

We suppose that for the uniform elliptic differential operator

$$L(v) = \sum_{i=1}^{n} \sum_{j=1}^{n} D_i(a_{ij} D_j v) + cv$$

for $a_{ij} = a_{ji} \in C^1(\overline{Q}), c \in C(\overline{Q}), c \leq 0$, there are known all eigenvalues $\{\lambda_i\}_{i \in \mathbb{N}}$ and eigenfunctions $\{v_i\}_{i \in \mathbb{N}}$ for the classical problem

$$L(v) + \lambda v = 0 \text{ on } Q \tag{11.44}$$

$$v|_{\partial Q} = 0. \tag{11.45}$$

Using the Fourier method of separation of variables construct the classical solution (supposing that there exists)

$$u \in C^2(Q \times (0, T)), \quad u, D_t u \in C(Q \times (0, T) \cup \overline{Q_0} \cup \Gamma_T)$$

with $F \in C(Q \times (0, T) \cup \overline{Q_0} \cup \Gamma_T)$ for mixed problem

$$D_t^2 u - L(u) = F \text{ on } Q \times (0, T) \tag{11.46}$$

with initial conditions

$$u|_{\overline{Q_0}} = f \in C(\overline{Q_0}), \tag{11.47}$$

$$D_t u|_{\overline{Q_0}} = g \in C(\overline{Q_0}), \tag{11.48}$$

with boundary condition

$$u|_{\Gamma_T} = r \in C(\Gamma_T), \tag{11.49}$$

as a series

$$u(x, t) = \sum_{i=1}^{\infty} T_i(t) v_i(x),$$

supposing the "good" convergence of the series, i.e., that we can exchange the order of the infinite sum and double differentiation with respect to t.

Solution. Since $u \in C(Q \times (0, T) \cup \overline{Q_0} \cup \Gamma_T)$ we can expand the function $x \mapsto u(x, t)$ into the Fourier series with respect to the eigenfunctions $\{v_i\}_{i \in \mathbb{N}}$ on Q for an arbitrary but fixed $t, 0 \leq t < T$,

$$u(x, t) = \sum_{i=1}^{\infty} T_i(t) v_i(x) \tag{11.50}$$

and this series converges in the space $L_2(Q)$ for every fixed t. Applying the operator L on (11.50) we obtain by (11.44)

$$(L(u))(x, t) = \sum_{i=1}^{\infty} T_i(t) L(v_i)(x) = - \sum_{i=1}^{\infty} T_i(t) \lambda_i v_i(x). \tag{11.51}$$

Applying the operator D_t^2 on (11.50) we obtain

$$D_t^2 u(x, t) = \sum_{i=1}^{\infty} T_i''(t) v_i(x). \tag{11.52}$$

We expand the function $x \mapsto F(x,t) \in C(\overline{Q})$ also in a Fourier series for an arbitrary but fixed $t, 0 \le t < T$,

$$F(x,t) = \sum_{i=1}^{\infty} a_i(t)v_i(x), \text{ where } a_i(t) = \int_Q F(x,t)\overline{v_i(x)}\, dx. \qquad (11.53)$$

Putting (11.51), (11.52) and (11.53) in (11.46) we obtain

$$\sum_{i=1}^{\infty} T_i''(t)v_i(x) + \sum_{i=1}^{\infty} \lambda_i T_i(t)v_i(x) = \sum_{i=1}^{\infty} a_i(t)v_i(x).$$

The preceding equality holds if and only if

$$T_i''(t) + \lambda_i T_i(t) = a_i(t) \qquad (t \in (0,T), i \in \mathbf{N}). \qquad (11.54)$$

We have by the initial conditions (11.47) and (11.48)

$$\sum_{i=1}^{\infty} T_i(0)v_i(x) = u(0,x) = f(x) = \sum_{i=1}^{\infty} b_i v_i(x),$$

$$\sum_{i=1}^{\infty} T_i'(0)v_i(x) = D_t u(0,x) = g(x) = \sum_{i=1}^{\infty} c_i v_i(x),$$

where we have taken the expansions of the functions f and g with respect to the eigenfunctions $\{v_i\}_{i \in \mathbf{N}}$. Hence

$$T_i(0) = b_i, \ T_i'(0) = c_i \qquad (i \in \mathbf{N}), \qquad (11.55)$$

where

$$b_i = \int_Q f\overline{v_i}\, dx \text{ and } c_i = \int_Q g\overline{v_i}\, dx.$$

The initial problem (11.54) and (11.55) for $\lambda = \lambda_i \ne 0$ has the following solution

$$T_i(t) = b_i \cos(\sqrt{\lambda_i}t) + \frac{c_i}{\sqrt{\lambda_i}} \sin(\sqrt{\lambda_i}t) + \frac{1}{\sqrt{\lambda_i}} \int_0^t a_i(\tau) \sin(\sqrt{\lambda_i}(t - \tau))\, d\tau \qquad (11.56)$$

and for $\lambda_1 = 0$

$$T_1(t) = b_1 + c_1 t + \int_0^t a_1(\tau)(t - \tau)\, d\tau. \qquad (11.57)$$

So we have obtained the classical solution of the mixed problem (11.46),(11.47),(11.48) and (11.49) as a series (11.50), where T_i is given by (11.56) and (11.57).

Example 11.36 *Let Q, Γ_T, Q_0, Q_T and L be same as in Example 11.35. Let $a_{ij} = a_{ji}, c \in C^1(\overline{Q}), c \le 0$ and $f, g \in L_2(Q), F \in L_2(Q \times (0,T))$. We shall call a function*

$u \in W^1(Q \times (0,T))$ the solution of the generalized mixed problem for the hyperbolic equation (11.46)- (11.49) for $r = 0$ if it satisfies

$$\int_{Q \times (0,T)} (\sum_{i=1}^{n} \sum_{j=1}^{n} a_{ij}(x)(D_j u)(D_i \overline{\varphi}) - (D_t u)(D_t \overline{\varphi}) - cu\overline{\varphi}) \, dx \, dt$$

$$= \int_{Q \times (0,T)} F\overline{\varphi} \, dx \, dt + \int_{\overline{Q_0}} g\overline{\varphi} \, dS, \qquad (11.58)$$

for every function $\varphi \in W^1(Q \times (0,T))$ whose traces on $\overline{Q_T}$ and $\overline{\Gamma_T}$ satisfy

$$\varphi|_{\overline{Q_T}} = 0 \text{ and } \varphi|_{\overline{\Gamma_T}} = 0, \qquad (11.59)$$

and

$$u|_{\overline{Q_T}} = f \text{ and } u|_{\overline{\Gamma_T}} = 0. \qquad (11.60)$$

Prove that

a) $a_i \in L_2(0,T)$ for $a_i(t) = \int_Q F(x,t)\overline{v_i(x)} \, dx$ $\quad (i \in \mathbf{N})$;

b) functions $T_i v_i$ belong to the space $W^1(Q \times (0,T))$ and

$$T_i v_i|_{\overline{Q_0}} = b_i v_i, \quad T_i v_i|_{\overline{\Gamma_T}} = 0 \text{ and } T_i(0) = c_i;$$

c) the function $T_i v_i$ satisfies the generalized mixed problem for $F = a_i v_i$, $f = b_i v_i$, $g = c_i v_i$ and $r = 0$;

d) the function $u_k = \sum_{i=1}^{k} T_i v_i$ $\quad (k \in \mathbf{N})$ satisfies the generalized mixed problem for

$$F_k = \sum_{i=1}^{k} a_i v_i, \quad f_k = \sum_{i=1}^{k} b_i v_i, \quad g_k = \sum_{i=1}^{k} c_i v_i \text{ and } r = 0;$$

e) if the series $\sum_{i=1}^{\infty} T_i v_i$ converges in the space $W^1(Q \times (0,T))$ to a function u, then u is the solution of the generalized mixed problem.

Solution.

a) Since

$$F \in L_2(Q \times (0,T)), \quad v_i \in L_2(Q) \text{ and } a_i(t) = \int_Q F(x,t)\overline{v_i(x)} \, dx \quad (i \in \mathbf{N}).$$

we have by Cauchy-Schwartz inequality

$$|a_i(t)| \le (\int_Q |F(x,t)|^2 \, dx)^{\frac{1}{2}} (\int_Q |v_i(x)|^2 \, dx)^{\frac{1}{2}}.$$

Then by Fubini theorem

$$\int_0^T |a_i(t)|^2 \, dt \le \int_{Q \times (0,T)} |F(x,t)|^2 \, dx \, dt \int_Q |v_i(x)|^2 \, dx,$$

and so $a_i \in L_2(0,T)$ for $i \in \mathbf{N}$.

b) We have for $\lambda_i \neq 0$ (see Example 11.35)

$$T_i(t) = b_i \cos(\sqrt{\lambda_i}t) + \frac{c_i}{\sqrt{\lambda_i}} \sin(\sqrt{\lambda_i}t) + \frac{1}{\sqrt{\lambda_i}} \int_0^t a_i(\tau) \sin(\sqrt{\lambda_i}(t-\tau)) \, d\tau, \quad (11.61)$$

and for $\lambda_1 = 0$

$$T_1(t) = b_1 + c_1 t + \int_0^t a_1(\tau)(t - \tau) \, d\tau. \quad (11.62)$$

Since $a_i \in L_2(0,T)$ there exists a sequence of functions $\{a_i^k\}_{k \in \mathbb{N}}$ from $C[0,T]$ such that $a_i^k \overset{L_2(0,T)}{\longrightarrow} a_i$ as $k \to \infty$. We have for $\lambda_i \neq 0$

$$T_{ik}(t) = b_i \cos(\sqrt{\lambda_i}t) + \frac{c_i}{\sqrt{\lambda_i}} \sin(\sqrt{\lambda_i}t) + \frac{1}{\sqrt{\lambda_i}} \int_0^t a_i^k(\tau) \sin(\sqrt{\lambda_i}(t - \tau)) \, d\tau,$$

and so $T_{ik} \in C^2[0,T]$, and for $\lambda_1 = 0$

$$T_{1k}(t) = b_1 + c_1 t + \int_0^t a_1^k(\tau)(t - \tau) \, d\tau \in C^2[0,T] \text{ and } T_{ik}(0) = b_i.$$

We have for $\lambda_i \neq 0$

$$T_{ik}'(t) = -b_i \sqrt{\lambda_i} \sin(\sqrt{\lambda_i}t) + c_i \cos(\sqrt{\lambda_i}t) + \int_0^t a_i^k(\tau) \cos(\sqrt{\lambda_i}(t - \tau)) \, d\tau,$$

and for $\lambda_1 = 0$

$$T_{1k}'(t) = c_1 + \int_0^t a_1^k(\tau) \, d\tau.$$

Applying once more the preceding procedure we obtain

$$T_{ik}''(t) = -b_i \sqrt{\lambda_i} \cos(\sqrt{\lambda_i}t) - c_i \sin(\sqrt{\lambda_i}t) - \sqrt{\lambda_i} \int_0^t a_i^k(\tau) \sin(\sqrt{\lambda_i}(t-\tau)) \, d\tau + a_i^k(t)$$

and for $\lambda_1 = 0$

$$T_{1k}''(t) = a_1^k(t).$$

Letting $k \to \infty$ in the preceding equalities and using the Cauchy-Schwartz inequality we obtain $T_i \in W^2(0,T)$ and

$$T_{ik} \overset{W^2(0,T)}{\longrightarrow} T_i \text{ as } k \to \infty \quad (i \in \mathbf{N}). \quad (11.63)$$

Hence for $\lambda_i \neq 0$

$$T_i'(t) = -b_i \lambda_i \sin(\sqrt{\lambda_i}t) + c_i \cos(\sqrt{\lambda_i}t) + \int_0^t a_i(\tau) \cos(\sqrt{\lambda_i}(t - \tau)) \, d\tau, \quad (11.64)$$

and for $\lambda_1 = 0$

$$T_1'(t) = c_1 + \int_0^t a_1(\tau) \, d\tau, \quad (11.65)$$

and

$$T_i''(t) = -b_i\lambda_i \cos(\sqrt{\lambda_i}t) - c_i\sqrt{\lambda_i}\sin(\sqrt{\lambda_i}t) - \sqrt{\lambda_i}\int_0^t a_i(\tau)\sin(\sqrt{\lambda_i}(t-\tau))\,d\tau + a_i(t)$$
(11.66)

and for $\lambda_1 = 0$

$$T_1''(t) = a_1(t). \tag{11.67}$$

Since for the function $v_i \in W^1(Q)$ there exists a sequence $\{v_{ik}\}_{k\in\mathbf{N}}$ from $C^1(\overline{Q})$ such that

$$v_{ik} \overset{W^1(Q)}{\longrightarrow} v_i \text{ as } k \to \infty,$$

we conclude by (11.63) that

$$T_i v_i \in W^1(Q \times (0,T)) \text{ and } T_{ik}v_{ik} \overset{W^1(Q\times(0,T))}{\longrightarrow} T_i v_i \text{ as } k \to \infty.$$

Since $T_{ik} \in C^2[0,T]$ and $T_{ik}(0) = b_i$ we have

$$T_{ik}v_{ik}|_{\overline{Q_0}} = b_i v_i.$$

We obtain in an analogous way

$$T_i v_i|_{\overline{\Gamma_T}} = 0.$$

Since $T_i \in W^2(0,T)$ we obtain by Example 9.39 $T_i \in C^1[0,T]$ and then by (11.64) and (11.65)

$$T_i'(0) = c_i. \tag{11.68}$$

c) Since $\{v_k\}_{k\in\mathbf{N}}$ are the generalized eigenvalues for the problem

$$Lv + \lambda v = 0 \text{ on } Q \text{ and } v|_{\partial Q} = 0,$$

we have (see Examples 11.31, 11.32) for every $\varphi \in \overset{\circ}{W}{}^k(Q)$

$$-\sum_{i=1}^n \sum_{j=1}^n \int_Q a_{ij}(D_j v_k)(D_i\overline{\varphi})\,dx + \int_Q cv_k\overline{\varphi}\,dx + \lambda\int_Q v_k\overline{\varphi}\,dx = 0. \tag{11.69}$$

On the other side we have by (11.66), (11.67), (11.61) and (11.62) for $T_i \in W^2(0,T)$

$$T_k'' = -\lambda_k T_k + a_k.$$

By partial integration (see Example 9.40) we obtain

$$\int_0^T (-\lambda_k T_k + a_k)\overline{\varphi}\,dt = \int_0^T T_k''\overline{\varphi}\,dt = T_k'(T)\varphi(T) - \int_0^T T_k'\overline{\varphi'}\,dt \quad (\varphi \in W^1(0,T)).$$
(11.70)

We take now a function $\varphi \in W^1((0,T) \times Q)$ such that it satisfies (11.59). By Remark 2 from Example 9.28 there exists a sequence $\{\varphi_m\}_{m \in \mathbf{N}}$ of functions from $C^1(\overline{(0,T) \times Q})$ such that

$$\varphi_m \xrightarrow{W^1(Q \times (0,T))} \varphi \text{ and } \varphi_m|_{\Gamma_T} = 0.$$

Since the trace of the function $x \mapsto \varphi_m(x,t)$, which belongs to $C^1(\overline{Q})$, on ∂Q is zero and the function $t \mapsto \varphi_m(x,t)$ belongs to $C^1[0,T]$ we have by (11.69) and (11.70)

$$I = \int_{(0,T) \times Q} \Big(\sum_{i=1}^n \sum_{j=1}^n \int_Q a_{ij} D_j(T_k v_k)(D_i \overline{\varphi_m}) - D_t(T_k v_k)(D_t \overline{\varphi_m})$$

$$- c T_k v_k \overline{\varphi_m} \Big) \, dx \, dt - \int_{Q_0} c_k v_k \overline{\varphi_m} \, dS$$

$$= \int_0^T \Big(T_k(t) \int_Q \big(\sum_{i=1}^n \sum_{j=1}^n a_{ij}(x) D_j v_k(x) D_i \overline{\varphi_m}(x,t) - c(x) v_k(x) \overline{\varphi_m}(x,t) \big) \, dx \Big) \, dt$$

$$- \int_Q \Big(v_k(x) \int_0^T T_k'(t) D_t \overline{\varphi_m}(x,t) \, dt \Big) \, dx - \int_{Q_0} c_k v_k(x) \overline{\varphi_m}(x,0) \, dS$$

$$= \lambda_k \int_0^T \Big(T_k(t) v_k \overline{\varphi_m}(x,t) \, dx \Big) \, dt - \int_Q v_k(x) \Big(\int_0^T (\lambda_k T_k(t) - a_k(t)) \overline{\varphi_m}(x,t)) \, dt$$

$$+ T_k'(T) \overline{\varphi_m}(x,t)) \, dx - T_k'(0) \overline{\varphi_m}(x,0) \Big) \, dx - \int_Q c_k v_k(x) \overline{\varphi_m}(x,0) \, dx.$$

By (11.68) we obtain

$$I = \int_{Q \times (0,T)} a_k v_k \overline{\varphi_m} \, dx \, dt - T_k'(T) \int_{\overline{Q \times (0,T)}} (1 \cdot v_k) \overline{\varphi_m} \, dS.$$

Taking $m \to \infty$ we obtain by $\varphi|_{\overline{Q_T}} = 0$ the desired conclusion of c).

d) Follows by c).

e) By the first example in this section the trace of the function

$$u = \sum_{i=1}^\infty T_i v_i$$

on the set $\overline{Q_0}$ is a limit of the sequence $\{u_k\}_{k \in \mathbf{N}}$ from d) as $k \to \infty$ in the space $L_2(\overline{Q_0})$. In an analogous way we have that $u|_{\overline{\Gamma_T}} = 0$. Since by d) the function u_k satisfies (11.58), letting $k \to \infty$ we obtain by the first example in this section and Cauchy-Schwartz inequality that u satisfies (11.58).

Example 11.37 *Prove that if in the generalized mixed problem $f \in \overset{\circ}{W}\!{}^1(Q)$, then*

a) *the series $\sum_{i=1}^{\infty} T_i(t)v_i(x)$ converges in the space $W^1(Q \times (0,T))$ to function which is the generalized solution of the generalized mixed problem (11.58), (11.60) for $r = 0$;*

b) *there exists $C > 0$ such that*

$$\|u\|^2_{W^1(Q \times (0,T))} \leq C(\|f\|^2_{W^1(Q)} + \|g\|_{L_2(Q)} + \|F\|_{L_2(Q)}). \tag{11.71}$$

Solution.

a) Since $g \in L_2(Q)$ we have for its Fourier series $\sum_{i=1}^{\infty} c_i v_i$ for $c_i = \int_Q g \bar{v}_i \, dx$ by the Parseval equality

$$\|g\|_{L_2(Q)} = \sum_{i=1}^{\infty} |c_i|^2. \tag{11.72}$$

We have for the Fourier series

$$F(x,t) = \sum_{i=1}^{\infty} a_i(t) v_i(x)$$

again by the Parseval equality that almost everywhere with respect to t from $[0,T]$

$$\sum_{i=1}^{\infty} |a_i(t)|^2 = \int_Q |F(x,t)|^2 \, dx.$$

Therefore we have by Beppo-Levi theorem

$$\sum_{i=1}^{\infty} \int_0^T |a_i(t)|^2 \, dt = \int_{Q \times (0,T)} |F(x,t)|^2 \, dx dt. \tag{11.73}$$

We shall use that $g \in \overset{\circ}{W}{}^k(Q)$ and Examples 11.31, 11.32 and that the system $\{v_i\}_{i \in \mathbf{N}}$ is orthogonal in the space $L_2(Q)$. Then the system

$$\{\frac{v_i}{(v_i|v_i)^{1/2}}\}_{i \in \mathbf{N}}$$

is orthonormal in the space $\overset{\circ}{W}{}^k(Q)$, where the scalar product $(v|w)$ is given by

$$(v|w) = \sum_{i=1}^{n} \sum_{j=1}^{n} \int_Q a_{ij}(D_j v)(D_i \bar{w}) \, dx - \int_Q c v \bar{w} \, dx \tag{11.74}$$

(see Examples 11.31, 11.32). The function f has the following Fourier series expansion in the space $W^1(Q)$

$$f = \sum_{i=1}^{\infty} b_i v_i = \sum_{i=1}^{\infty} b_i (v_i|v_i)^{1/2} \frac{v_i}{(v_i|v_i)^{1/2}}.$$

Hence by the Parseval equality

$$(f|f) = \sum_{i=1}^{\infty} |b_i(v_i|v_i)^{1/2}|^2.$$

Since by Example 11.32

$$(v_i|v_i) = \lambda_i, \tag{11.75}$$

we obtain

$$\sum_{i=1}^{\infty} |b_i|^2 \lambda_i = (f|f). \tag{11.76}$$

We introduce a new scalar product $[\cdot|\cdot]$ in the following way

$$[v|w] = \int_{Q\times(0,T)} \left(\sum_{i=1}^{n}\sum_{j=1}^{n} a_{ij}(D_j v)(D_i \overline{w}) - cv\overline{w} + (D_t v)(D_t \overline{w}) \right) dx dt, \tag{11.77}$$

which by the uniform ellipticity of the considered differential operator induces a norm which is equivalent to the usual norm in $W^1(Q \times (0,T))$. The scalar product (11.77) can be written also in the following form

$$\begin{aligned}
[v|w] &= \int_0^T (v(\cdot,t)|w(\cdot,t))\, dt \\
&+ \int_0^T \left(\int_Q v(x,t)\overline{w(x,t)} + D_t v(x,t) D_t \overline{w(x,t)}\, dx \right) dt.
\end{aligned}$$

The system of functions $\{v_i\}_{i\in\mathbf{N}}$ is orthogonal in the space $W^1(Q)$ with respect to the scalar product $(\cdot|\cdot)$ given by (11.74) and therefore we have for $i \neq j$

$$[T_i v_i | T_j v_j] = \int_0^T T_i(t)\overline{T_j(t)}(v_i|v_j)\, dt$$

$$+ \int_0^T (T_i(t)\overline{T_j(t)} + T_i'(t)\overline{T_j'(t)}) \cdot \left(\int_Q v_i \overline{v_j}\, dx \right) dt + \int_0^T T_i(t)\overline{T_j(t)} \cdot \left(\int_{\partial Q} v_i \overline{v_j}\, dx \right) dt = 0,$$

since $v_i|_{\partial Q} = 0$.

Therefore the system $\{T_i v_i\}$ is orthogonal with respect to the scalar product $[\cdot|\cdot]$ and the series

$$\sum_{i=1}^{\infty} T_i v_i$$

converges if and only if the series

$$\sum_{i=1}^{\infty} [T_i v_i | T_i v_i] \tag{11.78}$$

converges. By the definition of the scalar product $[\cdot|\cdot]$ we have

$$[T_iv_i|T_iv_i] = (v_i|v_i)\int_0^T |T_i(t)|^2\, dt + (\int_Q |v_i(x)|^2\, dx)(\int_0^T (|T_i(t)|^2 + |T_i'(t)|^2)\, dt).$$

This implies by $(v_i|v_i) = \lambda_i$ and the normdness of the system $\{v_i\}_{i\in\mathbb{N}}$ in the space $L_2(Q)$

$$[T_iv_i|T_iv_i] = (\lambda_i + 1)\int_0^T |T_i(t)|^2\, dt + \int_0^T |T_i'(t)|^2\, dt. \tag{11.79}$$

To prove the convergence of the series (11.78) we have to find by (11.79) the estimations for $|T_i|$ and $|T_i'|$.
By (11.61), (11.62), (11.63) and (11.64) and the Cauchy-Schwartz inequality we obtain first for $\lambda \neq 0$

$$|T_i(t)|^2 \leq 3\left(|b_i|^2 + \frac{|c_i|^2}{\lambda_i} + \frac{1}{\lambda_i}|\int_0^T a_i(\tau)\sin(\sqrt{\lambda_i}(t-\tau))\,d\tau|\right)$$

$$\leq 3\left(|b_i|^2 + \frac{|c_i|^2}{\lambda_i} + \frac{1}{\lambda_i}(\int_0^T |a_i(\tau)|^2\, d\tau)T\right),$$

and

$$|T_i'(t)|^2 \leq 3\left(|b_i|^2\lambda_i + |c_i|^2 + |\int_0^T a_i(\tau)\cos(\sqrt{\lambda_i}(t-\tau))\,d\tau|\right)$$

$$\leq 3\left(|b_i|^2 + |c_i|^2 + (\int_0^T |a_i(\tau)|^2\, d\tau)T\right).$$

We have for $\lambda_1 = 0$

$$|T_1(t)|^2 \leq 3\left(|b_1|^2 + |c_1|^2 T^2 + (\int_0^T |a_1(t)|^2\, dt)\frac{T^2}{2}\right),$$

$$|T_1'(t)|^2 \leq 2\left(|c_1|^2 + (\int_0^T |a_1(t)|^2\, dt)T\right).$$

Putting these estimations in (11.79) and using the fact that the sequence $\{1 - \frac{1}{\lambda_i}\}_{i\in\mathbb{N}}$ is bounded (since $\lambda_i \to +\infty$ as $i \to \infty$) and (11.72), (11.73) and (11.76) we obtain

$$\sum_{i=1}^\infty [T_iv_i|T_iv_i] = \sum_{i=1}^\infty \left((\lambda_i + 1)\int_0^T |T_i(t)|^2\, dt + \int_0^T |T_i'(t)|^2\, dt\right)$$

$$\leq C'\left(\sum_{i=1}^\infty |b_i|^2(\lambda_i + 1) + \sum_{i=1}^\infty |c_i|^2 + \sum_{i=1}^\infty \int_0^T |a_i(\tau)|^2\, d\tau\right)$$

$$\leq C''\left(\|f\|_{W^1(Q)}^2 + \|g\|_{L_2(Q)}^2 + \|F\|_{L_2(Q\times(0,T))}\right),$$

which implies the convergence of the series (11.78).

b) The inequality (11.71) follows by the last inequality in a), the Parseval equality and the inequality

$$\|u\|^2_{W^1((0,T)\times Q)} \le C_1 \sum_{i=1}^{\infty} [T_i v_i | T_i v_i].$$

11.3   Numerical Solutions of PDEs in the Framework of Functional Analysis

11.3.1   Preliminaries

Let B_1 and B_2 be two Banach spaces and $A : B_1 \to B_2$ a linear operator. We shall consider for $f \in B_2$ the equation

$$A(u) = f. \tag{11.80}$$

We introduce two another Banach spaces $B_{1,n}$ and $B_{2,n}$ $(n \in \mathbf{N})$, which are usually finite dimensional, and linear (discretization) operators

$$D_{1,n} : B_1 \to B_{1,n} \text{ and } D_{2,n} : B_2 \to B_{2,n}.$$

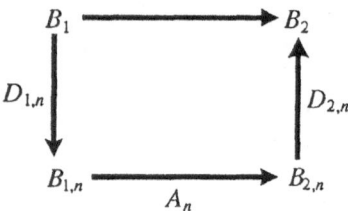

Figure 11.1

Then we reduce the equation (11.80) on the approximate equation

$$A_n u_n = f_n, \tag{11.81}$$

where $f_n = D_{2,n}(f)$. Supposing the existence and continuity of A_n^{-1} we can solve (11.81) $u_n = A_n^{-1}(f_n)$.

Definition 11.6   *(i) A and A_n are compatible if*

$$\lim_{n \to \infty} \|D_{2,n} A(u) - A_n(D_{1,n} u)\| = 0.$$

(ii) *The numerical procedure is it stable if the sequence $\{A_n^{-1}\}_{n \in \mathbf{N}}$ is uniformly bounded.*

(iii) *the sequence $\{u_n\}_{n \in \mathbf{N}}$ converges to u if*

$$\lim_{n \to \infty} \|D_{1,n} u - u_n\| = 0.$$

(iv) *the procedure is convergent for $f \in B_2$, if for every f and a convergent sequence $\{f_n\}_{n \in \mathbf{N}}$ the sequence $\{u_n\}_{n \in \mathbf{N}}$, where $u_n = A_n^{-1} f_n$, converges to u.*

11.3.2 Examples and Exercises

Example 11.38 *Prove that if A and A_n are compatible and the numerical procedure is stable, then the procedure is convergent.*

Solution. Since the procedure is stable there exists a constant $M > 0$ such that $\|A_n^{-1}\| < M$ $(n \in \mathbf{N})$. Therefore we have

$$
\begin{aligned}
\|D_{1,n} u - u_n\| &= \|A_n^{-1}(A_n(D_{1,n} u - u_n))\| \\
&\leq M \|A_n(D_{1,n} - u_n)\| \\
&\leq M \left(\|A_n(D_{1,n} u) - D_{2,n} A(u)\| + \|D_{2,n} f - f_n\| \right).
\end{aligned}
$$

Since the right side tends to zero as $n \to \infty$ we obtain the desired conclusion.

Exercise 11.39 *For the mixed problem for the heat equation*

$$\frac{\partial u}{\partial t} - \frac{\partial^2 u}{\partial x^2} = F \ \text{on} \ (0, \pi) \times (0, T)$$

with the conditions

$$u(x, 0) = f(x) \qquad (x \in [0, \pi])$$
$$u(0, t) = u(\pi, t) = 0 \qquad (t \in [0, T]),$$

give the approximate solution by the notions from Preliminaries for

$$B_1 = C_0([0, \pi] \times [0, T]), \ B_2 = C([0, T] \times [0, T]) \times C[0, \pi],$$

$B_{1,n} = \mathbf{R}^s$, *where s is the number of the points (mh, nk) of the lattice M for the parallelogram $[0, \pi] \times [0, T]$ and $B_{2,n} = \mathbf{R}^{ss} \times \mathbf{R}^{s_1}$, where s_1 is the number of the points of the lattice M in $[0, \pi]$.*

Hints. Take $A = (\frac{\partial}{\partial t} - \frac{\partial^2}{\partial x^2}, I)$ for $I(u) = u(x, 0)$ and $A_n = (T_{h,k}, I_h)$, where

$$T_{h,k}(u) = \frac{1}{k}(u(x, t+k) - ((1 - \frac{2k}{h^2})u(x, t) + \frac{k}{h^2}u(x+h, t) + \frac{k}{h^2}u(x-h, t)))$$

and $I_h(u) = u(x, 0)$, $x = mh$.

Exercise 11.40 *Prove that for $\frac{k}{h^2} \leq \frac{1}{2}$ the procedure from Exercise 11.39 converges.*

Hint. Prove the compatibility and stableness of the preceding procedure and then apply Example 11.38.

Example 11.41 *Let $D_{2,n}$ $(n \in \mathbf{N})$ be continuous operators. If*

$$\lim_{n \to \infty} \|D_{2,n}u\| = \|u\| (u \in B_2), \tag{11.82}$$

then there exists a constant $M > 0$ such that

$$\|D_{2,n}u\| \leq M\|u\| (u \in B_2, n \in \mathbf{N}).$$

Solution. By (11.82) we obtain the pointwise boundedness of the family of functionals $\{\|D_{2,n}\|\}_{n \in \mathbf{N}}$. Then by theorem on uniform boundedness we obtain the desired conclusion.

11.4   Miscellaneous

11.4.1   Preliminaries

Let Q be a bounded region in $\mathbf{R}^n$. We denote by $C^\lambda(\overline{Q})$ the Banach space of all Hölder continuous functions on $\overline{Q}$, i.e.,

$$|D^\alpha u(x) - D^\alpha u(y)| \leq H|x - y|^\lambda$$

for $H > 0$, $0 < \lambda \leq 1$, $|\alpha| = 1$, endowed with the norm

$$\|u\|_\lambda = \sup_{x,y \in Q, x \neq y} \frac{|u(x) - u(y)|}{|x - y|^\lambda}.$$

We introduce another Banach space of functions $C^{2+\lambda}(\overline{Q})$, which contains all functions from $C^2(Q)$ for which the second derivative belongs to $C^\lambda(\overline{Q})$. This space is endowed with the norm

$$\|u\|_{2+\lambda} = \|u\|_{C(Q)} + \max_{1 \leq j \leq n} \|D_j u\|_{C(Q)} + \max_{1 \leq i,j \leq n} \|D_i D_j u\|_{C(Q)} + \max_{1 \leq i,j \leq n} \|D_i D_j u\|_\lambda,$$

where

$$\|u\|_{C(Q)} = \sup_{x \in Q} |u(x)|.$$

We consider the problem

$$L(u) = F \text{ on } Q \text{ and } u|_{\partial Q} = 0, \tag{11.83}$$

where

$$L(u) = \sum_{i=1}^{n}\sum_{j=1}^{n} a_{ij}(x)\frac{\partial^2 u}{\partial x_i \partial x_j} + \sum_{i=1}^{n} a_i(x)\frac{\partial u}{\partial x_i} + a(x)u$$

for $a_{ij}, a_i, a \in C(\overline{Q}), a(x) \le 0$ $(x \in Q)$ and L is uniformly elliptic, i.e., there exists $c' > 0$ such that for all $z \in \mathbf{R}^n$

$$\sum_{i=1}^{n}\sum_{j=1}^{n} a_{ij}(x)z_i z_j \ge c'|z|^2.$$

For the solution of this problem in $C^{2+\lambda}(Q)$ the following *Schauder a-priori estimation* holds

$$\|u\|_{2+\lambda} \le C\|F\|_\lambda, \tag{11.84}$$

where $C > 0$.
The special case $L = \Delta$ is uniquely solvable in $C^{2+\lambda}(Q)$.

Let $\{S(t)\}_{t\ge 0}$ be a family of linear continuous operators defined on a Banach space X and with values in the same Banach space.

Definition 11.7 *A family $\{S(t)\}_{t\ge 0}$ is a semigroup of operators if it satisfies:*

(i) $S(0) = I$;

(ii) $S(t_1 + t_2) = S(t_1)S(t_2) = S(t_2)S(t_1)$ for every $t_1, t_2 \ge 0$.

The semigroup $\{S(t)\}_{t\ge 0}$ is strongly continuous at zero , or C_0–semigroup if for every $x \in X$ we have

(iii) $\lim_{t\to 0+0} \|S(t)x - x\| = 0$.

11.4.2 Examples and Exercises

Example 11.42 *Consider the equation*

$$L_a(u) = (1-a)\Delta(u) + aL(u) = F, \tag{11.85}$$

where $0 \le a \le 1$, with the boundary condition $u|_{\partial Q} = 0$.
Specially for $a = 0$ we obtain the problem

$$\Delta u = F \text{ on } Q \text{ and } u|_{\partial Q} = 0.$$

Let

$$A = \{a|\, 0 \le a \le 1\},\ F \in C^\lambda(\overline{Q})$$

implies that $u \in C^{2+\lambda}(\overline{Q})$ is the solution of (11.85). Prove that the set A is

a) closed set;

b) open set.

Solution.

a) Let $\{a_i\}_{i\in\mathbb{N}}$ be an arbitrary convergent sequence from A. Denote by a the limit of this sequence. We have to prove that $a \in A$. We denote by u_i the solution of (11.85) which corresponds to a_i. The estimation (11.84) implies

$$\|u_i\|_{2+\lambda} \leq c\|F\|_{\lambda} \qquad (i \in \mathbb{N}).$$

Hence the sequences $\{u_i\}_{i\in\mathbb{N}}, \{Du_i\}_{i\in\mathbb{N}}$ and $\{D^2u_i\}_{i\in\mathbb{N}}$ are equicontinuous. Then by the Arzela–Ascoli theorem there exists an uniformly convergent subsequence
$\{u_{i_j}\}_{j\in\mathbb{N}}$ of $\{u_i\}_{i\in\mathbb{N}}$ for which also the sequences of first and second derivatives are uniformly convergent. Denote the limit by $u \in C^{2+\lambda}(\overline{Q})$. Therefore we have on Q

$$F = \lim_{j\to\infty} L_{a_j}(u_{i_j}) = L_a(u) \text{ and } u|_{\partial Q} = 0.$$

This implies $a \in A$.

b) We shall prove that A is the neighborhood of every its point. For that purpose we have to find for an arbitrary but fixed $a_0 \in A$ a number $\varepsilon > 0$ such that $(a_0 - \varepsilon, a_0 + \varepsilon) \subset A$. We introduce the family $\{\Psi_a\}_{a\in[0,1]}$ of functions $\Psi_a : C^{2+\lambda}(\overline{Q}) \to C^{2+\lambda}(\overline{Q})$ defined by

$$\Psi_a(u) = v \qquad (u \in C^{2+\lambda}(\overline{Q})),$$

where v is the unique solution of the problem

$$L_{a_0}(v) = (a - a_0)(\Delta u - L(u)) + F \text{ on } Q \text{ and } v|_{\partial Q} = 0. \qquad (11.86)$$

If we prove that Ψ_a for $a \in (a_0 - \varepsilon, a_0 + \varepsilon)$ has a fixed point, i.e., $\Psi_a(u_a) = u_a$ for some a, then it would be $u_a|_{\partial Q} = 0$ and (11.86) would imply

$$L_a(u_a) = (a - a_0)(\Delta u_a - L(u_a)) + F \text{ on } Q.$$

Then it would follow $L_a(u_a) = F$, i.e., the fixed point u_a would be the solution of (11.85).
To prove that Ψ_a has a fixed point we shall find $\varepsilon > 0$ such that for $a \in (a_0 - \varepsilon, a_0 + \varepsilon)$ the map Ψ_a is a contraction and then we can apply Banach fixed point theorem .
Let

$$u_1, u_2 \in C^{2+\lambda}(\overline{Q}), \quad v_1 = \Psi_a(u_1) \text{ and } v_2 = \Psi_a(u_2),$$

where v_1 and v_2 are the corresponding unique solutions of (11.85). Then we have for their difference

$$L_a(v_1 - v_2) = (a - a_0)(\Delta(u_1 - u_2) - L(u_1 - u_2)).$$

Therefore by Schauder estimation (11.84) we obtain

$$
\begin{aligned}
\|\Psi_a(u_1) - \Psi_a(u_2)\|_{2+\lambda} &= \|v_1 - v_2\|_{2+\lambda} \\
&\leq C|a - a_0|\|\Delta(u_1 - u_2) - L(u_1 - u_2)\|_{\lambda} \\
&\leq CM|a - a_0|\|u_1 - u_2\|_{2+\lambda},
\end{aligned}
$$

where $M > 0$ is a constant independent of u_1, u_2 and C. Taking $|a - a_0| < \varepsilon$ for $\varepsilon = \frac{1}{2CM}$ we obtain

$$
\|\Psi_a(u_1) - \Psi_a(u_2)\|_{2+\lambda} < \frac{1}{2}\|u_1 - u_2\|_{2+\lambda}.
$$

Hence Ψ_a is a contraction for $|a - a_0| < \varepsilon$. Therefore Ψ_a by Banach fixed point theorem has a fixed point u_a for every $a \in (a_0 - \varepsilon, a_0 + \varepsilon)$, and this is the solution of the problem (11.85). Therefore $(a_0 - \varepsilon, a_0 + \varepsilon) \subset A$, i.e., A is an open set.

Example 11.43 *Prove that for every $F \in C^\lambda(Q)$ the problem (11.83) is uniquely solvable.*

Solution. Consider instead of the equation $L(u) = F$ the perturbed equation (11.85), i.e.,

$$
L_a(u) = (1 - a)\Delta(u) + aL(u) = F,
$$

where $0 \leq a \leq 1, u|_{\partial Q} = 0$. Take the set

$$
A = \{a \,|\, 0 \leq a \leq 1\}.
$$

$F \in C^\lambda(\overline{Q})$ implies that $u \in C^{2+\lambda}(\overline{Q})$ is the solution of (11.85). The set A is by Example 11.42 open and closed set. Therefore $A = [0, 1]$. Hence $1 \in A$, what implies the desired conclusion.

Exercise 11.44 *Every C_0-semigroup $\{S(t)\}_{t\geq 0}$ is continuous for any $t > 0$.*

Hints. The right continuity follows easily by the definition.
For the left continuity prove first the pointwise boundedness and then use theorem on uniform boundedness .

Exercise 11.45 *For the mixed type problem for heat equation*

$$
\frac{\partial u}{\partial t} = \frac{\partial^2 u}{\partial x^2} \qquad (0 < x < \pi, t > 0),
$$

with boundary conditions

$$
u(0, t) = u(\pi, t) = 0 \qquad (t > 0),
$$

and initial condition

$$u(0, x) = f(x) \qquad (0 < x < \pi)$$

for $f \in L_2(0, \pi)$ consider the family of operators $\{S(t)\}_{t \geq 0}$ as a map $t \mapsto S(t)$ with the domain $[0, +\infty)$ and range $L(L_2(0, \pi), L_2(0, \pi))$ given by

$$S(t)f(x) = u(x, t),$$

where u is the solution of the considered mixed problem obtained by the Fourier method of separation of variables (see Chapter 6) given by

$$u(x, t) = \sum_{n=1}^{\infty} c_n e^{-n^2 t} \sin nx,$$

where c_n are the Fourier coefficients of the function f.
Prove:

a) *$\{S(t)\}_{t \geq 0}$ is a semigroup of operators;*

b) *$\{S(t)\}_{t \geq 0}$ is a C_0–semigroup.*

Exercise 11.46 *Prove that for a C_0–semigroup $\{S(t)\}_{t \geq 0}$ there exist two real constants $M > 0$ and $\omega > 0$ such that*

$$\|S(t)\| \leq M e^{\omega t} \qquad (t \in [0, +\infty)).$$

Chapter 12

Distributions in the theory of PDEs

12.1 Basic Properties

12.1.1 Preliminaries

In this chapter O denotes an open set in $\mathbf{R}^n$.

The *support* of a continuous function $\varphi : O \to \mathbf{C}$, denoted by $\operatorname{supp} \varphi$, is the closed set defined by

$$\operatorname{supp} \varphi = \overline{\{x \in O \mid \varphi(x) \neq 0\}}.$$

An infinitely differentiable function $\varphi : O \to \mathbf{C}$ is in the set C_0^∞ if it has a compact support. Then we shortly say that φ is a *test function*. The space $\mathcal{D}(O)$ is the set $C_0^\infty(O)$ endowed with the convergence defined below.

A sequence $\{\varphi_j\}_{j \in \mathbf{N}}$ from $\mathcal{D}(O)$ converges to the zero function $\varphi = 0$ if

(i) there exists a compact set $K \subset O$ such that for all $j \in \mathbf{N}$ it holds $\operatorname{supp} \varphi_j \subset K$;

(ii) for every multiindex $\alpha \in \mathbf{Z}_+^n$ and every $x \in K$ it holds $\lim\limits_{j \to +\infty} \dfrac{\partial^\alpha}{\partial x^\alpha} \varphi_j(x) = 0$.

Definition 12.1 *A distribution T on O is a linear continuous functional on $\mathcal{D}(O)$, where the continuity of T means that for every sequence $\{\varphi_j\}_{j \in \mathbf{N}}$ which converges to zero in $\mathcal{D}(O)$ it holds*

$$\lim_{j \to +\infty} T(\varphi_j) = 0.$$

The set of distributions on O will be denoted by $\mathcal{D}'(O)$.

The set of distributions is a vector space. The value $T(\varphi)$ of a distribution T at a test function φ is also denoted by $\langle T, \varphi \rangle$.

Definition 12.2 *A sequence of distributions* $\{T_j\}_{j \in \mathbf{N}}$ *from* $\mathcal{D}'(O)$ *converges to an element* $T \in \mathcal{D}'(O)$ *if for every* $\varphi \in \mathcal{D}(O)$ *it holds*

$$\lim_{j \to +\infty} \langle T_j, \varphi \rangle = \langle T, \varphi \rangle. \tag{12.1}$$

Then we say that T is a *weak limit* of the sequence $\{T_j\}_{j \in \mathbf{N}}$.

Every locally integrable function f on O defines a unique $T_f \in \mathcal{D}'(O)$, such that

$$\langle T_f, \varphi \rangle = \int_O f(x)\varphi(x)\, dx \quad (\varphi \in \mathcal{D}(O)). \tag{12.2}$$

Such distributions are called *regular*. However, the functional δ_a, $a \in O$, given by

$$\langle \delta_a, \varphi \rangle = \varphi(a) \quad (\varphi \in \mathcal{D}(O)), \tag{12.3}$$

is *not* regular. If in (12.3) $a = 0$, it will be denoted simply by δ; this is the well known *delta distribution* ("delta function").

The *support of a distribution* T is the smallest closed set $K \subset O$ such that for every $\varphi \in \mathcal{D}(O)$, with support in $O \setminus K$ it holds $\langle T, \varphi \rangle = 0$. Clearly, the support of δ_a is the single point a.

Let $\alpha = (\alpha_1, \alpha_2, \ldots, \alpha_n) \neq (0, 0, \ldots, 0)$ be a multiindex from $\mathbf{Z}_+^n$. In order to define the distributional partial derivative of a distribution T, let us assume first that a function $f = f(x_1, \ldots, x_n)$ has a continuous partial derivative in x_1 on $\mathbf{R}^n$. Then for $\varphi \in \mathcal{D}(\mathbf{R}^n)$ it holds

$$\int_{\mathbf{R}^n} \frac{\partial f(x)}{\partial x_1} \varphi(x)\, dx = f(x)\, \varphi(x)\Big|_{-\infty}^{\infty} - \int_{\mathbf{R}^n} f(x)\frac{\partial \varphi(x)}{\partial x_1}\, dx = -\int_{\mathbf{R}^n} f(x)\frac{\partial \varphi(x)}{\partial x_1}\, dx.$$

(See also Section 9.1.) Since both f and its derivative $\dfrac{\partial f}{\partial x_1}$ define unique regular distributions T_f and $T_{\frac{\partial f}{\partial x_1}}$, the obtained equality can be written as

$$\langle D^{(1,0,\ldots,0)} T_f, \varphi \rangle = -\left\langle T_f, \frac{\partial \varphi}{\partial x_1} \right\rangle.$$

where $D^{(1,0,\ldots,0)} T_f$ is, in fact, the distributional partial derivative in x_1 of the distribution T_f. Thus for an arbitrary element $T \in \mathcal{D}'(O)$ and a multiindex $\alpha \in \mathbf{Z}_+^n$, the α–th *distributional derivative* of T, denoted by $D^\alpha T$, is defined by

$$\langle D^\alpha T, \varphi \rangle = (-1)^{|\alpha|} \left\langle T, \frac{\partial^\alpha \varphi}{\partial x^\alpha} \right\rangle \quad (\varphi \in \mathcal{D}(O)). \tag{12.4}$$

Since $\varphi \in C^\infty(O)$, we get the essential property of the space of distributions, namely that every distribution has a distributional derivative of arbitrary order. For the relation between the distributional and "classical" derivatives, see Example 12.9.

Definition 12.3 *The space* $\mathcal{S}(\mathbf{R}^n)$ *of* rapidly decreasing functions *on* $\mathbf{R}^n$ *is the set of infinitely differentiable functions on* $\mathbf{R}^n$ *such that for all multiindices* α *and* β *it holds*

$$\lim_{|x|\to+\infty}\left|x^\alpha\frac{\partial^\beta}{\partial x^\beta}\varphi(x)\right|=0,$$

endowed with the following convergence:

A sequence $\{\varphi_j\}_{j\in\mathbf{N}}$ *converges in* $\mathcal{S}(\mathbf{R}^n)$ *to the zero function* $\varphi=0$ *iff for all multiindices* α *and* β *and every* $x\in\mathbf{R}^n$ *it holds*

$$\lim_{j\to+\infty}x^\alpha\frac{\partial^\beta}{\partial x^\beta}\varphi_j(x)=0. \qquad (12.5)$$

The space of *tempered distributions* $\mathcal{S}'(\mathbf{R}^n)$ is the space of linear continuous functionals on $\mathcal{S}(\mathbf{R}^n)$, the continuity being defined analogously to (12.1). The space $\mathcal{D}(\mathbf{R}^n)$ is dense in $\mathcal{S}(\mathbf{R}^n)$. Thus the space $\mathcal{S}'(\mathbf{R}^n)$ can be considered as a subspace of the space of distributions $\mathcal{D}'(\mathbf{R}^n)$. In fact, it holds

$$\mathcal{D}(\mathbf{R}^n)\subset\mathcal{S}(\mathbf{R}^n)\subset\mathcal{S}'(\mathbf{R}^n)\subset\mathcal{D}'(\mathbf{R}^n).$$

The important property of $\mathcal{S}(\mathbf{R}^n)$ is that the Fourier transformation is a topological isomorphism of $\mathcal{S}(\mathbf{R}^n)$ (see Chapter 8). Thus, in view of the Parseval equality, the *distributional Fourier transform* $\mathcal{F}T$ of a tempered distribution T is defined by

$$\langle\mathcal{F}T,\varphi\rangle=\langle T,\mathcal{F}\varphi\rangle\quad(\varphi\in\mathcal{S}(\mathbf{R}^n)),$$

and it is also a tempered distribution.

Let f and g be locally integrable functions on $\mathbf{R}^n$ such that the improper integral

$$\int\limits_{-\infty}^{\infty}|f(\tau)\,g(x-\tau)|\,d\tau$$

converges for almost all $x\in\mathbf{R}^n$ and defines a locally integrable function on $\mathbf{R}^n$. Then the *convolution* $f*g$ is defined by

$$(f*g)(x)=\int\limits_{\mathbf{R}^n}f(\tau)\,g(x-\tau)\,d\tau\quad(x\in\mathbf{R}^n). \qquad (12.6)$$

One can prove that the function $f*g$ is locally integrable and the convolution (12.6) is commutative.

In the following section, we shall have to deal with the convolution in the space of distributions. To that end, let us take $\varphi\in\mathcal{D}(\mathbf{R}^n)$ and then, using the Fubini theorem, calculate the following integral:

$$\int_{\mathbf{R}^n} (f * g)(x)\, \varphi(x)\, dx \;=\; \int_{\mathbf{R}^n_x} \int_{\mathbf{R}^n_\tau} f(\tau)\, g(x - \tau)\, \varphi(x)\, d\tau\, dx$$

$$=\; \int_{\mathbf{R}^n_\tau} g(\tau) \int_{\mathbf{R}^n_x} f(x - \tau)\, \varphi(x)\, dx\, d\tau$$

$$=\; \int_{\mathbf{R}^n_\tau} \int_{\mathbf{R}^n_x} f(x)\, g(\tau)\varphi(x + \tau)\, dx\, d\tau\,.$$

Let now f and g be two distributions on $\mathbf{R}^n$. Then the upper calculation suggests us to define the convolution of f and g by

$$\langle f * g, \varphi \rangle = \langle f(x), \langle g(\tau), \varphi(x + \tau) \rangle \rangle \quad (\varphi \in \mathcal{D}'(\mathbf{R}^n)), \tag{12.7}$$

provided this relation defines an element from $\mathcal{D}'(\mathbf{R}^n)$. The problem of existence of the convolution of two distributions is rather involved. Let us just say that (12.7) exists if at least one of the distributions f and g has a compact support. In particular, if $g = \delta$, then it holds

$$f * \delta = f$$

for every $f \in \mathcal{D}'(\mathbf{R}^n)$.

(Note that $\operatorname{supp} \delta = \{0\}$, hence a compact set.)

12.1.2   Examples and Exercises

Example 12.1 *Prove that the following sequences converge to the delta distribution* δ *(given by (12.3) for $a = 0$,) in $\mathcal{D}'(\mathbf{R})$:*

a) $\left\{ \dfrac{j}{\pi(1 + j^2 x^2)} \right\}_{j \in \mathbf{N}}$;

b) $\left\{ \dfrac{j}{2} e^{-j|x|} \right\}_{j \in \mathbf{N}}$.

Solutions.

a) Let $\varphi \in \mathcal{D}(\mathbf{R})$. For every $j \in \mathbf{N}$, the function $\dfrac{j}{\pi(1 + j^2 x^2)}$ is locally integrable on $\mathbf{R}$ (in fact, it is infinitely differentiable on $\mathbf{R}$), hence it defines a unique regular distribution via the formula (12.2). Thus we have

$$\left\langle \frac{j}{\pi(1 + j^2 x^2)}, \varphi \right\rangle = \frac{1}{\pi} \int_{-\infty}^{\infty} \frac{j\, \varphi(x)}{1 + j^2 x^2}\, dx = \frac{1}{\pi} \int_{-\infty}^{\infty} \frac{\varphi(t/j)}{1 + t^2}\, dt.$$

By supposition, the support of φ is contained in some interval $[-L, L]$, which implies that the last integral is equal to

$$\frac{1}{\pi} \int_{-jL}^{jL} \frac{\varphi(t/j)}{1+t^2}\, dt = 2\frac{\varphi(0)}{\pi} \cdot \arctan(jL) + \frac{1}{\pi} \int_{-jL}^{jL} \frac{\varphi(t/j) - \varphi(0)}{1+t^2}\, dt. \qquad (12.8)$$

Since it holds

$$\lim_{j\to\infty} 2\frac{\varphi(0)}{\pi} \cdot \arctan(jL) = \varphi(0) = \langle \delta, \varphi \rangle,$$

we have yet to prove that the last integral in (12.8) tends to zero as $j \to \infty$. To that end, we use the mean value theorem and obtain

$$\left| \frac{1}{\pi} \int_{-jL}^{jL} \frac{\varphi(t/j) - \varphi(0)}{1+t^2}\, dt \right| \le \frac{2}{\pi} \cdot \max_{|\xi| \le jL} |\varphi'(\xi)| \cdot \int_{0}^{jL} \frac{t\, dt}{j(1+t^2)}$$

$$= \frac{1}{\pi} \cdot \max_{|t| \le L} |\varphi'(t)| \; \frac{\ln(1+j^2 L^2)}{j}.$$

Since $\lim\limits_{j\to\infty} \dfrac{\ln j}{j} = 0$, the last expression tends to zero as $j \to \infty$.

b) Left to the reader.

Remark 12.1.1 One can prove that *every* distribution can be obtained as a weak limit of a sequence of test functions.

Exercise 12.2 *Construct the sequences $\{f_j\}_{j\in\mathbf{N}}$ and $\{g_j\}_{j\in\mathbf{N}}$ of locally integrable functions, which both converge almost everywhere (a.e.) to zero, and the first converges to the delta distribution δ in $\mathcal{D}'(\mathbf{R})$, while the other does not converge at all in $\mathcal{D}'(\mathbf{R})$.*

Answer. Let us put, e.g.,

$$f_j(x) = \begin{cases} j/2 & \text{if } |x| \le \dfrac{1}{j}, \\ 0 & \text{otherwise,} \end{cases} \qquad \text{and} \qquad g_j(x) = \begin{cases} j^2 & \text{if } |x| \le \dfrac{1}{j}, \\ 0 & \text{otherwise.} \end{cases}$$

We have for every $x \ne 0$:

$$\lim_{j\to\infty} f_j(x) = \lim_{j\to\infty} g_j(x) = 0$$

which means that these two sequences converge a.e. to zero.

If $\varphi \in \mathcal{D}'(\mathbf{R})$, then, using the mean value theorem for definite integrals, we obtain

$$\langle f_j, \varphi \rangle = \frac{j}{2} \int_{-1/j}^{1/j} \varphi(x)\,dx = \frac{j}{2} \cdot \frac{2}{j} \cdot \varphi(\xi_j) = \varphi(\xi_j),$$

where $\xi_j \in [-1/j, 1/j]$. This implies

$$\langle f_j, \varphi \rangle = \lim_{j \to \infty} \varphi(\xi_j) = \varphi(0) = \langle \delta, \varphi \rangle.$$

Further on, if $\varphi \in \mathcal{D}'(\mathbf{R})$, then

$$\langle g_j, \varphi \rangle = j^2 \int_{-1/j}^{1/j} \varphi(x)\,dx. \tag{12.9}$$

Assume that φ is identically equal to 1 in some neighbourhood of zero; then the expression on the right–hand side of (12.9) does not converge as $j \to \infty$.

Example 12.3 *The functional x^{-1} on $\mathcal{D}(\mathbf{R})$ is defined by*

$$\langle x^{-1}, \varphi \rangle = \int_{-\infty}^{\infty} \frac{\varphi(x) - \varphi(0) \cdot H(1-x)}{x}\,dx \quad (\varphi \in \mathcal{D}), \tag{12.10}$$

where H is the Heaviside function given by

$$H(x) = \begin{cases} 1 & \text{if } x > 0, \\ 0 & \text{if } x \le 0. \end{cases}$$

Prove that x^{-1} is a distribution on $\mathcal{D}(\mathbf{R})$ with the property

$$x^{-1} \cdot x = 1, \tag{12.11}$$

where the last equality is in the sense of $\mathcal{D}'(\mathbf{R})$.

Remark 12.3.1 In general, there does not exist a definition of the product of arbitrary two distributions, which would generalize the usual product of continuous functions and would also preserve the commutative and the associative law. However, it is possible to define the product $f \cdot g$ of a distribution f and an infinitely differentiable function g by

$$\langle f \cdot g, \varphi \rangle = \langle f, g \cdot \varphi \rangle \quad (\varphi \in \mathcal{D}(\mathbf{R})).$$

Then $f \cdot g$ is also a distribution.

Solution. The linearity of x^{-1} is obvious. If a sequence of functions $\{\varphi_j\}_{j \in \mathbf{N}}$ from

$\mathcal{D}(\mathbf{R})$ tends to zero in the sense of $\mathcal{D}(\mathbf{R})$, then, by definition, there exists a compact set $K \subset \mathbf{R}$ such that for every $j \in \mathbf{N}$ it holds $\operatorname{supp}\varphi_j \subset K$. Then we have

$$\left|\langle x^{-1},\varphi_j\rangle\right| \le \int\limits_{K\cap\{x<1\}} \frac{|\varphi_j(x)-\varphi_j(0)|}{|x|}\,dx + \int\limits_{K\cap\{x>1\}} \frac{|\varphi_j(x)|}{|x|}\,dx.$$

Thus we get

$$\left|\langle x^{-1},\varphi_j\rangle\right| \le \left(\max_{x\in K}|\varphi_j'(x)| + \max_{x\in K}|\varphi_j(x)|\right)\cdot m(K),$$

where $m(K)$ is the measure of the compact set K. Hence, by the the definition of the convergence in $\mathcal{D}(\mathbf{R})$, the right–hand side tends to zero as $j \to \infty$.

Let us prove now the equality (12.11). If $\varphi \in \mathcal{D}(\mathbf{R})$, then from (12.10) it follows

$$\langle x^{-1}\cdot x,\varphi\rangle = \langle x^{-1}, x\cdot\varphi(x)\rangle$$

$$= \int\limits_{-\infty}^{\infty} \frac{x\varphi(x)-(x\varphi(x))(0)\cdot H(1-x)}{x}\,dx$$

$$= \int\limits_{-\infty}^{\infty} \varphi(x)\,dx = \langle 1,\varphi\rangle.$$

Example 12.4 *Show that the following two distributional products exist:*

$$(x^{-1}\cdot x)\cdot\delta \quad and \quad x^{-1}\cdot(x\cdot\delta), \tag{12.12}$$

but are nonequal. In (12.12), the distribution x^{-1} is given by (12.10), and the delta distribution δ is given by (12.3) (for $a=0$).

Solution. Firstly, let us calculate the distributional product $g\cdot\delta$, for $g \in C^\infty(\mathbf{R})$. If $\varphi \in \mathcal{D}'(\mathbf{R})$, then it holds

$$\langle g\cdot\delta,\varphi(x)\rangle = \langle\delta,g(x)\,\varphi(x)\rangle = \langle\delta,(g\cdot\varphi)\rangle$$

$$= (g\,\varphi)(0) = g(0)\,\varphi(0) = g(0)\cdot\langle\delta,\varphi\rangle,$$

which means that

$$g\cdot\delta = g(0)\,\delta. \tag{12.13}$$

in the distributional sense.

The function $g(x) = x$ is in $C^\infty(\mathbf{R})$, hence by (12.13) it holds

$$x\cdot\delta = x|_{x=0}\cdot\delta = 0.$$

Since for any distribution T it holds $0 \cdot T = 0$, we have

$$x^{-1} \cdot (x \cdot \delta) = x^{-1} \cdot 0 = 0. \tag{12.14}$$

In view of relation (12.11), see Example 12.3, and equation (12.13) (for $g(x) = 1$), it holds

$$(x^{-1} \cdot x) \cdot \delta = 1 \cdot \delta = \delta. \tag{12.15}$$

Hence from (12.15) and (12.14) we obtain the inequality

$$(x^{-1} \cdot x) \cdot \delta \neq x^{-1} \cdot (x \cdot \delta). \tag{12.16}$$

Remark 12.4.1 The inequality (12.16) shows that the associative law does not always hold in $\mathcal{D}'(\mathbf{R})$. In fact, if the multiplicative product in $\mathcal{D}'(\mathbf{R})$ is defined as a generalization of the usual product of continuous functions, the space of distributions cannot be an algebra.

Example 12.5 *Find the distributional products*

a) $x^p \cdot \delta^{(q)}$ $(p, q \in \mathbf{N})$;   b) $e^{ax} \cdot \delta^{(q)}$ $(a \in \mathbf{R}, q \in \mathbf{N})$,

where $\delta^{(q)}$ is the $q-$th distributional derivative of δ for $q \in \mathbf{N}$.

Solutions.

a) For $\varphi \in \mathcal{D}(\mathbf{R})$ it holds

$$\langle x^p \cdot \delta^{(q)}, \varphi \rangle \;=\; \langle \delta^{(q)}, x^p \varphi(x) \rangle = (-1)^q \langle \delta, (x^p \varphi(x))^{(q)} \rangle.$$

Assume first $p > q$. Then it holds

$$(x^p \varphi(x))^{(q)} \;=\; \sum_{j=0}^{q} \binom{q}{j} (x^p)^{(j)} \varphi^{(q-j)}(x)$$

$$\tag{12.17}$$

$$=\; \sum_{j=0}^{q} \binom{q}{j} p(p-1) \cdots (p-j+1) x^{p-j} \varphi^{(q-j)}(x).$$

So we have

$$\langle x^p \cdot \delta^{(q)}, \varphi \rangle \;=\; (-1)^q \sum_{j=0}^{q} \left\langle \delta, \binom{q}{j} p(p-1) \cdots (p-j+1) x^{p-j} \varphi^{(q-j)}(x) \right\rangle$$

$$=\; (-1)^q \sum_{j=0}^{q} \binom{q}{j} p(p-1) \cdots (p-j+1) \left. \left(x^{p-j} \varphi^{(q-j)}(x) \right) \right|_{x=0}$$

$$=\; 0 = \langle 0, \varphi \rangle.$$

Assume next $p \leq q$. Then for all $x \in \mathbf{R}$ it holds for

$$(x^p)^{(j)} = \begin{cases} 0 & \text{for} \quad j > p, \\ \\ p! & \text{for} \quad j = p, \end{cases}$$

and therefore we have from (12.17)

$$\langle x^p \cdot \delta^{(q)}, \varphi \rangle = (-1)^q \left\langle \delta, (x^p \varphi(x))^{(q)} \right\rangle$$

$$= (-1)^q \left\langle \delta, \sum_{j=0}^{p-1} \binom{q}{j} p(p-1) \cdots (p-j+1) x^{p-j} \varphi^{(q-j)}(x) \right\rangle$$

$$+ (-1)^q \left\langle \delta, \binom{q}{p} p! \, \varphi^{(q-p)}(x) \right\rangle$$

$$+ (-1^q) \left\langle \delta, \sum_{j=p+1}^{q} \binom{q}{j} (x^p)^{(j)} \varphi^{(q-j)}(x) \right\rangle$$

$$= 0 + (-1)^q q(q-1) \cdots (q-p+1) \left\langle \delta, \varphi^{(q-p)} \right\rangle + 0$$

$$= q(q-1) \cdots (q-p+1)(-1)^p \langle \delta^{(q-p)}, \varphi \rangle.$$

b) Similarly as in a), we get

$$e^{ax} \cdot \delta^{(q)} = \sum_{j=0}^{q} (-a)^j \delta^{(q-j)}.$$

Exercise 12.6 *Prove that*

$$\rho(x) \cdot \delta' = \rho'(0)\,\delta + \rho(0)\,\delta',$$

where ρ is an arbitrary continuously differentiable function on $\mathbf{R}$. In particular, prove that in $\mathcal{D}'(\mathbf{R})$ it holds

$$H'(x) = \delta.$$

Remark 12.6.1 Assume additionally $\rho(0) = 0$ and $\rho'(0) \neq 0$. Then note that the function ρ is equal to zero on the support $\{0\}$ of the distribution δ, but still their product is nonzero.

Example 12.7 *Let ρ be an infinitely differentiable function on $\mathbf{R}$ with simple zeros $a_1, a_2, \ldots, a_m$. Prove that the equation with the unknown distribution T*

$$\rho(x) \cdot T = 0 \tag{12.18}$$

has the same solutions as the equation

$$\left(\prod_{j=1}^{m}(x-a_j)\right)\cdot T = 0. \tag{12.19}$$

Moreover, the solutions of (12.19) (hence also of (12.18)) are of the form

$$T = \sum_{j=1}^{m} C_j\,\delta_{a_j},$$

see (12.3), where C_j, $j = 1, 2, \ldots, m$, are real constants.

Solution. Clearly, it is enough to analyze the case when ρ has only one simple zero at some point a. In that case, we have to prove that equation (12.18) is equivalent with equation

$$(x-a)\cdot T = 0, \tag{12.20}$$

and its solution is the distribution $T = A\,\delta_a$, for some constant A.

Let us introduce the function ρ_1 by

$$\rho(x) = (x-a)\,\rho_1(x). \tag{12.21}$$

Then it holds $\rho_1(a) \neq 0$ and the mapping $\varphi \mapsto \psi = \rho_1\varphi$ is a bijection from $\mathcal{D}(\mathbf{R})$ onto itself. Putting (12.21) into (12.18) we obtain

$$\langle \rho(x)\cdot T, \varphi(x)\rangle = \langle T, (x-a)\,\rho_1(x)\,\varphi(x)\rangle = \langle T, (x-a)\,\psi(x)\rangle$$

$$= \langle(x-a)\cdot T, \psi(x)\rangle.$$

which implies the equivalence of the equations (12.18) and (12.20).

Let us find now the solution of (12.20). To that end, note that the mapping $\psi \mapsto (x-a)\psi$ from the set

$$\mathcal{A} = \{\psi \in \mathcal{D}(\mathbf{R})|\ a \notin \operatorname{supp}\psi\}$$

into itself is, in fact, a bijection. Thus for every test function φ with the property $a \notin \operatorname{supp}\varphi$ it holds

$$\langle T, \varphi\rangle = 0,$$

which is equivalent with the statement $\operatorname{supp} T = \{a\}$. Any distribution T whose support is a single point a is necessarily of the form

$$T = A\,\delta_a + \sum_{k=1}^{p} A_k\,\delta_a^{(k)}, \tag{12.22}$$

for some constants A and A_k, $k = 1, 2, \ldots, p$. Choose now $k \in \{1, 2, \ldots, p\}$. Then for $\varphi \in \mathcal{D}(\mathbf{R})$ it holds:

$$
\begin{aligned}
\langle (x - a) \cdot \delta_a^{(k)}, \varphi \rangle &= \langle \delta_a^{(k)}, (x - a)\,\varphi(x) \rangle \\
&= (-1)^k \left\langle \delta_a, ((x - a)\,\varphi(x))^{(k)} \right\rangle \\
&= (-1)^k \left\langle \delta_a, (x - a)\,\varphi^{(k)}(x) + k\,\varphi^{(k-1)}(x) \right\rangle \\
&= (-1)^k \left. \left((x - a)\,\varphi^{(k)}(x) + k\,\varphi^{(k-1)}(x) \right) \right|_{x=a} \\
&= (-1)^k\, k\,\varphi^{(k-1)}(a).
\end{aligned}
$$

Since there exists a $\varphi \in \mathcal{D}(\mathbf{R})$ such that $\varphi^{(k-1)}(a) \neq 0$, it follows that the distribution $\delta_a^{(k)}$ is not a solution of (12.20), hence also not of (12.18). Thus T from (12.22) is a solution of (12.18) (for $m = 1$ and $a_1 = a$) if and only if $A_1 = A_2 = \ldots = A_p = 0$, which finally gives us the solution

$$
T = A\,\delta_a \quad \text{for some constant } A.
$$

Example 12.8 *Let us denote*

$$
e_j(x) = \exp(2j\pi\imath x) \quad (j \in \mathbf{Z}),
$$

D^α *the derivation operator in the sense of $\mathcal{D}'(\mathbf{R})$ and assume that for the sequence $\{c_j\}_{j \in \mathbf{Z}}$ of complex numbers there exist a positive constant A and a natural number k such that*

$$
|c_j| \leq A \cdot j^k \quad (j \in \mathbf{Z}). \tag{12.23}
$$

a) *Prove that the sequence of functions*

$$
f_m(x) = \sum_{j=-m}^{m} c_j e_j(x) \quad (m \in \mathbf{N}),
$$

converges in $\mathcal{D}'(\mathbf{R})$ as $m \to \infty$ to the distribution

$$
f(x) = \sum_{j \in \mathbf{Z}} c_j e_j(x). \tag{12.24}
$$

b) *Prove that for $\alpha \in \mathbf{N}$ and f from (12.24) it holds*

$$
D^\alpha f(x) = \sum_{j \in \mathbf{Z}} (2j\pi\imath)^\alpha c_j e_j(x), \tag{12.25}
$$

where the convergence in (12.25) is in the sense of $\mathcal{D}'(\mathbf{R})$.

c) *Find the sum of (12.24) in $\mathcal{D}'(\mathbf{R})$, if $c_j = 1$ for all $j \in \mathbf{Z}$.*

Solutions.

a) Let us us start from the sequence of functions $\{f_{m,k+2}(x)\}_{m \in \mathbf{N}}$ (k from (12.23)), where
$$f_{m,k+2}(x) = \sum_{\substack{0 \neq j = -m}}^{m} \frac{c_j}{(2j\pi \imath)^{k+2}} e_j(x) \quad (m \in \mathbf{N}).$$

In view of (12.23), this sequence uniformly converges on every compact set $K \subset \mathbf{R}$, hence its limit is a continuous function on $\mathbf{R}$; let us denote it by F_{k+2}. The distributional derivative of order $k+2$ of the function F_{k+2} is
$$D^{k+2} F_{k+2}(x) = \lim_{m \to \infty} f_{m,k+2}(x) = f(x) - a_0.$$

In other words, the distribution f, given by (12.24), is the limit of the sequence $\{a_0 + f_{m,k+2}(x)\}_{m \in \mathbf{N}}$ in $\mathcal{D}'(\mathbf{R})$.

b) Since for $\alpha \in \mathbf{N}$ and $m \in \mathbf{Z}_+$ it holds
$$D^\alpha f_m(x) = \sum_{j=-m}^{m} (2j\pi \imath)^\alpha e_j(x),$$

part a) implies that the sequence of functions $\{D^\alpha f_m\}_{m \in \mathbf{N}}$ converges in $\mathcal{D}'(\mathbf{R})$ to the distribution $D^\alpha f$ and thus (12.25) holds.

c) From part a) it follows that the sequence $\left\{ \sum_{j=-m}^{m} e_j(x) \right\}_{m \in \mathbf{N}}$ converges in $\mathcal{D}'(\mathbf{R})$ to a distribution which we denote by g. Then we have in the sense of $\mathcal{D}'(\mathbf{R})$:
$$(1 - e_1(x)) g(x) = \lim_{m \to \infty} (1 - e_1(x)) \sum_{j=-m}^{m} e_j(x) = \lim_{m \to \infty} (e_{-m}(x) - e_{m+1}(x)).$$
(12.26)

Let us prove next that the last limit is equal to 0 in $\mathcal{D}'(\mathbf{R})$. To that end, let us analyze the difference $e_{-m}(x) - e_{m+1}(x)$:
$$\langle (e_{-m}(x) - e_{m+1}(x)), \varphi(x) \rangle = \int_{\mathbf{R}} e^{-2\pi m \imath x} \varphi(x)\, dx - \int_{\mathbf{R}} e^{2\pi (m+1)\imath x} \varphi(x)\, dx$$
$$= \frac{-1}{2\pi m \imath} \int_{\mathbf{R}} e^{-2\pi m \imath x} \varphi'(x)\, dx + \frac{1}{2\pi (m+1)\imath} \int_{\mathbf{R}} e^{2\pi (m+1)\imath x} \varphi'(x)\, dx$$

Thus we have
$$|\langle e_{-m}(x) - e_{m+1}(x), \varphi(x) \rangle| \leq \frac{1}{2\pi m} \int_{\mathbf{R}} |\varphi'(x)|\, dx + \frac{1}{2\pi (m+1)} \int_{\mathbf{R}} |\varphi'(x)|\, dx \leq \frac{C}{m},$$

for some constant $C = C(\varphi)$. The last right–hand side tends to 0 as $m \to \infty$, which implies that the right–hand side of (12.26) tends to zero in $\mathcal{D}'(\mathbf{R})$ as $m \to \infty$. Thus we obtained that the sought after distribution g is the solution of the equation

$$\left(1 - e^{2\pi \imath x}\right) \cdot g(x) = 0. \tag{12.27}$$

For $|x| < m$, the solutions of the equation $e^{-2\pi \imath x} = 1$ are the integers j such that $|j| < m$. Example 12.7 tells us that the solutions of (12.27) in $\mathcal{D}'(-m, m)$ are exactly the solutions of the following equation

$$\left(\prod_{j=-m+1}^{m-1} (x - j)\right) \cdot g(x) = 0. \tag{12.28}$$

The same example gives us the solution of (12.28) in $\mathcal{D}'(-m, m)$

$$g(x) = \sum_{j=-m+1}^{m-1} A_j\, \delta_j(x),$$

while in $\mathcal{D}'(\mathbf{R})$ the solution of (12.27) is

$$g(x) = \sum_{j \in \mathbf{Z}} A_j\, \delta_j(x).$$

We next show that all constants A_j, $j \in \mathbf{Z}$, are equal to a single constant C. To that end, note that for a test function φ_m such that

$$\operatorname{supp} \varphi_m \subset (m - 2/3, m + 2/3) \quad \text{and} \quad \varphi(x) = 1$$

on the interval $(m - 1/3, m + 1/3)$, it holds

$$\langle g, \varphi_m \rangle = \left\langle \sum_{j \in \mathbf{Z}} A_j\, \delta_j\, \varphi_m \right\rangle = \langle A_m \delta_m, \varphi_m \rangle = A_m. \tag{12.29}$$

From the other hand, g is 1–periodic, i.e.,

$$\langle g, \varphi \rangle = \langle g(x), \varphi(x - 1) \rangle$$

for every test function φ. But then it follows from (12.29):

$$A_m = A_{m-1} = C \quad \text{for every} \quad m \in \mathbf{Z}.$$

So we get

$$g(x) = C \cdot \sum_{j \in \mathbf{Z}} \delta_j(x).$$

Example 12.9

a) *Let f be a continuous function on $\mathbf{R}\backslash\{a\}$, which is also continuously differentiable on the intervals $(-\infty, a]$ and $[a, +\infty)$. Prove that*

$$Df = T_{f'} + [f]_a\, \delta_a,$$

where Df denotes the distributional derivative of the function f, while $T_{f'}$ is the regular distribution defined by the classical derivative f' of f, see equation (12.2). As usual,

$$[f]_a = f(a+) - f(a-)$$

is the jump of f at the point a.

b) *If f is a piecewise continuously differentiable function on $\mathbf{R}$ with isolated discontinuities at the points a_j, $j \in J$, J a finite or infinite subset of $\mathbf{N}$, then*

$$Df = T_{f'} + \sum_{j \in J} [f]_{a_j}\, \delta_{a_j}.$$

Remark 12.9.1 In a), by assumption, f' exists and is continuous on the set $\mathbf{R}\backslash\{a\}$, but not in the point a. Since a point is a set of measure zero, the classical derivative f' of the function f defines a locally integrable function on $\mathbf{R}$.

Solution. Clearly it is enough to prove part a). To that end, for $\varphi \in \mathcal{D}(\mathbf{R})$ we have using (12.4)

$$
\begin{aligned}
\langle Df, \varphi \rangle &= -\langle f, \varphi' \rangle = - \int_{-\infty}^{a} f(x)\,\varphi'(x)\,dx - \int_{a}^{+\infty} f(x)\,\varphi'(x)\,dx \\[2mm]
&= -f(x)\,\varphi(x)\Big|_{-\infty}^{a-} + \int_{-\infty}^{a} f'(x)\,\varphi(x)\,dx - f(x)\,\varphi(x)\Big|_{a+}^{\infty} + \int_{a}^{+\infty} f'(x)\,\varphi(x)\,dx \\[2mm]
&= \langle T_{f'}, \varphi \rangle + (f(a-) - f(a-))\,\varphi(a).
\end{aligned}
$$

Since $\varphi(a) = \langle \delta_a, \varphi(x) \rangle$, we obtain the statement.

Example 12.10 *Find all distributions y such that*

a) $y' = 0$;   b) $y^{(m)} = 0, \quad m \in \mathbf{N}$.

Solution. a) Let $y \in \mathcal{D}'(\mathbf{R})$ be a solution of the given equation, i.e.,

$$\langle y', \varphi \rangle = 0 \quad \text{for every} \quad \varphi \in \mathcal{D}(\mathbf{R}). \tag{12.30}$$

Let χ be a fixed test function with the property

$$\int\limits_{-\infty}^{+\infty} \chi(x)\,dx = 1.$$

Now every $\varphi \in \mathcal{D}(\mathbf{R})$ can be written in the form

$$\varphi(x) = \chi(x) \cdot \int\limits_{-\infty}^{+\infty} \varphi(x)\,dx + \psi'(x) \quad (x \in \mathbf{R})$$

for some test function ψ, depending on φ (prove that). Then we have

$$\langle y, \varphi \rangle = \left\langle y, \chi(x) \cdot \int\limits_{-\infty}^{+\infty} \varphi(x)\,dx + \psi'(x) \right\rangle = \int\limits_{-\infty}^{+\infty} \varphi(x)\,dx \cdot \langle y, \chi \rangle + \langle y, \psi' \rangle.$$

By assumption we have

$$0 = -\langle y', \psi \rangle = \langle y, \psi' \rangle;$$

then putting $C = \int\limits_{-\infty}^{+\infty} \psi(x)\,dx$, we obtain

$$\langle y, \varphi \rangle = C \cdot \int\limits_{-\infty}^{+\infty} \varphi(x)\,dx = \langle C, \varphi \rangle$$

for every $\varphi \in \mathcal{D}'(\mathbf{R})$. Thus the sought after solution is

$$y = C,$$

where C is an arbitrary constant.

Answer. b) $y(x) = C_0 + C_1 x + \cdots + C_{m-1} x^{m-1}$,   where C_1, $C_2, \ldots, C_n$ are arbitrary constants.

Exercise 12.11 *Show that the linear first order ODE*

$$y'(x) + p(x)\,y = q(x),$$

where p and q are infinitely differentiable functions on an interval (a, b) has only classical solutions in $\mathcal{D}'(a, b)$, which are also infinitely differentiable functions.

Exercise 12.12 *Let $A(x) = [a_{i,j}(x)]_{i,j=1}^{n}$ denote an infinitely differentiable matrix on $\mathbf{R}$, i.e., for every $i, j \in \{1, 2, \ldots, n\}$ the function $a_{i,j} = a_{i,j}(x)$, $x \in \mathbf{R}$, is in $C^{\infty}(\mathbf{R})$. Prove then that the system of ODEs*

$$\frac{d\mathbf{y}}{dx} = A(x)\,\mathbf{y}(x), \tag{12.31}$$

where $\mathbf{y}(x) = (y_1(x), y_2(x), \ldots, y_n(x))$ is the unknown column–vector, has only classical solutions in the space of distributions $\mathcal{D}'(\mathbf{R})$.

Exercise 12.13 *Prove that the general solution of the equation*

$$x^n y^{(m)} = 0, \quad n > m, \tag{12.32}$$

in $\mathcal{D}'(\mathbf{R})$ is the distribution

$$y(x) = \sum_{j=0}^{m-1} \left(a_j H(x) x^{m-j-1} + b_j x^j \right) + \sum_{k=m}^{n-1} c_k H(x) \delta^{m-k},$$

where a_j, b_j, $j = 0,1,\ldots,m-1$, and c_k, $k = m, m+1, \ldots, n-1$, are arbitrary constants.

Example 12.14 *Prove the following equalities for every $f \in \mathcal{D}'(\mathbf{R})$:*

a) $\delta_a * f = f$ $(a \in \mathbf{R})$;   b) $D^m \delta * f = D^{(m)} f, \quad m \in \mathbf{N}$.

Remark 12.14.1 One often meets the equality in a) in the (formal) form

$$\int_{-\infty}^{+\infty} f(y)\, \delta(x-y)\, dy = f(x) \quad (x \in \mathbf{R}).$$

Solutions.

a) Let $\varphi \in \mathcal{D}(\mathbf{R})$. Then it holds

$$\langle \delta * f, \varphi \rangle = \langle f(x), \langle \delta(\tau), \varphi(x+\tau) \rangle \rangle = \langle f, \varphi \rangle.$$

b) For $\varphi \in \mathcal{D}(\mathbf{R})$ and $m \in \mathbf{N}$ it holds

$$\langle D^m \delta * f, \varphi \rangle = \langle f(x), \langle D^m \delta(\tau), \varphi(x+\tau) \rangle \rangle = (-1)^m \langle f(x), \varphi^{(m)}(x) \rangle = \langle D^m f, \varphi \rangle.$$

Example 12.15 *Let $P(x,D)$ be the linear differential operator*

$$P(x,D) = \sum_{j=0}^{m} a_j(x) D^j, \tag{12.33}$$

where $a_j \in C^\infty(\mathbf{R})$, $j = 0,1,\ldots,m$, and D is the distributional derivation operator.

a) *Prove that the solution of the following equation in $\mathcal{D}'(\mathbf{R})$*

$$P(x,D)\delta * u = \left(\sum_{j=0}^{m} a_j(x) D^j \delta \right) * u = \delta$$

is the function

$$u(x) = H(x) \cdot v(x),$$

where H is the Heaviside function, while $v \in C^m(\mathbf{R})$ is the solution of the initial value problem

$$P(x,D)v = 0, \quad v(0) = v'(0) = \ldots = v^{(m-2)}(0) = 0, \ v^{(m-1)}(0) = 1.$$

b) *Suppose that each a_j, $j = 0, 1, \ldots, m$, is a constant; then for the operator given by (12.33) we put simply $P(D)$. Assume F is a distribution with support in $[0, \infty)$. Prove then that the solution of the equation*

$$P(D)\delta * u = F,$$

is the distribution

$$u = (H \cdot v) * F,$$

where v is the function defined in part a).

Solution.

a) Firstly we find the distributional derivatives of $u = E = Hv$.

$$
\begin{aligned}
E'(t) &= H(t)v'(t) + (H(0+) - H(0-))\delta(t) = H(t)v'(t), \\
E''(t) &= H(t)v''(t), \\
&\;\;\vdots \\
E^{(m-1)}(t) &= H(t)v^{(m-1)}(t), \\
E^{(m)}(t) &= H(t)v^{(m)}(t) + \delta(t).
\end{aligned}
$$

Therefore

$$P(D)E = H(t)P(D)v + \delta(t) = \delta(t).$$

Specially for $m = 1$ we have

$$E(t) = H(t)e^{-a_0 t/a_1}.$$

b) Left to the reader.

Exercise 12.16 *Let a cylindrical, homogeneous isotropic rod of length ℓ with insulated sides be on initial temperature $0°C$. For times $t > 0$, there is located a cross section $x = b$ $(0 < b < \ell)$, a plane heat source of constant strength q. If both ends $(x = 0$ and $x = \ell)$ of the rod are kept at the temperature $0°C$, find the temperature $u = u(x, t)$ in the rod for $t > 0$, $0 < x < \ell$.*

Hint. The physical assumptions give the following mathematical problem:

$$a^2 \frac{\partial^2 u}{\partial x^2} = \frac{\partial u}{\partial t} - q_1\, \delta_b \quad (0 < x < \ell,\ t > 0),$$

$$u(0, t) = u(\ell, t) = 0 \quad (t > 0), \tag{12.34}$$

$$u(x, 0) = 0 \quad (0 < x < \ell).$$

where δ_b is the delta function with support in b. The solution of this problem is

$$u(x, t) = \frac{2q_1\ell}{a^2\pi^2} \sum_{n=1}^{\infty} \frac{1 - \exp(-\dfrac{n^2\pi^2 a^2 t}{\ell})}{n^2} \cdot \sin\frac{n\pi b}{\ell} \sin\frac{n\pi x}{\ell},$$

where $q_1 = K\,q$, K a constant depending on the material the rod was made from.

12.2   Fundamental Solutions

12.2.1   Preliminaries

A distribution E is the *fundamental solution* of a differential operator $P(x, D)$, where

$$P(x, D) = \sum_{|\alpha| \leq m} a_\alpha(x) D^\alpha,$$

if it satisfies the equation

$$P(x, D)E = \delta.$$

12.2.2   Examples and Exercises

Example 12.17 *Let the function $f : \mathbf{R}^2 \to \mathbf{R}$ be given for $(x, y) = x + \imath y$ by*

$$f(x + \imath y) = \frac{1}{\pi} \cdot \frac{1}{x + \imath y}.$$

Prove that

a) *the function f is locally integrable on $\mathbf{R}^2$, and so it defines a distribution on $\mathbf{R}^2$;*

b) *the function f is the fundamental solution for the Cauchy–Riemann operator*

$$\bar{\partial} = \frac{1}{2}\left(\frac{\partial}{\partial x} + \imath \frac{\partial}{\partial y}\right).$$

Solution.

a) Since we have $|f(z)| = \dfrac{1}{\pi} \cdot \dfrac{1}{|z|}$ for $z = x + \imath y$, it follows that $f \in L^1_{loc}(\mathbf{R}^2)$.

b) We have to prove that

$$< \bar{\partial} f, \varphi >= \varphi(0). \tag{12.35}$$

To that end, we start from the equality

$$< \bar{\partial}, \varphi >= - < f, \bar{\partial}\varphi >= -\frac{1}{2\pi} \int_{\mathbf{R}^2} \left(\frac{\partial \varphi}{\partial x} + \imath \frac{\partial \varphi}{\partial y}\right) \frac{dx\,dy}{x + \imath y}.$$

Changing the variables $x = r\cos t$, $y = r\sin t$, $\varphi_1(r, t) = \varphi(r\cos t, r\sin t)$, we obtain

$$< \bar{\partial} f, \varphi >= -\frac{1}{2\pi} \int_0^{2\pi} \int_0^\infty \left(\frac{\partial \varphi_1}{\partial r} - \frac{\imath}{r} \cdot \frac{\partial \varphi_1}{\partial t}\right) dr\,dt.$$

Therefore we have

$$< \bar{\partial} f, \varphi >= -\frac{1}{2\pi} \int_0^{2\pi} (\varphi(\infty, t) - \varphi_1(0, t))\,dt + \frac{\imath}{2\pi} \int_0^{2\pi} (\varphi(r, 2\pi) - \varphi_1(r, 0))\,dr.$$

By the compactness of the support of the function φ and hence also of φ_1 and the periodicity of φ_1, we obtain (12.35).

Example 12.18 *Prove that the function $E : \mathbf{R} \times \mathbf{R} \to \mathbf{R}$ given by*

$$E(x,t) = \frac{1}{2}H(t - |x|) = \begin{cases} \dfrac{1}{2} & \text{if } t > |x|, \\ 0 & \text{if } t < |x|, \end{cases} \tag{12.36}$$

is the fundamental solution of the one–dimensional wave equation. In other words, it holds

$$\frac{\partial^2 E}{\partial t^2} - \frac{\partial^2 E}{\partial x^2} = \delta.$$

As before, H is the Heaviside's function, i.e., the characteristic function of the interval $(0, +\infty)$.

Solution. Let us note first that E from (12.36) is a locally integrable function on the set $O = \{(x,t) \mid x \in \mathbf{R}, t > 0\}$, and thus defines a regular distribution which we simply denote also by E. For any $\varphi \in \mathcal{D}(\mathbf{R})$ it holds

$$\left\langle \frac{\partial^2 E}{\partial t^2} - \frac{\partial^2 E}{\partial x^2}, \varphi \right\rangle = \frac{1}{2} \int_{-\infty}^{+\infty} \int_{|x|}^{+\infty} \frac{\partial^2 \varphi}{\partial t^2} \, dt \, dx - \frac{1}{2} \int_{0}^{+\infty} \int_{-t}^{t} \frac{\partial^2 \varphi}{\partial x^2} \, dx \, dt$$

$$= \frac{1}{2} \int_{-\infty}^{+\infty} \left(\frac{\partial \varphi(x,t)}{\partial t} \Big|_{t=|x|}^{t=\infty} \right) dx - \frac{1}{2} \int_{0}^{+\infty} \left(\frac{\partial \varphi(x,t)}{\partial x} \Big|_{x=-t}^{x=t} \right) dt$$

$$= -\frac{1}{2} \int_{0}^{+\infty} \frac{\partial \varphi(x,x)}{\partial t} \, dx - \frac{1}{2} \int_{-\infty}^{0} \frac{\partial \varphi(x,-x)}{\partial t} \, dx$$

$$\quad - \frac{1}{2} \int_{0}^{+\infty} \frac{\partial \varphi(t,t)}{\partial x} \, dt + \frac{1}{2} \int_{0}^{+\infty} \frac{\partial \varphi(-t,t)}{\partial x} \, dt.$$

Putting $y = -x$ in the second integral and taking y instead of x in the other three integrals of the last right-hand side, we get

$$\left\langle \frac{\partial^2 E}{\partial t^2} - \frac{\partial^2 E}{\partial x^2}, \varphi \right\rangle = -\frac{1}{2} \int_{0}^{+\infty} \frac{d}{dy} \left(\varphi(y,y) \right) dy - \frac{1}{2} \int_{0}^{+\infty} \frac{d}{dy} \left(\varphi(-y,y) \right) dy. \tag{12.37}$$

From (12.37) it finally follows

$$\left\langle \frac{\partial^2 E}{\partial t^2} - \frac{\partial^2 E}{\partial x^2}, \varphi \right\rangle = \frac{1}{2}\varphi(0,0) + \frac{1}{2}\varphi(0,0) = \langle \delta, \varphi \rangle.$$

Exercise 12.19 *Find the fundamental solution of the two–dimensional wave equation.*

Hint. Take the function

$$E_2(x,t) = \frac{H(t - |x|)}{2\pi\sqrt{t^2 - |x|^2}},$$

where for $x = (x_1, x_2) \in \mathbf{R}^2$, we put $|x|^2 = x_1^2 + x_2^2$.

Example 12.20 *Let*

$$E(x) = \frac{1}{\sigma_n(2-n)|x|^{n-2}} \quad for \ n \geq 3,$$

where $\sigma_n = 2\pi^{n/2}/\Gamma(n/2)$.

    Prove that

a) *E is a locally integrable function;*

b) *E is the fundamental solution for the Laplace operator Δ.*

Solution.

a) There exist $M > 0$ and $a < n$ such that

$$|E(x)| \leq \frac{M}{|x|^a},$$

where the integral $\displaystyle\int\limits_{B(0,s)} \frac{dx}{x^a}$ converges for arbitrary $s > 0$, since

$$\int\limits_{B(0,s)} \frac{dx}{x^a} = \int\limits_0^s \left(\int\limits_{\partial B(0,r)} \frac{dS_x}{x^a} \right) dr = \int\limits_0^s \left(\int\limits_{\partial B(0,r)} dS \right) dr$$

$$= \sigma_n \int\limits_0^s r^{n-1-a}\, dr = \sigma_n \frac{s^{n-a}}{n-a}.$$

b) By definition of the distributional derivative and Lebesgue convergence theorem we have for every $\varphi \in \mathcal{D}$ and $0 < \varepsilon < s$

$$< \Delta E, \varphi > = < E, \Delta\varphi > = \int\limits_{\mathbf{R}^n} E\Delta\varphi\, dx = \lim_{\varepsilon \to 0+} \int\limits_{\varepsilon \leq |x| \leq s} E(x)\Delta\varepsilon(x)\, dx$$

and supp $\varphi \subset B(0, s)$. Applying the classical symmetric Green formula on the region $B(0, s) \setminus \overline{B(0, \varepsilon)}$, since $E \in C^\infty(\{x \,|\, |x| \geq \varepsilon\})$, we obtain

$$\int\limits_{\varepsilon \leq |x| \leq s} (E\Delta\varphi - \varphi\Delta E)\, dx = \int\limits_{\partial B(0,s) \cup \partial B(0,\varepsilon)} \left(E\frac{\partial\varphi}{\partial \mathbf{n}} - \varphi\frac{\partial E}{\partial \mathbf{n}} \right) dS$$

$$= \int\limits_{\partial B(0,\varepsilon)} \left(E\frac{\partial\varphi}{\partial \mathbf{n}} - \varphi\frac{\partial E}{\partial \mathbf{n}} \right) dS.$$

Therefore we have

$$\int_{\epsilon \le |x| \le s} E \Delta \varphi \, dx = \int_{\epsilon \le |x| \le s} \varphi \Delta E \, dx + \int_{\partial B(0,\epsilon)} \left(E \frac{\partial \varphi}{\partial n} - \varphi \frac{\partial E}{\partial n} \right) dS, \qquad (12.38)$$

since $\varphi = 0$ on $\partial B(0, s)$. Let now $|x| = r$. Then

$$\frac{\partial}{\partial x_i} \left(\frac{1}{r^{n-2}} \right) = (2 - n) x_i r^{-n}$$

and

$$\frac{\partial^2}{\partial x_i^2} \left(\frac{1}{r^{n-2}} \right) = (2 - n) r^{-n} + (2 - n) x_i (-\frac{n}{2}) 2 x_i r^{-n-2}.$$

Therefore we have

$$\Delta((2 - n) \sigma_n E) = (2 - n) n r^{-n} - n(2 - n)(\sum_{i=1}^n x_i^2) r^{-n-2} = 0.$$

Then we obtain by (12.38)

$$< \Delta E, \varphi > = \lim_{\epsilon \to 0+} \int_{\partial B(0,\epsilon)} \left(E \frac{\partial \varphi}{\partial r} - \varphi \frac{\partial E}{\partial r} \right) dS_\epsilon.$$

We shall prove that the last limit is equal $\varphi(0)$, i.e., $< \delta, \varphi >$. Using the polar coordinates we have that $dS_\epsilon = \epsilon^{n-1} \, dS_1$ and so

$$\int_{\partial B(0,\epsilon)} \left(E \frac{\partial \varphi}{\partial r} - \varphi \frac{\partial E}{\partial r} \right) dS_\epsilon = \int_{r=\epsilon} \epsilon^{2-n} \frac{\partial \varphi}{\partial r} \epsilon^{n-1} \, dS_1$$

$$- \int_{r=\epsilon} \varphi(\epsilon, \theta_1, \ldots, \theta_{n-1})(2 - n) \epsilon^{1-n} \cdot \frac{1}{\epsilon^{n-1}} \, dS_1$$

$$= \int_{r=\epsilon} \epsilon \frac{\partial \varphi}{\partial r} \, dS_1 + (n - 2) \int_{r=\epsilon} \varphi(\epsilon, \theta_1, \ldots, \theta_{n-1}) \, dS_1. \qquad (12.39)$$

Since

$$|\frac{\partial \varphi}{\partial r}| \le \sum_{i=1}^n \sup |\frac{\partial \varphi}{\partial x_i}| < M,$$

we obtain $\int_{r=\epsilon} \epsilon \frac{\partial \varphi}{\partial r} \, dS_1 \to 0$ as $\epsilon \to 0 +$. We have by Lebesgue theorem on convergence

$$\int_{r=\epsilon} \varphi(\epsilon, \theta_1, \ldots, \theta_{n-1}) \, dS_1 \to \sigma_n \varphi(0)$$

as $\epsilon \to 0 +$. Therefore letting $\epsilon \to 0+$ in (12.39) we obtain

$$< \Delta E, \varphi > = \lim_{\epsilon \to 0+} \int_{\partial B(0,\epsilon)} \left(E \frac{\partial \varphi}{\partial r} - \varphi \frac{\partial E}{\partial r} \right) dS_\epsilon = \varphi(0) = < \delta, \varphi > .$$

Exercise 12.21 *Prove that*

a) $\dfrac{H(t)}{(2\pi)^n}\displaystyle\int_{\mathbf{R}^n} \exp(-a^2|z|^2 t + \imath z x)\,dz = \dfrac{H(t)}{(2a\sqrt{\pi t})^n}\exp\left(-\dfrac{|x|^2}{4a^2 t}\right);$

b) $E(x,t) = \dfrac{H(t)}{(2a\sqrt{\pi t})^n}\exp\left(-\dfrac{|x|^2}{4a^2 t}\right);$

is the fundamental solution for the heat equation, i.e., it satisfies the equation

$$\frac{\partial u}{\partial t} - a^2\Delta u = \delta(x,t).$$

Hint. Apply the distributional Fourier transform, which gives the equation

$$\frac{\widehat{\partial u}}{\partial t} - a^2\widehat{\Delta u} = \hat{\delta}E.$$

Using the equality $\delta(x,t) = \delta(x)\delta(t)$, the form of a solution of the obtained ordinary differential equation in the space of the distributions (see Example 12.15), the inverse Fourier and a) deduce the desired form for E.

Exercise 12.22 *Assume that $\hat{\lambda}_j(\xi')$, $j = 1,...,k$, are continuous real functions in $\xi' = (\xi_1,...,\xi_{n-1}) \in \mathbf{R}^{n-1}$. Prove that there exists a measurable function $\Phi : \mathbf{R}^{n-1} \to [-k-1, k+1]$, such that*

$$\min_{1\le j\le k}\{|\Phi(\xi') - \hat{\lambda}_j(\xi')|\} \ge 1 \ \text{ for all } \xi' \in \mathbf{R}^{n-1}.$$

Example 12.23 (Malgrange – Ehrenpreiss) *Every linear partial differential operator with constant coefficients P(D) has a fundamental solution, i.e., there exists a solution of the equation*

$$P(D)u = \delta.$$

Solution. Let us define

$$K(x) = \int_{\mathbf{R}^{n-1}} \int_{\Im(\xi_n)=\Phi(\xi')} \exp(2\pi\imath x \cdot \xi)(P(\xi))^{-1}\,d\xi_n\,d\xi',$$

where, as before, $\xi = (\xi_1,\dots,\xi_{n-1},\xi_n) = (\xi',\xi_n)$, while Φ was defined in Example 12.22.

We shall show that K is well defined, and, moreover, is the desired fundamental solution of P. To that end, put first

$$P_N(\xi) = P(\xi)\left(1 + 4\pi^2\sum_{j=1}^n \xi_j^2\right),$$

where N is a (sufficiently big) natural number. Next, put

$$K_N(x) = \int_{\mathbf{R}^{n-1}} \int_{\Im(\xi_n)=\Phi(\xi')} \exp(2\pi\imath x \cdot \xi)(P_N(\xi))^{-1}\, d\xi_n\, d\xi',$$

where the function Φ was chosen as in Example 12.22. Since then on the domain of integration it holds $|P_N(\xi)| \geq C\cdot(1+|\xi|^2)^N$, the last integral converges for $N > n/2$.

Next, let us prove

$$P_N(D)K_N = \delta.$$

Assume $\varphi \in \mathcal{D}(\mathbf{R}^n)$. Since the adjoint operator to $P_N(D)$ is $P_N(-D)$, it holds

$$
\begin{aligned}
\langle P_N(D)K_N, \varphi \rangle &= \langle K_N, P_N(D)\varphi \rangle \\[2mm]
&= \int_{\mathbf{R}^n} \int_{\mathbf{R}^{n-1}} \int_{\Im(\xi_n)=\Phi(\xi')} \exp(2\pi\imath x \cdot \xi)\frac{P_N(-D)\varphi}{(P(\xi))}\, d\xi_n\, d\xi'\, dx \\[2mm]
&= \int_{\mathbf{R}^{n-1}} \int_{\Im(\xi_n)=\Phi(\xi')} (P_N(-x\imath))^{-1} \\[2mm]
&\qquad \cdot \int_{\mathbf{R}^n} \exp(2\pi\imath x \cdot \xi)(P_N(-D))\varphi(x)\, dx\, d\xi_n\, d\xi \\[2mm]
&= \int_{\mathbf{R}^{n-1}} \int_{\Im(\xi_n)=\Phi(\xi')} \mathcal{F}\varphi(-\xi)\, d\xi_n\, d\xi \\[2mm]
&= \int_{\mathbf{R}^n} \mathcal{F}\varphi(-\xi)\, d\xi = \varphi(0) = \langle \delta, \varphi \rangle.
\end{aligned}
$$

Note that we used the equality

$$\varphi(0) = \int_{\mathbf{R}^n} \mathcal{F}\varphi(\xi)\, d\xi,$$

which follows using the inverse Fourier transformation.

So we obtain

$$\delta = P_N(D)K_N = P_N(D)(1 - \Delta)^N K_N.$$

Now $K = (1 - \Delta)^N K_N$, hence K is the fundamental solution of the operator $P(D)$.

Exercise 12.24 (L. Nirenberg) *Let*

$$P(D) = \sum_{|\alpha|\leq m} a_\alpha D^\alpha \tag{12.40}$$

be a linear partial differential operator with constant coefficients. Then for any given function f from C_0^∞ there exists a solution u in C^∞ of the equation

$$P(u) = f.$$

Answer. The solution u is given by the convolution

$$u = K * f,$$

where K is the fundamental solution of the partial differential operator $P(D)$ from Example 12.23.

Bibliography

[1] Andrews, L. C., *Elementary Partial Differential Equations,* Academic Press College Division, Orlando, Florida, 1986.

[2] Bitzadze A. V., *Equations of Mathematical Physics,* Moscow, 1982 (in Russian).

[3] Budak, B. M., Samarski, A. A., Tikhonov A., *The Exercises in Mathematical Physics,* Nauka, Moscow, 1972 (in Russian).

[4] Carol, R. C., *Abstract Methods in Partial Differential Equations,* Happer and Row, 1969.

[5] Colombeau, J. F., *Elementary Introduction to New Generalized Functions,* North Holland, 113, 1985.

[6] Courant, R., Hilbert, D., *Methods of Mathematical Physics I, II,* Interscience Publishers, 1953, 1962.

[7] Epstein, B., *Partial Differential Equations,* McGraw– Hill, New York 1962.

[8] Friedman, A., *Partial Differential Equations,* Holt, Reinhart and Winston, 1969.

[9] Garabedian, P. R., *Partial Differential Equations,* John Wiley & Sohns, Inc., 1964.

[10] Gelfand, I. M., Shilov, G. E., *Generalized Functions I, II, III,* Fizmatgiz, 1958 (in Russian).

[11] Gilberg, D., Trudinger, N. S., *Elliptic Partial Differential Equations of Second Order,* Springer–Verlag, Berlin, 1977.

[12] Greenspan, D., *Introduction to Partial Differential Equations,* McGraw–Hill, New York, 1961.

[13] Hellwig, G., *Partielle Differentialgleichungen,* Teubner Verlag, Stuttgart, 1960.

[14] Hörmander, L., *The Analysis of Linear Partial Differential Operators I-IV*, Springer–Verlag, Berlin, 1983-1985.

[15] Hubson, V. C. L., Pym, J. S., *Applications of Functional Analysis and Operator Theory*, Academic Press, 1980.

[16] John, F., *Partial Differential Equations*, Springer–Verlag, 1982.

[17] Ladyzenskaya, O. A., Uraltseva N. N., *Linear and Quasilinear Elliptic Equations*, Academic Press, New York, 1968.

[18] Lakshmikantham, V., Leela, S., *Differential and Integral Inequalities*, Academic Press, New York, 1969.

[19] Mihajlov, V. P., *Partial Differential Equations*, Nauka, Moscow, 1983 (in Russian).

[20] Pap, E., *Null–Additive Set Functions*, Kluwer Academic Publishers, Dordrecht, 1995.

[21] Pap, E., *Complex Analysis through Examples and Exercises*, Kluwer Academic Publishers, Dordrecht, in print.

[22] Pinski, A. M., *Partial Differential Equations and Boundary Values with Applications*, McGraw–Hill, New York 1991.

[23] Renardy, M., Rogers, R. C., *An Introduction to Partial Differential Equations*, Springer–Verlag, 1993.

[24] Rubinstein, Z., *A Course in Ordinary and Partial Differential Equations*, Academic Press, 1969.

[25] Schmeelk, J., Takači, Dj., Takači, A., *Elementary Analysis through Examples and Exercises*, Kluwer Academic Publishers, Dordrecht 1995.

[26] Schwartz L., *Théorie des distributions I, II*, Paris, 1950–1951.

[27] Smoller, J., *Shock Waves and Reaction–Diffusion Equations*, Springer–Verlag, 1983.

[28] Sobolev, S. L., *The Equations of the Mathematical Physics*, Nauka, Moscow, 1966 (in Russian).

[29] Stepanov, V. V., *Ordinary Differential Equations*, Fizmatgiz, 1959 (in Russian).

[30] Strang G., *Introduction to Applied Mathematics*, Wellesley–Cambridge Press, 1992.

[31] Taylor M.E., *Partial Differential Equations I,II, III*, Springer–Verlag, 1996.

[32] Tikhonov A., Samarski A. A., *Equations of Mathematical Physics*, Pergamon Press, New York, 1963.

[33] Treves, F., *Basic Linear Partial Differential Equations*, Academic Press, 1975.

[34] Vvedensky, D., *Partial Differential Equations with Mathematica*, Addison–Wesley, 1993.

[35] Vladimirov, V., S.,*Equations of Mathematical Physics*, Nauka, Moscow, 1976.

[36] Wloka, J., *Partielle Differentialgleichungen*, Teubner, Stuttgart, 1982.

[37] Zeidler E., *Nonlinear Functional Analysis and its Applications, Vol. IV: Applications to Mathematical Physics*, Springer–Verlag, New York, 1988.

Index

C_0–semigroup, 369

Adjoint Theorem, 324

ball function, 108
Banach fixed point theorem, 323
Banach space
    reflexive, 322
Banach-Steinhaus theorem, 322
Beppo-Levi theorem, 249
Bessel
    equation, 108
    functions, 109
bilinear form
    coercitive, 330
boundary value problem, 2
    first, 143
    second, 143
    third, 144

canonical form
    for a quasi–linear second order PDE,
        51
    for a second order PDE, 51
Cauchy integral
    for a first order PDE, 17
Cauchy problem, 2
    n–dimensional, 81
    for a first order PDE, 17
    for a second order PDE, 52
    one–dimensional, 71
    parabolic equation, 183
Cauchy–Kowalevskaya theorem, 12
Cauchy–Riemann equations, 5
change of variables, 252
characteristic curve

for a second order PDE, 52
of a first order PDE, 18
Closed Graph Theorem
    on Hilbert space, 305, 307
    on normed space, 322
compact support, 256
complete solution
    of first order PDEs, 35
convergences pointwise, 90
convolution
    of distributions, 376
    of functions, 268, 375

D'Alambert's formula, 71
Darboux equation, 84
delta distribution, 374
delta net, 256
Diagonal Theorem, 306
differential operator
    hypoelliptic, 340
    of order 2k, 329
dipheomorphism, 260
Dirichlet boundary condition, 69
Dirichlet problem, 143
    exterior, 163
    on cylinder, 161
distribution, 373
    tempered, 375
distributional
    derivative, 374
    Fourier transform, 375
divergence theorem, 80

eigenfunctions, 96
eigenvalue, 96, 312

eigenvector, 312
elliptic second order PDE, 49
energy integral, 137
Euler movement equation,, 16
Euler-Poisson-Darboux equation, 85
exchange formula, 268

first integral
    of a first order PDE, 18
first order PDE, 17
    characteristic curve, 18
    first integral, 18
    linear, 17
    quasi–linear, 17
Fourier
    coefficients, 90
    integral theorem, 107
    inverse transform, 268
    method, 92
    series, 90, 304
    transform, 267
Fredholm alternative, 313
    first, 314
    forth, 319
    second, 317
    third, 317
Fubini theorem, 250
fundamental solution, 390
    of Laplace equation, 163
    of the one–dimensional wave equation, 391
    of the two–dimensional wave equation, 391

Gårding
    hyperbolicity condition, 271
Gårding
    inequality, 330
Gauss-Ostrogradsky theorem, 80, 82
general solution
    of first order PDEs, 36
generalized
    Dirichlet problem, 330

mixed problem, 359
    solution, 345
    solution of mixed problem, 359
generalized derivative, 279
Green
    operators, 336
Green formula, 170
    first (antisymmetric), 82
    second (symmetric), 82
Green function, 167

Hörmander theorem, 340
Hadamard example, 9
harmonic function, 144
Harnack
    first theorem, 180
    second theorem, 180
heat equation, 70
    one–dimensional, 6
heat flux, 223
Heaviside's function, 391
Helmholtz equation, 151, 180
Hilbert space, 303
hyperbolic second order PDE, 49

integral
    Lebesgue, 249
    Lebesgue on surface, 261
integral surface, 17

Kirchoff formula, 87

Lagrange–Charpite method, 36
Laplace equation, 5, 69
Lax-Milgram theorem, 334
Lebesgue theorem on convergence, 250
Legendre
    equation, 107
    polynomial, 107, 161
linear first order PDE, 17
linear operator, 303, 321
    adjoint, 305, 321
    bounded, 304, 321
    closed, 304, 321

compact, 305
  symmetric, 312
linear second order PDE, 49
Liouville equation, 138
locally quadratic $(n - 1)$–dimensional
    part of surface, 260
locally quadratic region, 260

maximum principle
  elliptic equation, 163
  parabolic equation, 183
  strong, 163, 173
Maxwell equations, 16
mean value theorem, 174
mixed type problem, 2, 93, 193
mollifier, 256
multi–index, 1

Neumann boundary condition, 70
nonlinear first order PDE, 35
norm, 250
normed space, 251

order of a PDE, 1

parabolic second order PDE, 49
Parseval identity, 270, 304
partial differential equation, 1
PDE
  hyperbolic, 268
  linear, 2
  of first order, 17
  quasi–linear, 2
  uniformly elliptic, 329
Pfaff's equation, 32
piecewise continuous function, 89
Poincare's inequality, 135
Poisson equation, 69
Poisson formula, 88
  for wave equation, 81
Poisson integral formula, 144
potential equation, 69
Projection theorem, 342

quasi–linear
  first order PDE, 17
  second order PDE, 49, 64

rapidly decreasing function, 274, 375
real analytic function, 12
region, 1
  starshaped, 287
regular distribution, 374
Rellich theorem, 285
Riccati equation, 138
Riemann's method, 130
Riesz representation theorem, 304
Ritz approximate solution, 342

scalar product, 303
Schauder a-priori estimation, 369
Schrödinger equation, 16, 274
second order PDE
  elliptic, 49
  hyperbolic, 49
  linear, 49
  parabolic, 49
  quasi–linear, 49
semigroup of operators, 369
sine–Gordon equation, 139
singular solution
  of first order PDEs, 36
slowly increasing function, 275
Sobolev
  lemma, 285
  space, 285
solution of a PDE, 1
spheric
  coordinate system, 160
  function, 108
  inversion, 168, 180
Sturm-Liouville
  series, 106
Sturm-Liouville problem, 96, 106
  regular, 106
superposition principle, 97
support

of a continuous function, 256, 373
of a distribution, 374

test function, 373
trace, 297

Uniform Boundedness Theorem, 305,
        322, 328
Urysohn property, 308

Volterra's method, 137

wave equation
    n–dimensional, 4
    homogeneous, 71
    nonhomogeneous, 71
    one–dimensional, 4, 69, 71
weak convergence, 322
    in Hilbert space, 304
weak limit of a sequence, 374
Weierstrass test, 90
well–posed problem, 2
Weyl lemma, 341